Redesigning Organizations

Denise Feldner

Editor

Redesigning Organizations

Concepts for the Connected Society

 Springer

Editor
Denise Feldner
Berlin, Germany

ISBN 978-3-030-27959-2   ISBN 978-3-030-27957-8   (eBook)
https://doi.org/10.1007/978-3-030-27957-8

This Springer imprint is published by the registered company Springer Nature Switzerland AG
The registered company address is: Gewerbestrasse 11, 6330 Cham, Switzerland

Foreword by Prof. Dr. Michael Huether

Old Institutions for a New World

Digitalization is in everyone's hands. The mass use of smartphones triggered a skyrocketing level of real-time connectedness and left its mark on people's daily lives. Shopping opportunities after office hours and constant updates on the US President's mood and sensitivities have a huge outreach; however, digitalization has implications far beyond these simple virtual interactions between consumers and suppliers of goods, services and information on large (American) platforms.

Hence, a coherent analysis of institutional adaptations due to digitalization needs to address the *technical dimension* and its implications for the respective *digital business models*. The technical dimension can be separated into a product and a process level, as well as into a tangible and an intangible dimension. This categorization leads to the distinction between *smart fabric* (physical machines exploiting real-time communication), *smart products* (individualized remote data analysis of physical products), *smart operations* (machine learning-based optimization of production processes), and *smart services* (intangibly supplied services). In order to understand the disruptive potential of the current transformation, it is crucial to understand that digitalization is tantamount to virtual copies of tangible, physical processes, and products. What's more, the depicted smart economy is detrimentally based on its network characteristic and comes with specific needs, especially in the fields of infrastructure and contactless continual connectivity (*Internet of things*). Non-connectivity might trigger disproportionally high costs as production processes potentially break down as soon as machines lack information for follow-up production processes. Disproportional costs of connectivity loss challenge the principle of unrestrained network neutrality.

For most (retail) observers of the digital transformation, machine-to-machine communication is only a minor aspect. However, for a categorization of digital business models, it is useful to additionally zoom in on the specific digital potentials, risks, and needs within the business world. Finally, the consumer–producer nexus of digital interactions allows us to come up with a fourfold divided representation of the respective interaction interfaces:

The *business-to-business* (*B2B*) interface reflects the industrial perspective and describes interactions between companies in a highly digitalized network or along the value chain ("Industrie 4.0").

- The *business-to-consumer* (*B2C*) interface prominently describes large platform-based service provision by "traditional" Internet companies. Consumers profit from low search, assessment, and coordination costs.
- The *consumer-to-consumer* (*C2C*) interface includes the *sharing economy*. Consumers develop market clearing price mechanisms on virtual platforms without making use of "professional" retail sellers.
- The *consumer-to-business* (*C2B*) interface describes *big data*-based business models, such as smart home services where consumers usually pay with their personal data.

All four interaction interfaces are based on platforms digitally matching supply and demand. The categorization enables the specific qualification of institutional implications within the concerned areas due to heterogeneous effects on productivity, employment, and competition. Such a specific analysis along the interfaces is collected in Table 1.

Interestingly, the analysis reveals that an often-echoed claim for basic competition reform and stronger monopoly control can be dispelled. In fact, even big platforms are still contestable by new competitors, do not necessarily exploit market

Table 1 Economic policy implications

	B2B	B2C	C2C	C2B
Productivity effects	(1) Network neutrality (2) Standards (3) Data law (4) Data protection (5) SME politics	./.	Regulatory level playing field	(1) Consumer protection (2) Terms and conditions (3) Education
Employment effects	(1) Education (2) Advanced training (3) Secure skilled worker supply	(1) Individualization of social security (2) Secure skilled worker supply	(1) Company characteristics of individuals (2) Regulatory level playing field	./.
Competition effects	(1) Standards (2) Data ownership	(1) Market analysis (2) Control of abusive practices	(1) Company characteristics of individuals (2) Regulatory level playing field	(1) GDPR (2) Consumer protection

Own Source Huether (2018)

power on different competition layers, and even enable new downstream business models (*inverse scaling*).

Especially in the C2B (but also in the B2C) interface—characterized by a large asymmetry of power between giant profit-oriented companies and personal data supplying individuals—consumer protection standards have to be taken care of. Within the C2B interface, this goes along with the need for minimum "data literacy" education, as well as clear and transparent rules for terms and conditions.

In the C2C interface, the rules of the game must be addressed so that they equally apply to consumers and companies supplying equivalent services. Once a platform allows individuals to supply a semi-professional activity, the supervising agency has to guarantee a level playing field without regulatory or fiscal loopholes.

The B2B interface analysis reveals a significant demand for efforts in the labor market and the field of advanced training. Especially with regard to regional clusters, attention must be paid to structurally weak regions, in order to shape the digital transformation in a socially acceptable manner.

Further research needs to go beyond poking in the dark of the digital transformation and must combine the technical dimensions that enable digitalization for business models in the different interfaces. Special attention should be paid to the B2B interface, where current innovations such as AI, robotics, or machine learning will trigger significant changes with highly disruptive potentials.

Prof. Dr. Michael Huether
Director of the Institut der Deutschen Wirtschaft

Honorary Professor for Political
Economy at the EBS University
Cologne

Prof. Dr. Michael Huether has studied economics and history at the Justus Liebig University Giessen and the University of East Anglia, Norwich, UK, from 1982 to 1987. After completing his doctorate studies in economics in 1991, he became a Research Fellow at the German Council of Economic Experts where he was promoted to Secretary General in 1995. In 1999, he took the position of Chief Economist of the DekaBank and in 2001 was appointed Head of the Economics and Communications Department. He has been Honorary Professor of Political Economy at EBS Business School since August 2001. Since July 2004, he has been a Director and Member of the Presidium of the Cologne Institute for Economic Research. From 2016 to 2017, he was Gerda Henkel Adjunct Professor at Stanford University. He was awarded the Order of Merit of the Federal Republic of Germany by the Federal President. He is Member of the supervisory board of Allianz Global Investors Kapitalanlagegesellschaft mbH, SRH Holding, and TÜV Rheinland Berlin Brandenburg Pfalz e.V. and Member of the main board of Atlantik-Brücke. He is Member of the Board of Trustees of the Max Planck Institute for the Study of Societies and Member of the EU Commission's Refit Platform, which advises the Commission on the efficient, effective, and practical implementation of EU legislation.

Foreword by André Loesekrug-Pietri

How Government and Civil Society Can Invent the State of the Twenty-First Century—The Case of Innovation

Digital Acceleration is Putting Extreme Pressure on Liberal Democracies

In this fantastic acceleration of time brought about by the Fourth Industrial Revolution, nation-states and liberal democracies find themselves at the precipice of paradigm shifts, both internally—regarding societal cohesion—and externally, with fast-changing geopolitics that might make or break their prosperity and even their existence in the years to come. In a world of immense technological progress driven by big data collection and intelligent analysis of sensory data across all platforms of society, the nation-state of 2019 faces a fundamental challenge to its core value systems: the emergence of technologies that are disrupting bureaucratic structures inherited from the twentieth century. On the waves of the globalizing market economy of the past decades, government and civil society long seemed to gain in interdependency, information became more open and more easily accessible with the advent of the Web, and new forms of government—like open democracy or global governance—seemed to be the future.

However, we see today that the increased generation and volatility of information have evolved from the anarchic online society of the late 1990s all the way to a platform economy run by the world's largest multi-national corporations. Today, we are witnessing an over-abundance of information and complexity that goes beyond the human capabilities of data analysis and opens the door to all kinds of manipulations and fake news. This world—in which the pioneers of emerging technologies rapidly hold monopolistic access to billions of users, in which the winner takes all—is both an immense challenge to the ability of the nation-state to act in time and an enormous opportunity for civil society to step up and complement what government is not able to achieve fast enough.

In different parts of the world, the current alterations of different state philosophies, or software, run on the late global market hardware of capitalism. Additionally, the geopolitical shifts in influence in the twentieth century show the paradox of balancing the long-term strategy planning necessary to prepare economies to win the competition of technological dominance with other states while keeping up an intense agility to react on a moment's notice to the accelerated developments of the world. The launch of the Sputnik satellite—and the associated sudden perception that the entire American continent was within reach of Soviet ballistic missiles—uniquely exemplified for the US Government their fragility and showed that they were mistaken about their own economic, technological, and political dominance at that moment. The government's reaction was to ensure that never again would they be surprised by another nation's innovative developments. The single best way to achieve this was to always be on top of all the others in their own high-tech developments. This led to the creation of DARPA—the Defense Advanced Research Projects Agency—and its phenomenal list of achievements and breakthroughs. China realized the lessons of the Cold War and hence vehemently aims today to overarch the USA's advantage in established technologies by leapfrogging to the next age of artificial intelligence (AI) and cyberspace.

These two examples—though executed on vastly different value systems and from contrasting starting points—show the necessity for governments to imagine a long-term vision inspiring the entire society at a moment where the world is increasingly difficult to forecast. It also underlines the performance of highly agile technology ecosystems that embrace experimental innovation and with it the transformational change necessary to stay ahead in a winner-take-all world. Both platforms and autocratic systems seem to benefit from a massive advantage to be the new pacesetters in this world. Will liberal democracies be able to cope and reinvent themselves?

New Technologies are Becoming Massively Political and Carry Strong Values

It is crucial to fully understand the central importance of technological dominance in an accelerated, globalized political environment, in which the market capitalization of the largest platforms far exceeds the national budgets of entire nation-states, and the market penetration of privately developed apps—like Facebook, Gmail, or WeChat—can tap a consumer market far greater than whole national populations. Private technological development increasingly dominates the shaping of cultures and carries with it the intrinsic value systems of the environments in which these technologies have been produced—and these can be markets with far inferior privacy standards, misregulated security or hidden strategic interests of enterprises and states. This increases the growing dependency of some countries on technology and risks undermining the value systems of liberal democracies. Europe is already dependent on a digital software infrastructure largely based on American standards, whereas future technological revolutions, like the 5G-based infrastructure enabling the Internet of things (IoT) economy of the coming years, are being progressed by Chinese telecom companies. Despite having invented the GSM standard, Europe seems incapable of producing its own tech giants on which the normative power and security standards of the future depend. In the wake of the instant cultural exchange across the entire planet, and the accessibility to travel and do business all around the world enabled by revolutions in communication and transportation, it is easy to forget the underlying reality of competing spheres of influence carried out on the back of technological competition. The world has never been so fragmented and divided about political and value systems as on the eve of this third decade of the twenty-first century. The myth of the end of history is long gone, and the Big Game as well as national rivalries have returned, with attacks on all multilateral institutions like never before.

Thus, the resilience of liberal democracies is under threat from an accelerated technological adoption that does not stop before national boundaries or cultural hemispheres. This is true especially in Europe, since market-leading applications and current-edge technologies—be it in AI, blockchain, new materials, energy, space, cyber, or biotech—are now largely pioneered from research papers into prototypes elsewhere in the world. This shift to foreign security critical infrastructure has accelerated throughout the era of the Internet economy, with USA and Chinese market leaders now holding large majorities of European market share in cloud infrastructure, pushing the boundaries in emerging fields like gene-editing and racing ahead in the development of AI.

We are just beginning to realize how much these technologies carry values: Cyber is at the core of the resilience of our critical infrastructure, be it the power network, hospitals or voting systems. Undermining it would put societies under immediate danger. Social media platforms are increasingly shaping public opinions and, despite all the positive aspects they bring—like connectivity, access to information and ability to lower barriers of entry—they have a proven tendency to increase polarizations of societies and thus undermine our democratic processes. Facial recognition, one of the most hyped technologies—at the convergence of big data and optics—is completely changing the social contract between the state and the individual.

It is therefore imperative for liberal democracies to reinvent the speed with which governments can face these new challenges coming from authoritarian states, but also non-state actors. One solution is to include civil society even more into decision-making between elections and to shape together the norms and technologies that will become critical in the years and decades to come. Experimentation must be embraced, and new ways continually be tried out as a way to reinvent, accelerate, and connect nations with speed while remaining true to their fundamental values of democracy, diversity, privacy, rule of law and human rights. Innovation is thus not only a benchmark of technological or economic prosperity, but inherently a guarantor of societal and political stability, and necessary for the resilience of liberal democracies.

Augmenting Leaders with Technology?

Beyond visionary technological pioneering, emerging technologies should have a massive influence on the way governments and institutions, as well as the market, make decisions. It is striking to analyze the growing contrast between an ever more complex world on the one hand and the administrative structure that has only marginally changed in most states and that has been less prone to disruptive changes—be it in the way they form their opinions, gather their data, or make informed decisions affecting the lives of millions. At a juncture where true "singularity" is happening—i.e., a world where the amount of daily received data, information, and messages has outstripped the cognitive capacity of any single individual—technology is potentially a gigantic source of improvement toward more rational, timely, and informed political decisions. Liberal democracies, again, face the challenge of being outrun by competing systems: The Chinese government, for example, leads the way with massive usage of big data and unified data lakes to attempt to fully control, direct, and influence the actions and decisions of 1.4 billion citizens, with the admitted goal of conducting and executing "state management" to perfection. And it is much better than George Orwell's *1984*, which is a comparison often heard. It looks much more like Aldous Huxley's *Brave New World*, where a generalized system of social credits will influence the "right" decisions of all citizens: Self-censorship is one of the most powerful tools ever invented. In a fully different register, large US platforms have a user base outstripping the largest countries on the planet, with databases going into the billions of users, which creates massive economies of scale while allowing individual targeting on a large scale. The commonality in both cases is that the rigorous and wide-scale analysis of all collectible data is central to their business models and, in turn, offers a great return on investment as evidenced in the economic success of these platforms. Liberal democracies and international institutions should take note from these lessons and implement new technologies in a swifter, privacy-concerned manner that, above all, benefits the citizen first. The deeply European concern for individual rights and privacy—beyond a regulatory framework like the General Data Protection Regulation (GDPR)—is a powerful engine for growth, value, and consumer engagement. European countries must ensure that they don't just end up being the best regulators, but also turn their values into prosperity.

Beyond Tech for Tech's Sake: Technology for a Purpose

Emerging technologies can and should be used to solve societies' most pressing challenges beyond economic growth—for example, to tackle environmental problems like air pollution, global warming, and the dependency on finite resources—by re-inventing and disrupting the many ways in which mankind has developed the established global economy on visionary ideas that were much too often based on unsustainable production processes. New materials, carbon sequestration, and energy extraction from newfound techniques and refined existing processes are all imperative in the struggle to keep the planet's climate tolerable. Likewise, only a purposeful and visionary human-centric digital transformation will allow humanity to achieve the full benefits from intelligent automation in virtually all aspects of life through AI and other assisting technologies. However, such a human-centric adoption requires the state and its institutions to engage monetarily and programmatically in the development of such technologies. In the field of AI, particular scrutiny is put on the question of what values and models are introduced into autonomous programs, and to whose benefit intelligent machines will execute their code in the end. Combining data science with the cutting edge of biotechnology, gene-editing and advances in agricultural technologies, emerging technologies can drastically and massively improve healthcare systems—from allowing medicine to be personalized to the very last individual patient (through advanced and instantly digitized DNA sequencing) to eventually renewing and repairing body functions beyond their natural lifespans. And lastly, the combined effort of state, civil society, and technological pioneers will enable a new exploration of the unknown—be it in the deep sea or outer space—that will give humanity a vision beyond the limited complexity of its earthly quarrels. Engaged citizens of liberal democracies should keep pressure on institutions and nations to leverage emerging technologies for prosperity, resilience, and transparency. They should push for an update of archaic policy-making mechanisms with technology-enhanced decision-making and evidence. And above all ensure that societies as a whole, and not just a happy few, benefit from new technologies.

A New Pact of Trust Between the Public Hand and Civil Society

The challenge to adapt liberal democracies to the pace of a globalized digitized world—preserving the fabric of societies by avoiding a digital divide, while withstanding the pressure from monopolistic challengers from overseas—will be massive for our societies.

Transformational change lies ahead, shaped by those who own the technologies of the future. Huge benefits can already be seen, for example, for artificial intelligence in industry as well as in everyday life—innovation is a critical asset for the strength of societies and needs to be considered as such in Europe. But at the same time, the consequences of unregulated digitalization are challenging the robustness of democratic processes, the horizontalization and polarization of social media are disrupting traditional political institutions, and future technologies are posing new security risks in both reality and cyberspace.

To achieve prosperity in the near- and long-term, along with political stability, it is necessary to forge a new pact of trust between the public and civil society to drive forward innovation—as an engine for inclusive change and progress. Only when the state—as an actor for the long term and driven by purpose—is connected with the technology ecosystem—that pushes innovative technologies forward—can humanity achieve both long-term vision and the agility that the current epoch commands. And give a clear sense of purpose to citizens, such as engaging on the topic of climate change, a human-centric digital transformation or massively improving healthcare. It is crucial to set long-term, visionary goals as guidelines of where society should be headed, and to navigate the short-term with agility and speed. Experimentation is hereby critical as it gives societies the continuing sense of urgency and feeling that nothing can be taken for granted.

With this new pact of trust between the pillars of European nations, a visionary government, and a pioneering civil society and entrepreneurial ecosystem, a new state model can be achieved in which transformative change is embraced, leading to a more prosperous, wealthy, fair, and secure society in the twenty-first century.

André Loesekrug-Pietri

Speaker of the Joint European
Disruptive Initiative (JEDI)

Entrepreneur

Technology Investor

Former Special Advisor to the French
Minister of Defence
Paris

André Loesekrug-Pietri is a Speaker for JEDI, Former Special Advisor to the French Minister of Defence, Founder of ACAPITAL, holds French–German nationalities, held leadership positions in private equity, government, industry, and as an entrepreneur. Starting as an Assistant to the CEO of Aerospatiale-Airbus, he then spent 15 years in private equity and venture capital, including 10 years in China, investing in European companies with global ambitions. In 2017, he took a break from the private sector to become Special Advisor to the French Minister of Defence, responsible in particular for European Defense policy as well as technology and innovation. He is currently Speaker of the Joint European Disruptive Initiative (JEDI), aiming to accelerate Europe's leadership in disruptive innovations. An HEC and Harvard Kennedy School graduate, he is a Lecturer at SciencesPo and Non-resident Fellow of the German Marshall Fund. He attended Sup-Aero aerospace engineering school, was named a Young Global Leader by the WEF, and is Private Pilot and Colonel with the French Air Force People's Reserve.

Foreword by Dr. Matthias Spielkamp

Algorithms—To Govern or Be Governed?[1]

It was a striking story. The headline read, "Machine Bias," and the teaser proclaimed: "There's software used across the country to predict future criminals. And it's biased against blacks." ProPublica, a Pulitzer Prize-winning non-profit news organization, had analyzed the risk assessment software known as COMPAS. The software is being used to forecast which criminals are most likely to reoffend. Guided by such forecasts, judges in courtrooms throughout the USA make decisions about the futures of defendants and convicts, determining everything from bail amounts to sentences. When ProPublica compared COMPAS's risk assessments for more than 10,000 people arrested in one Florida county with how often those people actually went on to reoffend, it discovered that the algorithm "correctly predicted recidivism for black and white defendants at roughly the same rate." But when the algorithm was wrong, it was wrong in different ways for blacks and whites. Specifically, the analysis found that "blacks are almost twice as likely as whites to be labeled a higher risk but not actually reoffend." And COMPAS tended to make the opposite mistake with whites: "They are much more likely than blacks to be labeled lower risk but go on to commit other crimes."

Whether it is appropriate to use systems like COMPAS is a question that goes beyond racial bias. Potential problems with other automated decision-making (ADM) systems exist outside the justice system, too. On the basis of online personality tests, ADMs are helping to determine whether someone is the right person for a job. Credit-scoring algorithms play an enormous role in whether you get a mortgage, a credit card, or even the most cost-effective cell phone deals.

There has been a lot of discussion about the potential and risk of artificial intelligence (AI) in recent years and with it comes a proliferation of terms like machine learning, self-learning systems, and deep neural networks. But today's AI

---

[1] This text is based on the article "Inspecting Algorithms for Bias," MIT Technology Review, Vol. 120, No. 4, pp. 96–98, https://www.technologyreview.com/s/607955/inspecting-algorithms-for-bias/; it has been substantially amended for this publication.

is tomorrow's off-the-shelf software solution, and the term automated decision-making (sometimes called algorithmic decision-making) better defines what societies are faced with: the automation of processes that were, so far, inaccessible to being automated—medical diagnostics, criminal justice, news curation, and more. Examples like the COMPAS risk assessment system show that these ADM systems can have a high impact without relying on the so-called artificial intelligence or machine learning, so algorithms and automation are at the core of the development.

It is important to note, though, that the term ADM has to be used with caution. When we talk about decisions being automated, what we are faced with are processes that automate a certain aspect of a decision (e.g., data gathering and analysis). This is different from human reasoning that results in an action a human has to take responsibility for. A machine can never bear responsibility. When it is argued that machines make decisions, what we really mean is that humans have devised a model, encoded in rules executable by a machine, to produce a certain result that is then used to prescribe a certain action.

It is not necessarily a bad idea to use ADM systems like COMPAS. In many cases, ADM systems can increase fairness. Human decision-making is, at times, so incoherent that it needs oversight to bring it in line with our standards of justice. But often, we do not know enough about how ADM systems work to know whether they are fairer than humans would be on their own. In part, because the systems make choices on the basis of underlying assumptions that are not clear even to the systems' designers, it can be extremely difficult to determine which systems are biased and which ones are not.

What should be done to get a better handle on ADMs? Democratic societies need more oversight than they have now over such systems. At the same time, technology should not be demonized undeservedly. What's important is that societies—and not only the makers of ADM systems—make the value judgments that determine the power of ADMs.

As shown in Fig. 1, these value judgments start ahead of the actual development of any executable code or data collection. The first (human) decision needed is about what processes to automate: Is it a good idea to have an automation system "decide" who will be incarcerated or receive a mortgage—or to have it "decide" who's going to get care based on a statistically projected life expectancy? Different societies will come up with very different answers to this question. Most of them have little, if anything, to do with the data gathered or the code executed, but with fundamental questions of how to distribute resources and opportunities fairly to the people affected. In this sense, ADM is but a lens to focus more clearly on the challenges we face as humans.

That does not mean there is no need to look at the details of the ADM processes themselves. COMPAS, for example, determines its risk scores from answers to a questionnaire that explores a defendant's criminal history and attitudes about crime. Does this produce biased results?

After ProPublica's investigation, Northpointe, the company that developed COMPAS, disputed the story, arguing that the journalists misinterpreted the data. So did three criminal justice researchers, including one from a justice reform

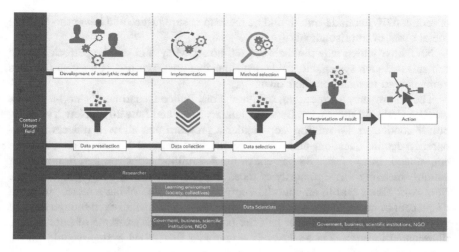

Fig. 1 A wide range of contributions to the development and deployment of ADM systems (© AlgorithmWatch, Creative Commons by 4.0)

organization. Who's right—the reporters or the researchers? Krishna Gummadi, Head of the Networked Systems Research Group at Max Planck Institute for Software Systems in Saarbrücken, Germany, offers a surprising answer: They all are. Gummadi, who has extensively researched fairness in algorithms, says ProPublica's results and Northpointe's results do not contradict each other. They differ because they use different measures of fairness.

Imagine you are designing a system to predict which criminals will reoffend. One option is to optimize for "true positives," meaning that you will identify as many people as possible who are at high risk of committing another crime. One problem with this approach is that it tends to increase the number of false positives: people who will be unjustly classified as likely reoffenders. The dial can be adjusted to deliver as few false positives as possible, but that tends to create more false negatives: likely reoffenders who slip through and get a more lenient treatment than warranted.

Raising the incidence of true positives and lowering the false positives are both ways to improve a statistical measure known as positive predictive value or PPV. That is the percentage of all positives that are true. As Gummadi points out, ProPublica compared false positive rates and false negative rates for blacks and whites and found them to be skewed in favor of whites. Northpointe, in contrast, compared the PPVs for different races and found them to be similar. In part, because the recidivism rates for blacks and whites do in fact differ, it is mathematically likely that the positive predictive values for people in each group will be similar, while the rates of false negatives are not.

One thing this tells us is that the broader society—lawmakers, the courts, an informed public—should decide what it wants such algorithms to prioritize. Are we primarily interested in taking as few chances as possible that someone will skip bail or reoffend? What trade-offs should be made to ensure justice and lower the massive social costs of imprisonment?

No matter which way the dials are set, no ADM system will be perfect. But we can still use such systems to guide decisions that are wiser and fairer than the ones humans tend to make on their own.

The controversy surrounding the New York Police Department's stop-and-frisk practices helps to show why. Between January 2004 and June 2012, New York City police conducted 4.4 million stops under a program that allowed officers to temporarily detain, question, and search people on the street for weapons and other contraband. But in fact, "88 percent of the 4.4 million stops resulted in no further action—meaning a vast majority of those stopped were doing nothing wrong," the *New York Times* said in an editorial decrying the practice. What's more, "In about 83 percent of cases, the person stopped was black or Hispanic, even though the two groups accounted for just over half the population." This example of human bias, illuminated through data analysis, is a reminder that ADM systems could play a positive role in criminal justice and a wide range of other fields.

But if we accept that ADM systems might make life fairer if they are well designed, how can we know whether they are so designed?

There have been numerous proposals made by researchers, lawmakers, and civil society organizations in recent months. One of the more popular is to create institutions similar to agencies routinely checking cars for safety, like the German TÜV or the MOT test stations in the UK. This idea has a certain appeal to it, because these institutions regularly are based on a combination of different properties needed to make the system work properly. First of all, there needs to be legal rules as the basis for imposing obligations on different actors—in this case, car makers and car owners. Then, there has to be an institution with the expertise to actually perform the checks. And last but not least, rules need to be enforced effectively, which in this case is done by police, who even have the power to take vehicles off the road.

But as convincing as it looks at first glance, the idea has major flaws. ADM systems are not just very complex, they are also very specific to certain tasks and contexts. An algorithm to curate movies for users of an online video platform may be very similar to one used to diagnose cancer. But the contexts in which the entire system operates—say, Netflix versus the NHS—are entirely different. ADM systems used in automated cars, credit scoring, medical diagnostics, criminal justice, financial markets, human resource management, and other purposes need to be dealt with differently from one another. The car safety analogy assumes that one agency could be responsible for—and capable of—dealing with all the challenges of automated processes. The fact that we have already developed very different regulatory approaches for these fields should be seen as an indication that with ADM processes, it should not be much different.

As a matter of fact, wherever we already have existing examples of regulating algorithmic processes, we see that they follow this sector-specific approach, for example, in high-frequency trading—The European Union's Markets in Financial Instruments Directive (DIRECTIVE 2014/65/EU, or MiFID 2) mentions the terms algorithm, algorithms and algorithmic 72 times—or for laws on automated driving in the USA and Germany. At the same time, we need to keep in mind that hard legal regulations in the form of laws and treaties have to remain the *ultima ratio* in liberal democracies: Only when stakeholders are unable to find ways to respect and obey fundamental values and principles of a given society—or are unwilling to do so— should the state act to curtail someone's actions.

But laws are only part of the regulatory toolbox that can and must be used to guide development and use of ADM systems. There are a number of important "soft law" activities being pursued at the moment: The Institute of Electrical and Electronics Engineers' (IEEE) Standards Association alone is currently developing a number of standards to provide ethical guidance for new technologies, seven of them directly addressing the context of ADM systems; the International Organization for Standardization (ISO), under its artificial intelligence initiative, has published two standards already and is in the process of developing another four; and there have been a number of declarations published by other different initiatives.

Societies need to support these approaches in order to find out which are sufficient and effective in keeping up with the fast-paced development of ADM systems. That can happen by public and private research funding in a wide range of fields: computer science and mathematics, of course, but equally important are approaches from law, sociology, philosophy, and other disciplines that deal with the wide range of consequences ADM systems produce. In addition, there needs to be a vigilant civil society keeping an eye on developments and raising the attention of the general public.

If society wants to benefit from advances in the development of ADM systems, we need to find ways to fend off the risks associated with them. To this aim, a true multi-disciplinary, multi-stakeholder approach must not remain an empty phrase.

Dr. Matthias Spielkamp
Founder and Executive Director of
AlgorithmWatch

Co-Founder and Publisher of iRights.info
Berlin

References

The European Union's Markets in Financial Instruments Directive (DIRECTIVE 2014/65/EU, or MiFID 2) https://eur-lex.europa.eu/legal-content/EN/TXT/HTML/?uri=CELEX:32014L0065&from=EN; accessed May 28, 2018

IEEE P7000 (Model Process for Addressing Ethical Concerns During System Design)

IEEE P7001 (Transparency of Autonomous Systems), IEEE P7003 (Algorithmic Bias Considerations)

IEEE P7006 (Standard for Personal Data Artificial Intelligence Agent), P7008 (Standard for Ethically Driven Nudging for Robotic, Intelligent and Autonomous Systems)

IEEE P7009 (Standard for Fail-Safe Design of Autonomous and Semi-Autonomous Systems)
IEEE P7010 (Wellbeing Metrics Standard for Ethical Artificial Intelligence and Autonomous
    Systems)
IEEE Standards Projects Provide Ethical Guidance for New Technologies, http://theinstitute.ieee.
    org/resources/standards/seven-ieee-standards-projects-provide-ethical-guidance-for-new-
    technologies; accessed May 28, 2018
*The IEEE Global Initiative on Ethics of Autonomous and Intelligent Systems Announces New
    Standards Projects*, http://standards.ieee.org/news/2017/ieee_global_initiative.html, accessed
    May 28, 2018
ISO/IEC JTC 1/SC 43 Artificial intelligence standards, https://www.iso.org/committee/6794475/x/
    catalogue/p/1/u/0/w/0/d/0; accessed May 28, 2018
ISO/IEC JTC 1/SC 43 Artificial intelligence standards under development, https://www.iso.org/
    committee/6794475/x/catalogue/p/0/u/1/w/0/d/0; accessed May 28, 2018
*List of Declarations on Automated Decision Making (ADM), Artificial Intelligence (AI), Machine
    Learning (ML)*, https://algorithmwatch.org/de/list-of-declarations-on-automated-decision-
    making-adm-artificial-intelligence-ai-machine-learning-ml/, accessed May 28, 2018

Dr. Matthias Spielkamp is the Founder and Executive Director of AlgorithmWatch, a non-profit advocacy and research organization focused on the consequences of algorithmic decision-making (ADM) on societies. He is co-founder and publisher of the online magazine iRigths.info, which in 2006 received the Grimme Online Award, Germany's premier award for online journalism. He testified before several committees of the German Bundestag, including on AI and robotics. Currently, he is a Bucerius Fellow of ZEIT Foundation; in 2015/16, he was a Fellow at Foundation Mercator. He serves on the governing board of the German section of Reporters Without Borders and the advisory councils of Foundation Warentest and the Whistleblower Network. In the steering committee of the German Internet Governance Forum (IGF-D), he acts as co-chair for the academia and civil society stakeholder groups. He has co-authored and edited several books on Internet governance, journalism, and copyright regulation and holds master's degrees in Journalism from the University of Colorado at Boulder and Philosophy from the Free University of Berlin. He has contributed to publications like the MIT Technology Review and Die Zeit, and he has been quoted by Engadget, Süddeutsche Zeitung, Wired, and many others. In 2017, he was named as one of 15 architects building the data-driven future by Silicon Republic.

Foreword by Prof. Dr. Holger Mey

A Major Milestone in the Process of Evolution

Today's technological megatrends—such as nanotechnology, biotechnology, robotics, and the so-called artificial intelligence—hold the potential to fundamentally and comprehensively change all societies. Such profound change is particularly likely when various new technologies merge. Predicting the concrete effects is difficult if not impossible, but the overall dimension of today's technological change seems no less than a major milestone in the process of evolution. Such technological transformations will certainly involve redesigning virtually all of society's institutions. No sphere of human activity, even the most personal, will remain unaffected.

Artificial intelligence (AI) is a dazzling term characterized by the lack of a precise, agreed-upon definition. Intelligence is hard to distinguish from terms like intellect, reason, brightness, or wisdom. Furthermore, it might be more appropriate to use the term "intelligence used by artificial organisms or machines."

Obviously, the human brain did not invent intelligence any more than human legs invented land or eyes invented light, as the German medical doctor and science author, the late Hoimar von Ditfurth, so eloquently described decades ago. Rather, legs are the evolutionary answer to the fact that there was land and, thus, the possibility of movement. Being able to walk or run offered an evolutionary advantage when fleeing from danger or finding food. The eyes did not invent light. Rather, there was the reflection of electromagnetic waves and, thus, the possibility of seeing. The development of a sensor and image generator allowed danger to be recognized and food to be found—a clear advantage in the fight for survival. Analogously, the brain did not invent intelligence, but the other way around. There was intelligence, and consequently, nature created an organ to develop better escape strategies, hunting and farming methods, and the use of tools.

Animals also use intelligence, only less than humans do. Computers can also use intelligence, and perhaps, one day more than humans can. Just because the electronic brain is not based on hydrocarbons, but on silicon or gallium arsenide, for example, does not mean that it cannot one day be intelligent. Why should the further development of intelligence stop where the human brain is today? Will the computers convince us that we are superfluous (R. Kurzweil)? Will humans continue to develop, becoming "homo deus" (Y. N. Harari)? Maybe humans will merge with computers and become cybernetic organisms or cyborgs. We are the Neanderthal of tomorrow. We exist today only so that, one day, the future can take place (H. von Ditfurth).

During rush hour in Washington, Paris, or any other big city, what do we see in the streets? Almost every intersection is blocked. People drive their cars into the intersection out of carelessness or ruthlessness and block all traffic. At the same time, we assume that, in the future, intelligent, networked traffic control of autonomous vehicles will prevent people from blocking traffic. But what happens if a pedestrian accidentally or intentionally steps onto the road while traffic is flowing? The perfectly functioning autonomous vehicle would brake immediately so as not to run over the pedestrian. Reckless pedestrians would thus be able to bring all traffic to a standstill at any time. What should be done? One might program autonomous cars to simply run over those pedestrians that step onto the road. Pedestrians (as individuals) would be cautious and circumspect and would therefore not obstruct traffic (as a collective). This might be quite conceivable in societies that traditionally or ideologically subordinate the individual to the society or the collective. For moral reasons, Western societies would deem this unacceptable. Western societies reject sacrificing the individual for the benefit of the collective.

So how do Western societies prevent gridlock? How do they ensure that an individual's careless behavior does not jeopardize the functioning of society? Control is necessary, in the sense of social institutions *employing technologies* to enforce rules. Video surveillance (CCTV) records the individual stepping onto the street, biometric face recognition identifies the person, and on his "mobile device," the jaywalker reads that a painfully high sum—corresponding to his income—has been deducted from his bank account. He might even read that he would have to pay five times the amount the next time, and if the interests of others and the community were violated again, he would have to go to prison for one month.

New York City was once an extremely dangerous city, and then at some point, it became one of the safest cities in the USA. In between, there was a mayor who increased police presence in the streets and introduced a "zero tolerance strategy" against infringement and major offences. Tolerance toward intolerance means the end of tolerance. Freedom for the enemies of freedom means the end of freedom. Those who do not understand this philosophical paradox expose societies to the ruthlessness of the individual.

The formal imposition of values like respect or tolerance is usually unnecessary. These are values that every individual and every group would define, at least initially, as meaningful and worth striving for. Whether waiting at a cash register or queuing for a taxi, pretty much any group of people anywhere would agree that it is not the person who cuts in line or using force to surpass others that should be served next, but the person whose turn it is. So why do people constantly cheat or even forcibly seek to gain advantages over others? The answer is quite simple: Because nobody prevents them from doing so. Disciplined waiting can be an element of culture and a result of good education. Being able to draw a number on a machine is another way of encouraging fairness, supporting justice, and preventing abuse. Ultimately, justice exists from being lived, protected and, if necessary, enforced. In the future, important social norms like civility, compassion, tolerance, and peacefulness will continue to depend on education, culture, and other aspects of socialization. Nevertheless, even in the future, humans will not escape their genetic heritage from the wilderness in the savanna. As such, social control—combined with consistent enforcement of the law—will be important for any human community that does not suffer under brutality and ruthlessness.

Time and again, it is said that the computer must ultimately be controlled by humans. However, this statement falls short of answering the question of who is in control. Should it be Adolf Hitler, Josef Stalin, Mao Zedong, or Pol Pot? The real challenge is to make sure the computer is controlled by the *right* humans, the "good humans." But who determines this? Humans have suffered most at the hands of other humans. The question is not whether we want, or will have, more or less supervision and control, but rather, who controls the controllers and who determines the values that are to be complied with.

Control allows for security, and it is not necessarily at the expense of freedom. On the contrary, security is a prerequisite for freedom. That is why freedom and internal security cannot be set against each other in the sense of a zero-sum game. Socialists understand this when it comes to social and material impoverishment (people who starve do not feel free), but it is not always understood when it comes to internal security.

Let us look at the big scheme of things: We are on the threshold of a new epoch, presumably, even a new evolutionary stage. The digitalization of the economy and society is in full swing and, with this, comes the need to critically re-assess all of society's institutions and to redesign many of them. The intelligence of machines ensures that they take on more and more activities and skills that were previously the privilege of people. At the same time, we should not forget that the intelligence of brutal power politicians and functionaries has done their societies no good. Present-day humanity is caught up in the tension between a bestial past in the wilderness and a capacity for critical reason integrated into one's respective culture or civilization.

Now that mankind is about to lose its presumably leading position in nature when it comes to intelligence vis-à-vis the machines, the question arises of what characterizes human beings apart from their intelligence. Does man also have, aside from intellectual capabilities, reason at his disposal, which allows him insights into the true essence of things? Man is considered to be the animal gifted with reason, but to be gifted alone does not ensure that man also acts reasonably. As we become more aware of what actually constitutes a human being, we might also become more conscious of the fact that it is reason that characterizes us. Perhaps, it is the intelligence of machines or artificial organisms that will help us to become human beings.

Prof. Dr. Holger Mey
Head of Advanced Concepts
AIRBUS Defence and Space in
Munich

Honorary Professor for Foreign
Policy, University of Cologne

Prof. Dr. Holger Mey is the Head of Advanced Concepts, Airbus Defence & Space, Munich, Germany. Before joining the former EADS Company in June 2004, he worked for 12 years as a self-employed security policy analyst and consultant in Bonn, Germany. During those years, he directed more than 30 studies for the German Ministry of Defense. Among many other roles, he served as President and CEO of the Institute for Strategic Analyses (ISA) in Bonn, Germany. He is an Honorary Professor at the University of Cologne, Germany. He began his professional career in 1986 as Research Associate at the Stiftung Wissenschaft und Politik (Foundation for Science and Politics) then at Ebenhausen, Germany. From 1990 to 1992, he served as a Security Policy Analyst in the Policy Planning Staff of the German Minister of Defense. From 1992 to 1994, already self-employed, he became Security Policy Advisor to the Chairman of the Defense Committee in the German Parliament. In 1992, he founded the ISA and became the founding Chairman and Director. He is Member of many international and national foreign and security policy associations, including the International Institute for Strategic Studies (IISS) in London and the Deutsche Gesellschaft für Auswärtige Politik (German Council on Foreign Relations, DGAP), Berlin. He published numerous articles in major security policy journals. He is Editor, Co-author, and Author of many books, including "Deutsche Sicherheitspolitik 2030", Frankfurt: Report Verlag, 2001 (English version: "German Security Policy in the 21st Century", New York/Oxford: Berghahn Books, 2004).

Preface

An Attempt to Drawing a Big Picture of Cyberspace

The climax of the digital transition's AI hype has been reached, while the potential of cyberspace—a virtual space consisting of zeros and ones in a global ecosystem of hardware equipment—continues to be intoxicating and enticing at the same time. Tech evangelists preach smart tech to increase demand power in this new market. At the same time, governments and non-state actors want more control, societies want progress and job security, but some are reticent about personal change. In addition, societal groups and philosophers have changed their minds about the unreflective use of digital tech in recent years and started to fight for basic rights in cyberspace. They warn against the effects of digital tech, uncontrolled use of personal data like tracking processes, tech dependence, and surveillance activities. These concerns are supported by think tanks, NGOs, multi-stakeholder groups, and in science, where resistance to the unreflective use of cyberspace and surveillance technologies is growing. However, a few experts have a conceptual idea of cyberspace as a realm of activity for humans. All operate on social media channels or in the environment of e-commerce corporations and use the Internet of Things (IoT). Hardly anyone can imagine what cyberspace looks like or how best to behave in cyberspace. Clumsy reactions to social media activities can be seen in politics but also in many other places. In order to enable a meaningful use of this new room for maneuver to the advantage of as many people as possible, it is key to visualize what has so far seemed rather chaotic. An attempt is made with this manual.

I. Challenges for Humankind and Industry in Cyberspace

The European World View is Subject to Change

Cyberspace is in transition from the Information Age into the Hybrid Age,[2] a new human arena for action provokes great enthusiasm: new business models, more effective processes, and cost efficiencies become possible for the first time. In the imagination of businessmen, a virgin space opens. Others have a growing fear of a dystopian future for society in this arena. However, Europe has realized that western societies are entering a new age—an age that could be more Asian than we can imagine (Khanna 2019). For Europeans, it will be an age determined by products and tech infrastructures that are not available yet in today's European industrial sector, but perhaps are in other markets, e.g. 5G technology from China. This recent situation in Europe is reminiscent of times in history when the western view of the world was already being questioned by scientific findings and technical revolutions. The world of today differs in almost all aspects from these worlds and from the world of Sigmund Freud. He was the first to name radical technical changes as narcissistic wounds of humanity. They were inflicted by science and the resulting technical progress in real life (Freud 1917). After Freud, others discovered more such wounds, and a few new ones seem to be unfolding now: *The first current threat to western societies is to lose their intellectual dominance in cyberspace.* Scientific theories and discoveries are also standing behind today's technological revolution. The first is the realization of artificial general intelligence (AGI). It is said to be superseding human intelligence, and this is clashing with the European view that the human is the most intelligent creature on earth. It questions the enlightened human's self-image as being in a position of intellectual supremacy. *The second threat to western societies is to lose the man-made system of order of a purely materialistic world.* The discovery challenging this existing self-image of societies in Europe is the quantum theory of Einstein: Some say a quantum computer may soon be ready for use. Not all questions about the famous physicist's theory have been answered. Einstein himself rejected aspects of his findings as basically impossible. His famous sentence *"He (God) does not play dice"*[3] referrs to these characteristics of physical systems. It concerns quantum entanglement. The phenomenon describes the ability of a physical particle to be detected simultaneously at different places on earth. Since the Enlightenment Europeans have been thinking from a purely materialistic world view. Quantum entanglement is opposed to this. It is hard to accept its existence. For most Europeans, there is only one world, the material world. Quantum mechanics and quantum effects fundamentally shake the enlightened Europeans' world view. If it was only theory, nobody would be affected in real life. Today, we see new products arriving based on these functionalities, e.g. quantum computer. Their applications pose an acute risk to

---

[2]The *Hybrid Age* is the transition period between the Information Age and the moment of technological singularity (when machines surpass human intelligence) that inventor Ray Kurzweil, author of The Singularity is Near, estimates we may reach by 2040, if not sooner.

[3]See Feldner, Part I, Chap. 1.

the European society's perception of the world in which it exists. From this mixture, flows the uncertainty of those people responsible for the wellbeing of their society and the need to develop a new way of thinking in politics and philosophy (see Maniam, Part III, Chap. 14).

The Fight for Dominant Designs in the Industry of the Future

The emerging hybrid age has more to offer than AGI and the quantum computer, new world views or dystopian future scenarios. Even companies like Tesla will not definitely dominate the German automotive industry, as some fear. Nevertheless, the company has opened an important global innovation battle for the next dominant design in the automotive sector. There are many other corporations and actors that are significant and have innovative ideas in the automotive sector, e.g., the American UBER or the Chinese Geely Automobile Holding. They are fighting for their places at the top of the global ranking and for tech supremacy. Only a few experts can predict today which technology will become the future leader in which industry sector. Those can even be non-digital as discussed in the fields of the mobility of the future and in the field of artificial general intelligence (AGI). Not all scientists believe that it is possible to reach human intelligence in the digital world of zeros and ones as those technologies offer a too low level of complexity compared to the requirements of the human brain. In the scientific field of neuromorphic computing, therefore, scientists are trying to build a brain with analog neuron circuits and neuromorphic hardware (Friedmann et al. 2016). *What is important is to devise a strategy about how to handle the uncertainty, how to monitor technological impacts and how to maintain western democratic societies in the future hybrid age.*

II. The Decade Ahead

Returning to what, in all likelihood will be key to the decade ahead; the pure process of digitization. In this process, information and communication technology (ICT) infrastructures are deployed on a technical level which enables higher connectivity of cyberspace through the IoT. ICT and the IoT create a more and more complex cyberspace. However, they can be applied in a way to prevent higher connectivity, e.g., through a ban of unmanned aircraft systems in the vicinity of Victoria Harbor, Hong Kong (see Thomson, Part III, Chap. 17). Digitization as a technical process includes all types of digital devices and technologies, such as AI, blockchain, GovTech, FinTech, smart homes, and e-mobility that enable digital business models, foster efficiencies, and bottom-up solutions to real-world problems. Digitization spans each topic that appears in the organizational processes of society. These are issues from the future of work, human rights, data ownership, access to the Internet, political opinion-making, state sovereignty, new technologies such as autonomous driving or drugs against cancer, automated warfare, drones or fighting robots. Their increasing ubiquity makes digital tech and devices the technical and communication instruments of the next decade which have a high potential to support human progress with significant effects or to harm humanity.

Any application for the benefit of humanity requires forward-looking intelligent regulation, a strong focus on a democratic cyber security and leadership that maximizes innovation, participation, and opportunity for humanity.

III. A Sense of Greater Complexity and Insecurity

Participation with ICT causes insecurity for humans and systems. As techniques they are not perfect. They arrive on the market and not all users know how to handle them. Societies worldwide experience what it means to deploy tech that comes with biases, bugs, that is overruled by better tech, that is misused on purpose or by ignorant people. This tech becomes the tools of espionage, data theft, propaganda, bot activities, cyberattacks, data protection violations, or election manipulation. Their use, creates a general sense of greater complexity and insecurity at all levels of society. However, it should not prevent people from using digital technologies as they promise comfort, greater efficiency, and human progress. *These connectivity effects on the economy, politics, and society (Part II of this manual) require management strategies and tailor-made policy measures that enable a secure cyberspace architecture (Part III of this manual) to maintain democracy and create trust (Part IV of this manual). A high quality learning culture, high quality education (Part V of this manual) together with sound digital and AI systems can advance democratic societies if they can be trusted.*

IV. Internet Governance

It is clear that none of the challenges have yet been addressed comprehensively by a national government or in society. *The reasons are:*

    I. *Experimental status. Digital technologies, which are intended to become disruptive products on the markets, are usually still in the experimental development phase when they are first applied.*

    II. *Cyber (in-)security. Digital products often find their way directly from the private research laboratories into application. This comes with challenges for trust building. Digital technologies and new business models are often not cyber-safe. Their code can be full of bugs, undisclosed value decisions and biases.*

    III. *Speed in development. Digital technology as a sector of the economy is evolving rapidly, leading to more uncertainty as digital products and business models are constantly evolving at a fast pace.*

    IV. *Undisclosed government structures. The governance of the networked world of the Internet of Things is not fully understood, as the cyberspace architecture itself is under construction and evolving rapidly. At the same time, the code is applying governance rules that change our current regulatory frameworks ("code is law").*

V. *Lack of common ideas.* *Societies have not yet decided what they define as safe in cyberspace, how they want to generate trust, what constitutes a human being and what values they are applying to the use of technology. In some fields strategies for policies, such as data policy and data ownership, are still missing.*

V. Formation of The Hybrid Age

Representatives of all blocks, from Luddites to digital utopians and tech evangelists, agree on the impact that digital tech will have on mankind. There are few inventions that will affect human life as much as machines that think and are interconnected across value chains and cultures, without respecting political, physical, and geographical boundaries. Only a few can foresee how intense their impact on society might be. *This point in time marks:*

I. *The start of the **formation of the Hybrid Age**.*

II. *The **once in a life-time chance** for countries to keep up or **leapfrog their competitors** that have been successful in the Third Industrial Revolution.*

III. *The right time for democracies and political parties to **rethink traditional habits** and societal organization forms of the past to keep up with other nations that move faster than westerners do. It is also the **once in a lifetime chance for non-state political systems** to take the leap forward.*

IV. *The best moment to think about the **crucial questions of being human in the hybrid age** since western societies are, at the moment, profoundly ill-prepared for future tech.*

VI. The Most Important Questions of Our Time

The Hybrid Age is the final part of the Fourth Industrial Revolution, in which man and machine combine in cyberspace with the help of smart tech. Some say that this point in time will be reached in 2040 if not earlier. After the biological and the cultural stages of human life, mankind is now entering its technological stage (Tegmark 2018). The late phase of this stage will involve—even if the question of when or how is highly controversial—the possibility that men will be outsmarted by AGI that eventually go beyond the human level (see I. Challenges for Humankind and Industry in Cyberspace). These future scenarios lead to questions for humanity that are so difficult that societies need to start discussing them now, so that they have the answers ready if AGI extends beyond the humal level. Many of the substantive quandaries raised by border-crossing information today could be resolved by a basic principle: conceiving of cyberspace as a distinct place (see Barber, Part II, Chap. 5). *Necessary intermediate steps toward an active understanding of cyberspace are:*

  I. *Separating from the world's physical and material framework, and developing a tangible picture of the hybrid world.*
 II. *Developing new framework conditions for management, politics, and law in the hybrid world.*
III. *Discovering how organizational systems and models of thought function in the traditional world and how the digital world can be linked with it.*
 IV. *Understanding how technological developments are represented over time and what this means for social adaptation.*
  V. *Understanding what it means for societies to merge into the hybrid age.*

The Idea Behind This Book—Drawing a Big Picture of the Cyberspace

This book is an attempt to provide a broad overview of humanity's new arena, the cyberspace. Chapters and articles in the book will provide an overview onto the most recent challenges posed on humans, on organizational structures, on the nation-state and on governing laws. The chapters collect and synthesize research, analysis and reports from practitioners, politicians, and scientists. They show where changes are about to take place, where the world as we know it must be redesigned and rethought. The four main topics of the manual *Redesigning Organizations— Concepts for the Connected Society* are connectivity (II), governance (III), trust (IV), and education (V). It brings together various cultural contexts since cyberspace is itself still an international field of inquiry. This is important although we are witnessing protectionism, trade wars, and the formation of a tripolar tech world (the US, China, Europe and their allies).

   In summary, the chapters of the book show the complexity of structures, viewpoints, and different speeds in cyberspace. Each author was asked to describe the developments related to cyberspace, taking into account fundamental aspects from a professional perspective. The authors were asked to identify challenges, to design and offer solutions. This required difficult decisions about what should or should not be included. Since the beginning of the construction of cyberspace the challenges of the experimental state of digital technologies have brought with them a moment of fundamental change. It therefore goes without saying that articles and analyses can only represent a snapshot of a time of fundamental change.

Berlin, Germany   Denise Feldner

Acknowledgements

I thank all contributors and authors who agreed to become part of this project and stayed with it until the end to make it an exciting journey for those who will read it. All authors are proven experts in their respective fields, highly respected, recognized thinkers and practitioners, founders, investors, current leaders, recognized artists, politicians—as well as young experts and future leaders.

I would like to thank Prof. Dr. Michael Huether, André Loesekrug-Pietri, Matthias Spielkamp, and Prof. Dr. Holger Mey for their valuable introductory words. They provide insights into their latest thoughts on the future of humanity, on how it is affected by machine biases and why we have to ask ourselves what actually constitutes a human being. They also talk about the great importance of creative business models and economists as well as how democracy can be preserved in the hybrid world. These are the key questions for a successful future for our democratic societies.

When I think of the book's creation, I would like to thank those who have contributed to polishing its texts, such as Bernadette Geyer, my American editor, and Tiberius Mitu, Bucharest, for his journalistic advice. I want to thank Prof. Dr. Thomas Osburg, Professor for Sustainable Marketing at Fresenius Business School in Munich, with whom I discussed the book's concept for the first time. My cordial thanks go to Barbara Bethke and Dr. Christian Rauscher, my book editors at Springer Nature in Heidelberg, for their unwavering support and endless patience. My gratitude goes to Alex Tuck and Ruth Barber, who supported the project with enthusiasm and professionalism. My very personal thanks go to Matthias M. Weber and Dr. Thomas Speckmann.

Denise Feldner

Contents

Part I   Chapter Zero

1   **Designing a Future Europe**   3
    Denise Feldner

Part II   In Search of Connectivity

2   **Cyber Commands—A Universal Solution to a Universal Cyber
    Security Challenge?**   27
    Siim Alatalu

3   **Digital Innovation Hubs and Their Position in the European,
    National and Regional Innovation Ecosystems**   45
    Maurits Butter, Govert Gijsbers, Arjen Goetheer
    and Kristina Karanikolova

4   **Transatlantic Privacies—Lessons from the NSA-Affair**   61
    Russell A. Miller

5   **Regulating the Internet—Necessary Evil or Squandered
    Opportunity?**   79
    Ruth Barber

6   **Data Ownership**   93
    Winfried Bullinger and Sophie Terker

7   **Redesigning Data Protection**   105
    Frederick Richter

8   **Erosion of Civil Rights in a Digital Society—Maintaining
    the Democratic Society**   113
    Jimmy Schulz

9   Transatlantic Cyber Forum—Cooperating on Borderless Cyber
     Security Challenges   123
     Sven Herpig and Julia Schuetze

10   Redesigning Corporate Responsibility How Digitalization
     Changes the Role Companies Need to Play for Positive
     Impacts on Society   137
     Nicolai Andersen

11   The Algorithmic Society   149
     Agnieszka M. Walorska

Part III   How Should One Respond to Political Risks?

12   Smart Cities and Smart Regions—The Future of Public
     Services—Solidarity and Economic Strength Through
     Smart Regions and Smart Cities   163
     Katherina Reiche

13   China's Authoritarian Internet and Digital Orientalism   177
     Maximilian Mayer

14   Digitalization and Public Policy—Conceptualizing
     a New Space   193
     Aaron Maniam

15   The Challenges of Digitalization for the (German) State   207
     Valentin Gauß

16   E-Estonia—"Europe's Silicon Valley" or a New "1984"?   215
     Florian Hartleb

17   Governance and Digital Transformation in Hong Kong   229
     Stephen Thomson

18   Blockchain—The Savior of Democracy?   239
     Alexander Braun

Part IV   In Tech We Trust?

19   Digital Propaganda—Russia or the Kid Next Door?   255
     Sarah Lohmann

20   Cyberspace as Military Domain: Monitoring Cyberweapons   267
     Thomas Reinhold

21   Trust in the Digital Age—The Case of the Chinese Social
     Credit System   279
     Peter Leibkuechler

22 Trust in the Functioning of Technology and Criminal Liability Based on the Example of Driving Automation 291
Nadine Zurkinden

23 Integral Corporate Cyber Security—Challenges and Chances for Showing the Way Towards Effective Cyber Governance 305
Hans-Wilhelm Duenn and Lukas W. Schaefer

24 Cyber Security… …by Design or by Counterplay?—Enabling and Accelerating Digital Transformation Through Managing Information Security Technology, Risk and Compliance at the Right Place 315
Thomas Hemker

Part V   The Future of Education and Work

25 Redesigning Traditional Education 329
Martina Francesca Ferracane

26 Mind the Gap!—Speed Matters in Education: Relating Technology to Human Capacities 345
Fré Ilgen

27 The (Post-)Digital University 357
Markus Deimann

28 Managing the Digital Change in Higher Education 365
Barbara Getto

29 The Shift from Stable Jobs to Dynamic Careers in Digital Manufacturing 373
Lina Huertas, Harald Egner and Martin Dury

Editor and Contributors

About the Editor

Denise Feldner is a Lawyer working in global investigations and an independent consultant for AI and digitization. She advises the Federal Foreign Office's RSF Hub, the German National Center for Applied Cybersecurity, investors, and founded *"The Post Digital Society"* a platform for fireside chats on the digital transformation. She is the strategic partnership lead at Crowdhelix Ltd., an Open Innovation platform founded at UC London. She specialized in innovation and science policy as Founding Manager at Germany's association of elite research universities and acted as Deputy Representative at the *Global Council of Research-Intensive University Networks*. Prior to that, she was Head of the president's staff at *Heidelberg University*. As Cooperation Manager at InnovationLab, a start-up founded by DAX companies and elite universities, she worked at the interface between science and the top management of DAX corporations. She headed the investigation team at a Berlin-based law boutique specialized in infrastructure for almost a decade. She is Invited Member of the Weizmann Institute's Young European Network, was named Young Leader of the British German Forum, the BDI, and the German Russian Young Leaders Conference. She is Fellow of the Helmholtz Academy and Member of the European Mission of the Hebrew University. Denise sits in the Advisory Board of the *Greater Bay Area AI and Society Institute* in Hong Kong and is Member of Atlantik-Brücke. She was born in East Germany, holds a master's degree in business law for tech corporations, and passed her management studies with honors at Prof. Malik Management, St. Gallen.

Contributors

Siim Alatalu EU CyberNet at the Estonian Information System Authority, Tallinn, Estonia

Nicolai Andersen EMEA Lead Innovation and Deloitte Garage at Deloitte, Hamburg, Germany

Ruth Barber London, Great Britain;
Berlin, Germany

Alexander Braun Creative Construction Heroes, Berlin, Germany

Winfried Bullinger CMS Hasche Sigle, Berlin, Germany

Maurits Butter The Netherlands Organization for Applied Scientific Research (TNO), The Hague, The Netherlands

Markus Deimann Teaching, FernUniversität in Hagen, Hagen, Germany

Hans-Wilhelm Duenn German Cyber Security Council e.V., Berlin, Germany

Martin Dury The Manufacturing Technology Centre (MTC), Coventry, UK

Harald Egner The Manufacturing Technology Centre (MTC), Coventry, UK

Denise Feldner Berlin, Germany

Martina Francesca Ferracane Oral3D, Milan, Italy

Valentin Gauß Ministry of Transport of the Federal State of Baden-Württemberg, Stuttgart, Germany

Barbara Getto Learning Lab, University Duisburg-Essen, Essen, Germany

Govert Gijsbers The Netherlands Organization for Applied Scientific Research (TNO), The Hague, The Netherlands

Arjen Goetheer The Netherlands Organization for Applied Scientific Research (TNO), The Hague, The Netherlands

Florian Hartleb Tallinn, Estonia

Thomas Hemker Security Strategy, EMEA CTO Office Symantec, Hamburg, Germany

Sven Herpig International Cyber Security Policy, Stiftung Neue Verantwortung (SNV), Berlin, Germany

Lina Huertas Technology Strategy for Digital Manufacturing, The Manufacturing Technology Centre (MTC), Coventry, UK

Fré Ilgen Berlin, Germany

Kristina Karanikolova The Netherlands Organization for Applied Scientific Research (TNO), The Hague, The Netherlands

Peter Leibkuechler Sino-German Institute for Legal Studies, Nanjing University, Nanjing, China

Sarah Lohmann American Institute for Contemporary German Studies, Johns Hopkins University, Washington, DC, USA

Aaron Maniam Blavatnik School of Government, University of Oxford, Oxford, Great Britain;
The Birthday Collective, Singapore, Singapore

Maximilian Mayer International Studies, University of Nottingham Ningbo China, Ningbo, China

Russell A. Miller Washington and Lee University School of Law, Lexington, USA

Katherina Reiche German Association of Local Public Utilities (Verband Kommunaler Unternehmen e.V.), Berlin, Germany

Thomas Reinhold Institute for Peace Research and Security Policy (IFSH), Hamburg, Germany

Frederick Richter German Foundation for Data Protection, Leipzig, Germany

Lukas W. Schaefer Research and Public Relations, German Cyber Security Council e.V., Berlin, Germany

Julia Schuetze Transatlantic Cyber Forum, Stiftung Neue Verantwortung (SNV), Berlin, Germany

Jimmy Schulz Committee on the Digital Agenda of the National Parliament of the Federal Republic of Germany, Berlin, Germany

Sophie Terker CMS Hasche Sigle, Berlin, Germany

Stephen Thomson School of Law, City University of Hong Kong, Hong Kong SAR, China;
Ombudsman, City of Hong Kong, Hong Kong SAR, China

Agnieszka M. Walorska Creative Construction Heroes, Berlin, Germany

Nadine Zurkinden Substantive and Procedural Criminal Law, University of Zurich, Faculty of Law, Zurich, Switzerland

Chapter Zero

Chapter 1
Designing a Future Europe

Denise Feldner

> *Without Security There is No Liberty.*
> (Wilhelm von Humboldt, Statesman, Academic, and Founder
> Alma mater Berolinensis)

A Strategic Mindset for The Single Market

We're facing another tech race between two superpowers, right? Not exactly. The race for tech and military supremacy is more global and complex than it seems at first glance because the modern internet tends to be agnostic to geopolitical borders and at the same time we see more nationalistic reactions to those developments. The race is an indication of the high potential and power that new technologies give investors, the industry, non-state actors but also states using cyber-statecraft to expel others from their economic and political positions in the global world order of post-war global world order. However, this does not mean that the world is shaking. The world is always volatile, uncertain due to natural development. There has never been a time when the world stood still. It means an integration of machines with digital technologies and devices. The project *Industry 4.0* is a German invention deeply rooted in Germany's industry management culture for the process of man-made digitization (see Reinhold, Part IV, Chap. 20). It is very much focused on technical risks, technical issues and investments that had to be made. In the last five years, the discussion started to shift and increasingly in the last two or three years. It stopped being only about the technical issues but more about the people (see Hemker, Part IV, Chap. 24). At the dawn of the 2020's, European societies are still amid the transitional phase to Industry 4.0. The transition also means the transition of politics, societies

D. Feldner (✉)
Berlin, Germany
e-mail: dfeldner@gmail.com

© Springer Nature Switzerland AG 2020
D. Feldner (ed.), *Redesigning Organizations*,
https://doi.org/10.1007/978-3-030-27957-8_1

and companies (see Huether, Forewords) into the hybrid age. The immense potential of this transition lies in increasing efficiency, reducing overall costs and starting new businesses. It is also helping with finding solutions to urgent crisis such as for health problems like cancer, fighting pollution and saving the planet's natural habitat, reducing resource consumptions or simply easing peoples work and health life. This transition for humans includes the opportunity to take a further step towards the advancement of the human condition (see Mey, Forewords) in the sense of the ideals of the Enlightenment (Pinker 2018). Those opportunities that result from technical disruptions make this moment the best point in time: to put humans in transition strategies and political concepts for the European single market first (see Huertas et al., Part V, Chap. 29). Strategic successes in business, politics, and society depend on the people. *It is the digital mindset, the culture of management and learning that has the power to take a society to the top, not global insecurity.*

New Alliances and Anti Alliances

This momentum of uncertainty is probably more difficult to change than a society's mental attitude to dealing with uncertainty and opportunities associated with technology (Acemoglu and Robinson 2017). An uncertain future can be met with the greatest possible freedom of thought and flexibility as things happen, but what societies can really control is the way that they respond to those things, including thinking beyond borders of the material world and considering the ambitions of third countries. This is an increasingly connected world. While the competition for tech leadership between the superpowers, United States (4.3% of the world's population) and China (18.5% of the world's population) has reached a new level, third countries that have close ties with them are forced to reposition themselves in the shifting world order. The Chinese government has set itself the goal of becoming the world's tech leader by 2030. The country is turning into something new, where billions of Renminbi are being continuously invested pushing the country to unimagined levels of growth. It is creating new markets and with them market defining rules. To achieve its goal, the Chinese government regards digital and AI tech as strategic for geopolitical, economic and security reasons (see Mayer, Part III, Chap. 13). These ambitions of the People's Republic of China come with a strong initiative, which can cause in western societies a future that will be more Asianized (Khanna 2019). This forecast reflects a crucial policy issue: *If western countries do not react in a focused manner and find their strategies and answers as democracies, they will be trapped between the two superpowers. This is their great challenge, which is to form new alliances and redesign old ones, such as the transatlantic partnership.*

Brain Drain

Turning heads toward Europe's 10% of the world's population today, there is the "old continent" suffering from a brain drain, especially in the significant field of AI tech, which in the past has lost tech talents mainly to the "Big Five" companies (Amazon, Facebook, Google, Microsoft, Apple) and research labs in the US. These "Big Five" seem to be the most powerful tech companies in the world, if not including their Chinese competitors' market power. Recent trends show that the brain drain is going in the direction where money is flowing. Where opportunities are promising, and tech ideas thrive. This includes opportunities arising from lower regulation, lower tech and lower ethical standards that allow unquestioning technological progress. This makes China interesting to potential participants and that doesn't mean one has to go to China to participate in its development. Companies like Huawei employ 80,000+ R&D employees (45% of the total workforce). To further increase their share of research personnel, they are building research labs in Europe, e.g. in Italy and France (Tao and Zhifeng 2018). This is a crucial point: *While European scientists are working for Chinese companies, Europe itself needs to adapt its policies and thought patterns to attract talented professionals, start-ups and venture capitalists to its own institutions.*

Europe's Death Valley and Global Tech Investments

Europe is still waiting for governmental initiatives to attract venture capital (VC) for high risk projects and to change its investment culture. Europe's *Death Valley* is the innovation gap in between a top-notch research sector, high potentials fleeing the continent, and patents that are transferred to products in foreign labs and companies. This gap can be closed with private money and initiatives, such as from the Public Group International Ltd. in London or J.E.D.I. in Paris. Public promotes European startups for creating GovTech for governments and administrations. J.E.D.I. tries to foster disruptive innovations and moonshot tech made in Europe. This kind of innovation, emerging from outside of public institutions, is urgently needed to support public institutions in maintaining democratic structures through tech made in Europe (see Butter et al., Part II, Chap. 3, Loesekrug-Pietri, Forewords). Europe as a political entity still lacks a major political push for a more innovative future. The present EU research framework, Horizon 2020, intends to foster cutting-edge research together with the industry, but that wasn't the success the European market requires. It is not enough to create major breakthroughs in science, these findings must make their way into products and real world workplaces. This must be a major goal for the €100 billion Horizon Europe, the next research and innovation framework. And although also Germany, a major research hub in Europe, feeds a unique innovation cluster that operates worldwide. It must achieve much more to fulfill the needs of a 21st century society. *A German contribution to a new European innovation strategy could be to integrate these institutions (e.g. German Academic Exchange Service, German Research Foundation, German Centers for Research and Innovation) into a strategically designed innovation value chain and European tech network together with private initiatives.*

Foreign players are investing in Europe. A prominent example is Softbank. The Japanese investment powerhouse is unrivaled in the global VC industry and reflects a recent trend. Many promising tech companies (unicorns) are owned by international investors, funds and banks. In times of zero interest rates the tech sector became a promising place to grow money stocks. As a result the sector is in a phase of over-funding. Europe's second disadvantage in this VC market is its "old money trap". The owners of old money, that can be found mainly in Europe, usually do not operate at the high-risk tech level. They are interested in receiving their money stocks. Much money flows into bricks and steel as many do not see an opportunity in technical dis-ruptions. In other regions of the world, there is more "new money"—such as in China or San Francisco where people got rich in the last decades, which often means less money per owner than in families who own "old money". These new investors have set themselves the goal of increasing money holdings and are therefore more willing to invest in high-risk projects that promise higher and riskier profits. They support the growth of the tech sector in their regions and are feeding the overfunding.

The US tech pioneers—belonging to the group of investors with the new money, the "Big Five" in the USA, haven't invested their cash in new tech for a while (2016: Apple $215.7b, Microsoft $102.6b, Google $73.1b) because there was no opportu-nity to invest in new tech in Silicon Valley. They have even been questioned about their status as tech companies. They should have become *de facto banks* because their money didn't pour into innovations but into balance sheet reserves. There was an effect on the US tech market, fresh (new) money and startups shifted to other innova-tive regions and projects in the US, pushing new developments. The money and ideas have moved from the Silicon Valley region to the Boston Area, Pittsburgh, Washing-ton D.C., Metro Area and Southern California's emerging technology ecosystems around Los Angeles, San Diego, Santa Barbara and San Bernadino. Later Google started creating new tech sectors, invested in e-cars, virtual reality glasses, and espe-cially in quantum computing, Amazon invested massive amounts of money in cloud tech and in the meantime became leader in R&D funding. Facebook invested in algorithms of any type that triggers social media correspondence and in cryptocur-rency. The "Big Five" companies helped to accelerate the recent global tech and AI hype in Europe. They built up research labs and attracted researchers. Those older companies are now losing their status as agile tech companies. A company that does not innovate ages and declines. In the 21st century that decline will presumably be as fast as the tech market accelerates. From this scenario derives the opinion of a few innovation experts that in a short time the platform economy will lose its market position and instead new business models based on new technologies will emerge (Charles-Edouard Bouée, former CEO of Roland Berger at the Axel Springer AI Summit 2018). Meanwhile Huawei a Chinese hightech company (Tao et al. 2017), is a controversial company. It is accused of being an active part of a Chinese borderless surveillance system. Its strategy of adapting to future challenges made the company as successful as it is today, e.g. in the 5G technology sector. In 2018, Huawei employed almost 45% of its staff (80,000+) in the R&D sector. The company maintains 40 research labs around the globe, some in Europe. It follows a strategy to invest 10% of the income in research on a regular basis (information collected 2019 in a personal interview at Huawei Headquarters in Shenzhen/China). As a response to these research and investment cultures, *Europe needs a new industry strategy and more investment in R&D correlated with a strong research management attitude in companies, more high-tech-focused investment strategies and high-risk projects in the tech sector. This must be supported by a future-oriented, start-up focused and open minded research community.*

Long-Term Public Investment in Infrastructures

In the US, the five new tech regions are old acquaintances when it comes to the research funding initiatives of the Advanced Research Projects Agency (DARPA) of the Department of Defense. As early as in 1969, the first DARPA-financed ARPANET hubs were in Utah, in the UC Los Angeles and in UC Santa Barbara. In 1970, Stanford University joined the group that would later become the parents of the Internet. In the same year, Harvard University and the Massachusetts Institute for Technology (MIT) on the Pacific Coast joined. In 1971, Carnegie Mellon University in Pittsburgh became a member of the club (DARPA 1978). Today, these regions are the regions where the "Big Five" were founded, where fresh money is invested in new tech, where new companies are founded, and business models emerge. This result translates into two fundamental public policy challenges for Europe: ***Long-term strategic public investment in (a) talented professionals, innovation and (b) infrastructures is required!*** ARPANET, the mother of the Internet, began growing fifty years ago, at the heart of what has become the world's most important technology center in recent decades, Silicon Valley. It can be deduced, that long-term strategic innovation management and public infrastructure investments are key to technological progress and economic prosperity in a society (see Reiche, Part III, Chap. 12). This is the objective in today's Europe, but not yet achieved (see Loesekrug-Pietri, Forwords, see Butter et al., Part II, Chap. 3). *This is what the Chinese government aims to accelerate through a variety of funding initiatives and strategies for research institutions, universities, industry or through the Belt and Road Initiative (OBOR) and the Asian Infrastructure Investment Bank: long-term supremacy in global infrastructures and in governance of digital technology.*

Shifting Tech Hubs

In terms of basic and applied research in AI—the current top-notch technology for the accleration of the digital transformation, the EU research hub is still said to be the world's most diverse and collaborative. The backbone of it is the European Research Area, which itself is underlined by political instruments that foster European innovation, e.g. the Horizon 2020, the EU Framework Programme for Research and Innovation. Europe's thriving research hubs are the UK, Switzerland, France, and Germany. Those regions are leaders in inventions, research and AI patent applications. Estonia is long known as the "European Silicon Valley" but offers a government and not privately driven model of digital transformation (see Hartleb, Part III, Chap. 16). A few others are catching up, such as Cluj-Napoca in Romania, that obtained the title "Silicon Valley of Eastern Europe" (De Man 2018). Europe's AI start-ups are clustered in London (see Exhibit. 1.1). Since the beginning of Brexit, new data shows a slight shift to Berlin, a city that has obtained international attention as start-up hub. Digital nomads from everywhere in Europe and Israel meet there. Even the US giant Amazon, that has heavily invested profits in new tech and talented professionals, moved its AI and Alexa research team to Berlin's vibrant city center. Paris has attracted more investment for start-ups in 2019 than Berlin.

Germany, the world's 4th biggest economy and Europe's industrial powerhouse, will most likely become the leader in manufacturing, robotics, and quantum computing. Almost half of the patents for autonomous driving have been successfully filed by German automotive companies. In 2018, the country was named the second most innovative country in the world (BCG 2019). *Despite this early lead in*

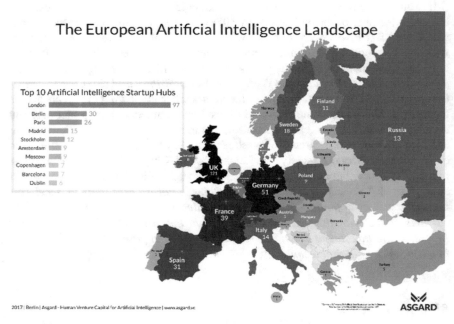

Exhibit. 1.1 The European artificial intelligence landscape, © Asgard 2017 (http://asgard.vc/wp-content/uploads/2017/07/European-Artificial-Intelligence-Hubs-and-Landscape-2017-by-Asgard-VC.png)

science and with patents, the country has not invested in products and in other fields that thrive and appears to be losing in sectors that comprise the new business world. This is a phenomenon we have seen in the West before, its leaders failed miserably on defining how important new technologies will become in the future, e.g. 5G. This scenario needs to be prevented. Europe must apply a long-term strategy focusing on European innovation infrastructures.

The Connectivity of a Leapfrogging Country

China's research has overtaken its key competitor, the United States, and is fast closing in on Europe's lead. China, the West's major tech rival is—still operating in relative isolation from the wider research and business community and is ready to become the global leader in tech and internet governance by 2030. China is seizing all possible opportunities to develop technology. Its large population (18.5% of the world's population) gives the country a unique advantage through sheer size. The Chinese government and industry leaders know how important strategic education and a high-quality learning culture are for the future prosperity of their society. Chinese politicians are considering the importance of the innovation value chain. This also means that Chinese companies and citizens by law must focus on tech investments when investing globally. An example was Ping An Insurance's investment in the Berlin-based FinTech company builder FinLeap in 2018. Even the internal administration of Chinese people (Social Scoring System) has been declared a testing ground for new tech for social stability reasons (see Leibkuechler, Part IV, Chap. 21, see Mayer, Part III, Chap. 13). The Chinese consider this governance systems as geared to stability and prosperity, not to control.

The Chinese digital market itself is protected on many levels; protected by the Golden Shield in cyberspace, and legally regulated for international corporations, especially for American companies. Chinese tech companies can have an impact at the same level as US companies through the third pillar of the One Belt and One Road Initiative (OBOR), the digital silk road. With the OBOR China promotes infrastructural projects, e.g., submerged cables in the Australian region, ICT infrastructures in neighboring countries such as Afghanistan and a smart city project in Duisburg, Germany (see Reiche, Part III, Chap. 12). The Chinese population is digitally connected worldwide. Companies expand globally through increasingly voracious customers of goods and digital services using Chinese devices and the Chinese platform economy. They are spreading tech developed in environments with less regulation and data inaccessible in Europe. *This marks the best point in time for European politicians to put tech, ethics and security on the political agenda with the highest priority in order to find a democratic response to changes in their economic partnership with China.*

Other Tech Hot Spots and Their Policy Language

Countries and governments in the EU that already treat digital tech as strategic
are Denmark, Sweden, Italy, Estonia, and Finland (see Hartleb for Estonia, Part III,
Chap. 16). Asia's other tech hot spots are Singapore, Taiwan, and Malaysia. Australia
and Canada are working hard on their tech leadership, and in the Middle East the
United Arab Emirates (UAE) invests heavily in digital tech. The UAE was the first
country with a minister for AI. Israel has top-notch knowledge in security tech and a
vibrant start-up scene not only in Tel Aviv but also in Jerusalem. These countries have
strategic strengths such as in finance (Singapore), in their learning culture (Taiwan)
or in technical and security terms (Israel). Their advantage is that politicians see
digital tech and AI as a relevant technology for their society's future. They focus on
it. That is what makes them leaders in digital tech and internet governance as they
allocate money and political measures in those fields. They are about to establish a
new policy language that fits with the world of the cyberspace, e.g., in Singapore,
and Estonia (see Maniam, Part III, Chap. 14). *For Europe there is still a lot of work
to do in the policy language field.*

New Colleagues and Other Surprises

As the status and speed of digitization varies from country to country and the western
world still thinks in categories such as the developed and the underdeveloped world,
leaps are not always reflected in official data and rankings. What is also not reflected
are developments that contradict expectations (to understand why expectations can be
wrong, see Rosling et al. 2018; Pinker 2018). What might be surprising is the status of
digitization in Hong Kong, despite being one of the Asian Tigers and a global leader
in FinTech, the city does not appear to be very open to transforming all sectors (see
Thomson, Part III, Chap. 17). This also applies to research in digital and AI tech on
Hong Kong Island. The AI and robotics scene in the Greater Bay Area Region of
the Pearl River Delta spanning northbound from Hong Kong Island, Shenzhen to
Guangzhou is much more vibrant. A development that should be closely observed
by Europeans. The pace of digitization is taking surprising paths. Basic and applied
research in the West became a job of private institutions that spent more money than
public institutions. As a result, artificial general intelligence (AGI) and quantum
computers may not be developed in a public but in a private lab.

In a "developing" country without any stable structural and economic legacies, it
may be easy to adopt recent tech developments and new technologies (see Exhibit.
1.2). This gives less developed countries the opportunity to leapfrog. The internet
in Africa is, for example, limited by a lower penetration rate than compared to the
rest of the world. The low penetration rate is attributed to weak connectivity, lack of
infrastructure and innovation. There is ample evidence that this is changing, and this

State and Pace	of Digitization
Countries Without strong organizational legacies with high innovation capacities *Leapfrogging opportunity high*	Countries With strong organizational legacies and without high innovation capacities *Leapfrogging opportunity low*
Countries striving for institutions *Leapfrogging opportunity depends on infrastructure and innovation capacities of the country/society*	Countries with developing institutions *Leapfrogging opportunity depends on infrastructure and innovation capacities of the country/society*

Exhibit. 1.2 State and pace of digitization. *Own Source* Feldner (2018)

gap is closing fast as more resources are being deployed in Africa to expand Africa's digital economy. This causes great opportunities for the continent.

The fabric of growing African connectivity are submerged cables that are deployed around Africa. The continent has been connected for a long time only by satellite or by fiber. Until 2009, the capacity was very low, and the costs have been high for connectivity. This changed in the last decade. Exhibit. 1.3 shows the state of connectivity within Africa and the Mediterranean in 2018. As a result of the increase in connectivity, an increase in mobile payments in Kenya and its neighboring Uganda took place. In 2007, with the launch of Vodafone's *M-Pesa,* a platform for mobile phone-based money transfer, financing, and micro-financing services, triggered an increase in FinTech. Today M-Pesa is the largest system in Kenya and Tanzania. It expanded to South Africa, Afghanistan and India, but also to Albania and Romania in the European Union.

Since 2009 the internet subscription in Africa grew from 4.5 million in 2000 to about 700 million in 2017. The average age of the African population is 19.5 years and will further push developments. Another driver for digital integration and connectivity is the e-healthcare sector. Actors are focusing on rural areas working on cutting-edge solutions to deliver healthcare in regions that are difficult to approach due to security, infrastructure, military or political reasons.

Key drivers are more than 300 African tech hubs that gather talented professionals. South Africa tops this list with 54 tech hubs followed by Egypt with 28. Tech nodes are in Kenya, Morocco, Nigeria (with 2.6% of the world's population and the world's fastest growing megacity), Ghana, Tunisia, and Uganda (Kamanthe 2018). These societies' future developments are affected positively by digital education products and Massive Open Online Courses delivered by educational companies such as Coursera, edX, Khan University, and Stanford OpenEdx that provide access to US and European education that wasn't available to Africans some years ago (Feldner 2018).

Another example of a rapidly developing digital market power is the Indian market with 17.9% of the world's population. The country will soon outperform China in terms of penetration rates and the sheer number of users that become future customers. What makes it easier for India to keep pace with developments is that it has neither

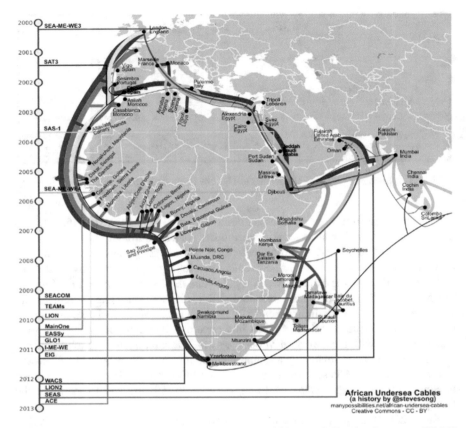

Exhibit. 1.3 African undersea cables (2018) Manypossibilities.net, © Creative Commons-CC-BY, https://manypossibilities.net/african-undersea-cables/

strong infrastructures nor binding institutional and organizational legacies in fields that are affected by digital technologies. Where everything is changing at the same time, this can assist a society to adopt new tech. Those countries aren't usually home to Luddites. It was an advantage for China as the country started at an agricultural level and jumped almost directly into the tech phase. It is likely that India will follow this path. *As a result, Europe will soon see another fast-growing economic power in Asia.*

From a Myth to Future Scenarios

These developments will provide access to basic legal and structural instruments in regions those populations today live without access to the (western-led) rule of

Digital Utopians in the second half of the 20th century	Digital Reality of the early 21st century
Cyber Law = Law of the Horse	Cyber Law necessary for different reasons and in areas such as criminal law, international law, IP law, patent law, data ownership(?)... labor law, certification systems

Exhibit. 1.4 Digital utopias and reality graph. *Own Source* Feldner (2018)

law (mainly in Africa and Asia). Digital infrastructures and devices can fundamentally change cyberspace as an arena of human activity, but also the real-world needs for legal infrastructures necessary to support and sort the complexity of a modern world economy (Hadfield 2018). The new buying power in Asia (60% of the world's population) will pose challenges for policymakers in their own country but also in the West to adapt traditional legal and economic systems. In contrast, the early years of cyberspace were characterized by the attitude of American scientists (having been involved in the creation of the ARPANET and the internet technologies) and Silicon Valley entrepreneurs who mainly spoke about powerful myths, stories, digital utopias and the great potential that digital technologies have for human progress, for democratization of communication as well as for their business ideas. Accordingly, it was common for early thinkers to assume that the internet and related tech did not need new regulation, as the internet was supposed to be a freely accessible and secure space for each user (see Schulz, Part II, Chap. 8). They declared any new regulation redundant (see Barber, Part II, Chap. 5). Even the former president of one of the major start-up cradles in the US, Stanford University, Professor Dr. Gerhard Casper, once said that a new law of cyberspace would be as effective as a law of the horse (Easterbrook 1996) see Exhibit. 1.4.

The Transfer of Scientific Culture into Business

This way of thinking in science concluded that there was no need for a realignment of the legal systems, for intellectual property rights, cyber security or data ownership (see current discussions Mayer-Schönberger and Ramge 2017). That led to the current state of laws regarding the data economy (see Bullinger and Terker, Part II, Chap. 6), privacy rights (see Miller, Part II, Chap. 4), and data protection (see Richter, Part II, Chap. 7). It also led to the current state of the economy and job sectors (see Huertas et al., Part V, Chap. 29). This kind of thinking changed in the last four to five years, in the connected world of the "Internet of Things (IoT)". The new world is where geography is an increasingly irrelevant factor and where job profiles appeared such as influencers, content moderators, where people are working for a "like" and a follow-up request on Twitter, and where the net incomes of most millennials are declining. Real world management has adopted the cyberspace culture: The first ARPANET project team at the University of California (including Vint Cerf, Steve Crocker) created the basic decision-making backbone for today's global cyberspace. It is operated by Internet Corporation for Assigned Names and Numbers an independent public private partnership in California. By that the net was built upon the decision making logic of scientists and has not changed in its role as the Internet's basic governance system. From there, the US scientific culture and mindset has evolved in the start-up company culture of Silicon Valley. Since most companies have been spin-offs from research and science, scientists went along with them. Thinking and cooperation cultures from science could expand (Isaacson 2014). This working culture in the sciences includes flat hierarchies. Corporations worldwide are now working in an agile management system and thus in loose project teams with flat hierarchies. The middle management has largely been abolished and stable jobs are becoming dynamic (see Huertas at al., Part V, Chap. 29). It presents serious policy challenges for labor markets that have been focused on hierarchical structures.

The Early Visionary and the Legal Systems

Lawrence "Larry" Lessig, an Internet pioneer and Harvard professor of law, wrote twenty years ago in a visionary article (Lessig 2000) that one day there will be regulation of the internet. What he called the "law of the code". He foresaw for the time when the internet would become an increasingly complex technology, and a military operation space (see Alatalu, Part II, Chap. 2). This where we have now arrived. Digital tech and AI give human life the potential to flourish like never before, or to destroy itself. If they are not managed with (a) ethical rules (see Spielkamp, Forewords; see Walorska, Part II, Chap. 11; see Mey, Forewords) or are not (b) be aligned with business needs (see Huether, Forewords) and (c) do not cover as much ground as they should, they do not match human goals. Digital tools are constrained by their coding. They are not free from biases and governance systems as products have been in the industrial revolutions. Digital tech is man-made and is interwoven with biases, thoughts, bugs and impressions of the humans that made it. Decision makers focus on one tech tool that is trustworthy and may assists with regaining trust in politicial decisions and institutions: the blockchain technology (see Braun, Part III, Chap. 18). Blockchain's basic technology beside the distributed ledger tech is smart contracts. The person that described them at first place was Nick Szabo (1994). He wrote an article that explained smart contracts as digital transaction protocol that execute terms. If those rules will govern our daily lives in the trustworthy blockchain world, the biggest question for the acceptance of this tech in democratic systems is *"Who will be the mastermind of these conditions?"*. Graduated lawyers are drafting in the analog world, controlling and executing contracts. In the digital world blockchain companies are here to circumvent notaries, lawyers and bankers, to dramatically reduce the overall costs and risks associated with those businesses. This reflects a fundamental challenge for the existing legal system, for trust in institutions and for governance systems (De Filippi and Wright 2018) and the basic demands of legal infrastructures will change fundamentally due to blockchain technologies (Baecker 2018) and the law of the code. This overall change began when the wall came down in the year 1989 and the World Wide Web. That was the moment when nearly forty percent of the world population accessed the western world's economic system (Hadfield 2017) and the emerging cyberspace. Now this young world faces an ever-growing IoT and is about to take another almost four billion people living in China, Russia, India, the Middle East or Africa (possibly another 60% of today's world's population) into the same legal and economic system and in cyberspace as an arena of human activity. The expectations of legal infrastructures are changing. Politics must not only deal with code law and smart contracts, but also with transnational legal challenges such as cyber security, tax systems and the cross-border flow of data, products and surveillance activities as well as with the people and their connectivity. *After several years of failed negotiations on an international cyber security treaty in the United Nations, one thing is clear so far: each country and nation will see this differently from its cultural background and will focus on its own political goals and power-building opportunities. The world is likely to be divided into several tech regions, driven by different policy guidelines. Europe must address its technological status and forge strategic tech and policy alliances.*

The Media's Love of Growing Insecurity

As a response to those developments, e.g. a growing perception and feeling of becoming more and more disconnected in Europe, the media coverage of tech applications such as AI, cyber security, election meddling (see Lohmann, Part IV, Chap. 19) or hacking activities has taken on a new form. The media is concerend about effects of the insecurity in cyberspace, from increasing connectivity and an overall growing insecurity. These developments were described as a negative advancement that went all the way to representations of a dystopian future for humanity. There were reasons for this new way of reporting. These reasons were incidents such as the NSA affair and Snowden disclosure in 2013, which caused a furore in Germany but not in the US (see Miller, Part I, Chap. 4). It was a spying activity from the US, backed by the "Five Eyes" alliance's intelligence services, against German politicians (see Herpig et al., Part II, Chap. 9).

Another incident that caused anger and insecurity among people, but this time across the transatlantic community, was the Cambridge Analytica scandal in 2015–2016. Scientists from the Psychometrics Centre at Cambridge University had signed a research cooperation agreement with Facebook. The data made accessible by the project were later misused for economic and political purposes. The project ended as an unethical and criminal project of a third party. Cambridge Analytica Ltd., a data analytics company, was alledged to have been misusing private data of millions of Facebook users for economic reasons and for the political benefit of the then US presidential candidate Trump.

Election meddling, deep fakes, hate speech, disinformation through social media platforms such as Facebook and YouTube instigated by Iran, North Korea, Russia and China or related groups, causes feelings of insecurity on a regular basis (see Lohmann, Part IV, Chap. 19). The only answer to those challenges is a political vision for internet governance including a culture of high-quality education that answers new questions that haven't existed before Cyberspace grew at such a fast pace. One key expression that comes to mind when experts speak about those developments and events in cyberspace is "Media Literacy". Although it is contested what "Media Literacy" might mean, democratic societies need an excellent education in media literacy to be able to maintain their democracy in the digital world. *The people must be empowered to understand and enabled to handle information flows in cyberspace, to build up capacities and cutting-edge knowledge in this field.*

A New Research Power in Cyberspace

Events such as the Cambridge Analytica scandal reflect the major struggles companies like Facebook are facing. In the years of its existence Facebook has had a clear focus on business development and on the development of new products. But after 14 years of existence, the revolutionary start-up became a sluggish company that struggles every day to reinvent and reposition itself. It was at this stage that Facebook and peers began a new phase in which they invested heavily in research and people to enable inventions and new products (see A.III, B.III). They also signed research cooperation contracts with well reputed research and technology organizations as they promised an increase of the company's reputation. The brightest minds from science moved to Facebook and developed research ideas for the company. At Professor John Martini's research lab in Santa Barbara Google maintains one of the most successful quantum computer teams in the world (Dönges 2014). *As stated earlier, important research and knowledge about critical technology is in private hands and no longer in the hands of governments or public research institutions. This needs to be rebalanced if democratic governments in Europe want to maintain political authority.*

Failures in Innovation Management

Although the platform companies triggered innovations and the current AI hype, Facebook, for example, has failed to adapt its economic advances in an ever growing IoT to its growing corporate responsibilities, to the expectations of society, its customers and to traditional compliance rules. After serious data scandals, Facebook and other American platform companies must answer questions about user and data privacy. The working conditions at Facebook, e.g. for content moderators, have already been criticized. There is a lack of powerful lobbyists for the rights of workers as they existed in the last industrial revolution (see Schulz, Part II, Chap. 7). A very smart solution to these problems that can help those companies to manage their innovation processes has come up from the business consulting world. The concept of "Corporate Digital Responsibility (CDR)" was invented and has emerged from traditional corporate social responsibility (see Andersen, Part II, Chap. 10).

Functional Sovereignty

Law scholars have investigated the US platform economy from a law and economics point of view. Some say that it has achieved "functional sovereignty". This term derives from a fundamental principle of international law (Schmitt 2017), which was developed in 1648 in the Treaty of Westphalia: state sovereignty. It describes the power of the political authority to act on behalf of citizens in relation to a specific (national) territory. Airbnb, for example, has developed market power to shape urban planning in smaller cities in the United States. Amazon has received offers from democratically elected mayors to assume political power when the company moves its headquarters to these cities. These companies gather more customers than countries like Estonia or Sweden have citizens (2.41 billion Facebook users worldwide in the second quarter of 2019). The result is a de-facto political influence that was reserved for elected representatives and represents a similar sovereignty in the hands of private companies. This power reflects a fundamental shift in the political power systems of western democracies.

The developments are turning companies into competitors for political authorities, which were traditionally responsible for organizing life in constituencies (see Reiche, Part III, Chap. 12). Amazon announced in early 2019 that it had encountered fierce opposition from local authorities and politicians from Long Island City in Queens. That kind of resistance was reason enough for Amazon not to establish the No. 2 headquarters in New York City. The incident is interesting for several reasons: New York is considered one of the world's most important centers for new technologies, and the city has always been the concrete jungle where dreams come true. The city's authorities have begun to take a critical look towards the tech world with its platform economy, as they did in Europe.

The Race to Dominate the Internet

Over the past decade we have seen a loss of political power in the elected authorities and their bodies (see Braun, Part III, Chap. 18) as a result of what happened within the platform economy. It was only recently that the authorities have begun to assume their role as regulators in cyberspace (see Barber, Part II, Chap. 5; Zurkinden, Part IV, Chap. 22). In the digital domain, confidence in political institutions is correspondingly at a record low—which is also reflected in the rise of influencers that can easily grab the prerogative of interpretation over political and social issues. This applies not only to Western authorities, but also to countries like China. The Chinese government reacted earlier and harsher to this deficit by introducing the questionable "Social Credit System" (see Leibkuechler, Part IV, Chap. 21). Other governments have hastily developed internet governance guidelines, created new ministries and developed AI guidelines to prove their ability to tackle these issues. After years of hesitation and what lawyers called the "wait and see approach", they are trying "sandboxing" models and are beginning to regulate, e.g. the EU General Data Protection Regulation (GDPR).

Today China seems to be the tech policy leader since its implementation of the digital strategies as part of the OBOR initiative, its Cybersecurity Law and with its holistic AI strategy from summer 2017. This poses a great challenge for its allies but especially for its traditional counterparts like the US. The US barely has an AI strategy. The Federal Government of Germany did not announce its AI strategy until November 2018. The Minister of State for Digitization in the Federal Chancellery, who took office in March 2018, acts as the "face" for internet governance in Germany. Responsibility for shaping policy lies with the head of the Federal Chancellery and his internet governance team. This fragmented authority is an expression of the fragmentation of authority in internet governance throughout the Federal Republic of Germany (Gauss, Part III, Chap. 15; Duenn and Schaefer, Part IV, Chap. 23).

In times of fundamental change, it is important to adapt adequate politics, policies, policy language and organizations to secure future prosperity and democracy. Regulatory and governance approaches in the West are currently primarily seen as risk minimization and security maximization. It led us to fail miserably on defining new critical technologies and this makes society ill-equipped for future tech and its effects on society and democracy. Europe needs to overcome existing prejudices and uncreative thought patterns that are not supportive. It needs a regulatory and governance mindset that will maximize knowledge for as many as possible; inventions and innovations to create European start-ups and opportunity. *This is a crucial point and the best point in time to redesign organizations that back democratic societies. For the connected and the post-digital society, suitable concepts are needed now, not traditional concepts.*

Spreading Tech Knowledge and Finding Opportunity in Disruption

Apart from political forces, AI tech is said to be the strongest driver of digital transformation, so it is important to understand the techniques and to develop them with

Actors in Cyberspace

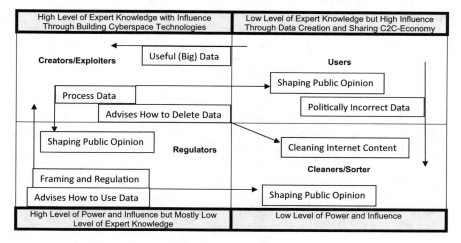

Exhibit. 1.5 Actors in cyberspace. *Own Source* Feldner (2018)

a vision of a prosperous future in a democracy. But the dramatic success in the AI research area "machine learning" have led in recent years to a flood of AI applications and devices that are not reflected in the European education system. An emerging status of ignorance and unawareness in society towards this development will lead to even more resistance to tech and create a long-lasting sense of insecurity or fear. This development has the potential to divide European societies (see Hartleb for Estonia, Part III, Chap. 16). If a society is digitally divided, there are a few who have access to knowledge and power and those who cannot catch up (see Exhibit 1.5). This level of insecurity and division has the potential to further weaken democracy and political decision makers.

There is a way to circumvent this scenario: the dissemination of knowledge through high-quality education for as many as possible. A prosperous future demands a high-quality learning culture. It is important to provide education for self-confident citizens. Not only IT literates have to find their way in Cyberspace, but the others as well.

The Remote Control of Our Life

A reason for this widespread feeling of uncertainty is that the 3rd generation mechanical AI learning models, applied in today's AI world, are obscure, unintuitive and even difficult for experts to understand. The fields of machine and deep learning are particularly cryptic for humans. While deep learning techniques are incredibly good at finding patterns in data, their complexity can make it impossible for people to understand how they come to their conclusions. The more that people depend on digital and AI applications, and the more tech affects human life, the more important it becomes that the technology is robust against the design process (see Walorska, Part II, Chap. 11) and becomes understandable. AI systems must be monitored on a permanent basis. For their safe use it is important that code is aligned to human goals and values. *It comes with the following questions:*

I.   *Should the research and development phase be subject to legal regulation, or to the monitoring of legal and certification professionals?*
II. *What does implementing cyber security and ethics in the coding process mean and will this serve human goals?*
III. *Who will be the one that decides what goals should be implemented in AI?*
IV. *Will the next "Big Five" global tech corporations be located in China, Singapore, Pakistan, India, in a post-Brexit UK or be in the hands of Russian investors?*

Enabling Democratic Goals

Since the 1990s, experts have been working to understand how technologies behind deep learning—neural networks—make decisions. The idea behind this work is that its results could facilitate the handling of the technology and minimize risks. A concept for this, the concept of explainable AI (XAI/ex AI), was introduced in 2004. Ex AI are AI systems whose actions are easy for people to understand and thus enjoy trust. In 2015, AI security research became mainstream in the US. **Until this year, critical discussions about AI risks were often misunderstood as the goal of hindering the AI process** (Tegmark 2018)! The year 2015 thus marks a very important date for the future of humanity, considering what AI can trigger for society, be it good or bad. In August 2016, the U.S. Defense Advanced Research Project Agency (DARPA) initiated the Explainable Artificial Intelligence Program (XAI) for military reasons. DARPA's political intent behind this research is to prevent agents and military personnel from blindly trusting an algorithm when using autonomous instruments like weapons or robots. The result of the program will be a toolkit library that can be used for the development of future systems. Upon completion of the program, these toolkits would also be available for further enhancement and migration to defense or business use in the US. In Europe, the High-Level Expert Group on Artificial Intelligence (HLEG on AI) presented its draft of an ethical guideline for trustworthy AI at the end of 2018. For the HLEG on AI, trustworthy AI means that general and abstract principles arising from human rights are underpinned by technical specifications in the design process for an algorithm. It is not yet clear what *trustworthy AI* means, since the meaning of the word has not been clarified (see Zurkinden, Part IV, Chap. 22). *A first step in Europe would be to clarify what is meant by trustworthy and how Europeans can put their ethics into practice by translating ethical principles into a code of conduct for technologies and companies. This leaves EU politicians with one question: How to handle actors, that are not subject to EU regulation?*

Don't Let People Be Outsmarted

The consequences of malfunctioning AI and digital tech for law enforcement, medicine, politics, critical infrastructures, in the media and for peace on earth can be serious, especially in democracies. The main risk might not be conscious intervention by people, but the lack of education in the population and a continuing loss of talents in Europe. This situation will lead us to new security concepts focusing on the individual and devices making up the IoT. It is foreseeable, that we will move to holistic security concepts and insurance policies for institutions and their individuals. As long as we do not see knowledge spreading in society, technology and tech companies are responsible for ensuring stability. An important step to counteract this technology-driven development is not regulation but education for all citizens (Getto, Part V, Chap. 28; Deimann, Part V, Chap. 27). This training must integrate findings from ex-AI- and cyber security research, media literacy and practical experiences of the physical world (see Ilgen, Part V, Chap. 26). It is necessary to teach this to children (see Ferracane, Part V, Chap. 25). If not, it will become more difficult for the younger generations to take on a critical position in the discussion about being human in the hybrid age. It will also become difficult for them to build a successful career as they are facing new actors and colleagues from around the globe (Feldner 2018). *Today is the best moment for liberal democracies to fight attacks by authoritarian regimes, non-state actors, and from their own ignorance, challenging developments with self-esteem, open-mindedness, an excellent education culture, and strong cyber defense skills.*

References

Articles

De Man, B. (2018). Cluj-Napoca: The Silicon Valley of Eastern Europe. *Smart Cities Council Europe*, 2018-08-14. https://eu.smartcitiescouncil.com/article/cluj-napoca-silicon-valley-eastern-europe.

Dönges, J. (2014). Quanten-KI: Google will eigenen Quantencomputer bauen, 08.09.2014, Spektrum. https://www.spektrum.de/news/google-will-eigenen-quantencomputer-bauen/1307826.

Easterbrook, F.-H. (1996). Cyberspace and the Law of the Horse. University of Chicago Law School, Chicago Unbound. https://chicagounbound.uchicago.edu/cgi/viewcontent.cgi?referer=&httpsredir=1&article=2147&context=journal_articles.

Feldner, D. (2018). How Is digitalization changing the future of work? June 16, 2018. https://www.theglobalist.com/artificial-intelligence-digitalization-education-facebook/.

Kamanthe, A. (2018). Tech Hubs in Africa, May 2, 2018, Atlas Corps. https://atlascorps.org/tech-hubs-africa/.

Lessig, L. (2000). Code is law. *Harvard Magazine*. https://harvardmagazine.com/2000/01/code-is-law-html.

Mayer-Schönberger, V., & Ramge, T. (2017). Die Daten gehören allen!, Hannoversche Zeitung. http://www.haz.de/Sonntag/Gastkommentar/Die-Daten-gehoeren-allen.

Szabo, N., (1994). Smart Contracts. http://www.fon.hum.uva.nl/rob/Courses/InformationInSpeech/CDROM/Literature/LOTwinterschool2006/szabo.best.vwh.net/smart.contracts.html.

Books

Acemoglu, D., & Robinson, J. A. (2017). Warum Nationen scheitern, Die Ursprünge von Macht, Wohlstand und Armut, Fischer.

Baecker, D. (2018). 4.0 oder Die Lücke die der Rechner lässt, Merve Verlag, pp. 185.

Bucher, T. (2018). *If...then, algorithmic power and politics*, Oxford University Press.

De Filippi, P., & Wright, A. (2018). *Blockchain and the law*. Harvard University Press.

Hadfield, G. K. (2017). *Rules for the flat world*. Oxford University Press.

Isaacson, W. (2014). *The innovators*. Bertelsmann.

Khanna, P. (2019). *The future is Asian*. Simon & Schuster.

Kurz, C., & Rieger, F. (2018). Cyberwar Die Gefahr aus dem Netz, Random House.

Pinker, S. (2018). *Enlightenment now, the case for reason, science, humanism, and progress*. Penguin Random House.

Rosling, H., Rosling, O., & Rosling, A. (2018). *Factfulness: Ten reasons we're wrong about the world- and why things are better than you think*. New York: Flatiron Books.

Schmitt, M. N. (2017). *Tallinn manual 2.0 on the international law applicable to cyber operations*, Second Edition, Cambridge University Press.

Tao, T., & Zhifeng, Y. (Ed.). (2018). Explorers huawei stories, Jichao, Zhang, 17. Riding the Microwave to Success, LID Publishing Limited.

Tao, T., De Cremer, D., & Chunbo, W. (2017). *Huawei leadership, culture, and connectivity*. Sage Publications.

Tegmark, M. (2018). Life 3.0, Penguin Random House.

Weber, S. (2001). Ist die Quantentheorie des Bewusstseins Humbug?, Telepolis 06. Februar 2001. https://www.heise.de/tp/features/Ist-die-Quantentheorie-des-Bewusstseins-Humbug-3443603.html.

Reports

DARPA, Heart, F., McKenzie, A., McQuillan, J., Walden, D. (1978). Completion Report 1978.

Studies

Fezer, K. H. (2018). Repräsentatives Dateneigentum, Studie der Konrad Adenauer Stiftung. https://www.kas.de/einzeltitel/-/content/repraesentatives-dateneigentum.

NACD Director's Handbook on Cyber-Risk Oversight, Directors Handbook Series, January 12, 2017. https://www.nacdonline.org/insights/publications.cfm?ItemNumber=10687.

Denise Feldner is an IT lawyer and innovation manager working in global investigation proceedings and is an independent consultant for digitization and cyber security. She founded *"The Post Digital Society"* a Berlin-based fireside chat platform on the digital transformation. Denise specialized in innovation and science politics as founding manager at the association of German elite research universities and acted as deputy representative at the *Global Council of Research-Intensive University Networks*. Prior to that, she was head of the president's team at *Heidelberg University*. As in-house counsel at InnovationLab GmbH, a start-up founded by DAX companies and elite universities, she worked at the interface of research in materials science and the top management of DAX corporations. Before that, she headed the investigation team at a Berlin-based law boutique specializing in infrastructures. Denise is an invited member of the Israeli Weizmann Institute's Young European Network, was named Young Leader of Wilton Park's British German Forum, the Federation of German Industries, and the German Russian Young Leaders Conference. She is fellow of the Helmholtz Leadership Academy and member of the European Mission to the Hebrew University Jerusalem. She is member of the Advisory Board of the *Greater Bay Area AI & Society Institute*, Hong Kong and Atlantik-Brücke. Denise was born in East Germany, studied law in Belgium, Germany, Greece and Hungary and qualified at the Kammergericht. She received a master's degree in business law for tech corporations from a Technical University in 2008 and passed her management studies with honors at Malik Management, St. Gallen.

In Search of Connectivity

Chapter 2
Cyber Commands—A Universal Solution to a Universal Cyber Security Challenge?

Siim Alatalu

Introduction

Digitization has thoroughly changed the ways of life around the world. Taking place back-to-back with what has become known as Globalization 3.0, it has empowered both individuals, enterprises and states with unprecedented leverage over and access to data that today in turn has unprecedented influence over how societies work. As characterized by Friedman, since the days of the first publicly available Internet browser Netscape, the connectivity between both people as well as applications has only grown, fueled by phenomena of the global economy like outsourcing, offshoring, open-sourcing, insourcing, supply-chaining, informing and lastly, new forms of collaboration like wireless access and Voice-over-Internet-Protocols—allowing one to do "any one of them, from anywhere, with any device" (Friedman 2005).

Effectively a part and parcel of digitization and the global connectivity offered by the Internet, cyber security too has established itself as a essential feature of life: "Internet, together with the information communications technology (ICT) that underpins it, is a critical national resource for governments, a vital part of national infrastructures, and a key driver of socio-economic growth and development" (Klimburg 2012, p. 2). At the same time, one should also consider the potential risks from cyberspace that are inherently global in nature. The World Economic Forum's (WEF) Global Risks Report 2018 featured cyber security risks as second only to environmental risks and noted *inter alia* that a "growing trend is the use of cyberattacks to target critical infrastructure and strategic industrial sectors, raising fears that, in a worst-case scenario, attackers could trigger a breakdown in the systems that keep societies functioning" (World Economic Forum 2018, p. 6).

Cyber security considerations have played both an encouraging, as well as discouraging role, leading some nations to a more open and proactive stance towards

S. Alatalu (✉)
EU CyberNet at the Estonian Information System Authority, Tallinn, Estonia
e-mail: siim.alatalu@ria.ee

© Springer Nature Switzerland AG 2020
D. Feldner (ed.), *Redesigning Organizations*,
https://doi.org/10.1007/978-3-030-27957-8_2

digitization, online provision of their public services, open access to and the freedom of the Internet, etc. Others, on the other hand, have opted for a strict control of the Internet on their territories, which paradoxically would mean very limited benefits from the global network for their citizens and societies but, rhetorically, had better defense against cyber threats. While the People's Republic of China might be the country that is best known for limiting Internet traffic, some authors also highlight that there are others and the tradecraft is particularly advanced in the Commonwealth of Independent States (CIS) and Russia (Deibert and Rohozinski 2010b, pp. 15–16).

Either way, a structured tackling of cyber threats (from detecting to thwarting and response) seem to have become an integral part of government functions, in order to safeguard their societies against threats from within cyberspace. Philosophically, this quest is a logical development of affairs. As suggested by Heidegger already in 1977, our "will to mastery [of technology] becomes all the more urgent the more technology threatens to slip from human control" (Heidegger 1977, p. 2).

Cyberspace can be defined as something "more than the Internet, including not only hardware, software and information systems, but also people and social inter-action within these networks" (Klimburg 2012). In the same article, Klimburg also refers to formal definitions of cyberspace by the International Telecommunications Union (ITU) ("systems and services connected either directly to or indirectly to the Internet, telecommunications and computer networks".), the International Orga-nization for Standardization (ISO) ("the complex environment resulting from the interaction of people, software and services on the Internet by means of technology devices and networks connected to it, which does not exist in any physical form".) and the United Kingdom ("all forms of networked, digital activities; this includes the content of and actions conducted through digital networks".) (Klimburg 2012, p. 8). Another noteworthy definition of cyberspace by The Tallinn Manual 2.0 defines it as the "environment formed by physical and non-physical components to store, modify, and exchange data using computer networks" (Schmitt 2017, p. 564).

With cyber security thus becoming relevant to national security, one will need to ask about the role of the military in safeguarding the nation in cyberspace. Will the nowadays global trend of establishing national military cyber commands become a theater of competition (if not conflict) for controlling cyberspace or actually a solution to the growing problem of threats from it?

This article seeks to analyze how to interpret the universal problem—the growing threats from cyberspace to the societies—from a national security perspective, what response options the states have and whether the practice of establishing military cyber commands will help nations cope with the emergent concerns.

National Cyber Security as a Matrix of Matrices

According to Collier, there are conceptually at least two ways that the risks of disrup-tive or destructive cyberattacks can further increase: first, the theoretically limitless amount of software in digital systems (and thereby the availability of a limitless amount of attack sequences and vectors), and second, the diffusion of cyberspace that supports the empowerment of new threat actors (Collier 2016, p. 2). The challenge that this poses to the states is based on the fact that no state today can fully control cyberspace (neither by having sovereignty over it nor avoiding imposing it), nor is it possible to have a civil-military differentiation of the nature of threats in cyberspace. Cyberspace remains by definition a "human-made domain in constant flux based on the ingenuity and participation of users themselves" (Deibert and Rohozinski 2010a, p. 16).

While it is widely acknowledged that, for example, NATO's role is the defense of NATO's networks (and it is subject to discussion whether these might also include the national networks of individual Allies), it is a matter of fact that the global infrastructure on which everything works has different and often private ownership. As acknowledged in 2011 by the then-commander of the U.S. Cyber Command, the vast majority of [the U.S.] military's information was being transferred on privately owned (i.e., commercial) infrastructure (Alexander 2011, p. 10). As a result, "cyberspace has emergent properties […] that elude state control" (Deibert and Rohozinski 2010a, p. 16).

To deal with risks to national security from cyberspace, it has become essential to have a national strategy for it. In the European Union member states, having a national cyber security strategy became mandatory when the Network and Information Security (NIS) Directive came into force in 2016, requiring the introduction of a "national strategy on the security of network and information systems" and the appointment of a National Competent Authority for cyber issues ("to monitor the application of the NIS Directive") (European Commission 2018). Interestingly, one of the reasons that make the definition of a national strategy complicated is that according to Bartholomees, "the word has a military heritage, and classic theory considered it a purely wartime military activity—how generals employed their forces to win wars" (Bartholomees 2006, p. 79).

Choices as to what a national cyber security strategy actually is rest with the nations. There is a paradox, however, that "national policies […] address an environment based on both infrastructure and functioning logic that has no regard for national boundaries" (Klimburg 2012, p. X). To make it even more complex, it is also no longer an option to not deal with the challenge due to its pervasiveness. As exemplified by Deibert and Rohozinski, "cyberspace is the domain through which electronic clearances take place, irrigation systems are controlled, hospitals and educational systems interconnect, and governments and private industries of all types function. It can be found aboard nuclear submarines and bicycles, watches and air traffic control systems—it is ubiquitous and pervasive, and is most acutely felt when it is absent" (Deibert and Rohozinski 2010a, pp. 18–19). Therefore, having in place a plan for safeguarding the nation in cyberspace is a necessity, rather than an option.

Further to the definitions offered above, cyberspace can be characterized as matrix of matrices where in the broadest sense technology meets with national security, strategy, operations, law and international relations. On one hand, cyberspace is a combination of technological layers that consists of the *physical network layer* (hardware, software and support infrastructure), the *logical layer* (the relationship between nodes in the physical layer, such as programs) and the *cyber-persona layer* (digital representation of an individual or entity in cyberspace, enabled by users) (Department of the Army HQ 2017, pp. 1–13). Manipulating one of them can lead to malfunctioning of the others. On the other hand, cyberspace reflects the nature of the society with its different, yet interoperable institutional and sectoral networks working together and across borders. Another way to structure challenges within and from cyberspace is to define them on the international, national and sub-national (e.g., defense sector) levels. For a European country, those challenges can be characterized as follows.

The International Level

(a) European Union—while the EU's original function is an economic union of sovereign member states, defense and cyber security issues have gradually gained ground in its deliberations. A landmark example of the proliferation of these topics on the pan-European level has been the recent Estonian Presidency of the Council of the EU in the 2nd half of 2017, which was notably rich in policy deliverables in those two areas. Concerning cyber security, the so-called Cybersecurity Package proposed by the EU Commission in September 2017 included a new cyber security strategy for the Union, as well as a proposal for an EU wide legislation to include an enhanced mandate for the European Network and Information Security Agency (ENISA) and for a EU-wide framework for cyber security certification. Earlier in 2017, the Council of the EU had also agreed on the so-called Cyber Diplomacy Toolbox (Härmä and Minarik 2017), which the Estonian presidency steered forward with an agreement on implementation guidelines. In 2018, the EU Network and Information Security (NIS) Directive has been transposed into Member States' legislation with the aim of enhancing cyber resilience Union-wide.

(b) NATO—compared to the EU, NATO has a longer record of accomplishment in cyber defense issues having reached its first Summit-level decisions for cyber defense already in 2002. In 2014, NATO declared cyber defense to be part of its collective defense framework, and in 2016, it went on to declare cyberspace a domain of operations. Compared to the EU, NATO has been less known for setting any cyber security requirements on its member states, bar the Cyber Pledge of 2016, where the heads of state and government *inter alia* committed to developing "the fullest range of capabilities to defend our national infrastructures and networks" (NATO 2016).

(c) Cyberspace in itself has become a source of globally spreading threats such as e.g. malware ["software that may be stored and executed in other software, firmware or hardware that is designed to adversely affect the performance of a computer system" (Schmitt 2017, p. 566)] that could be assessed to also threaten national security, as well as specific sectors. Cross-border and cross-domain by nature, it lacks physical borders and that allows malicious actors to operate freely. To make them accountable, a complex attribution process is needed to overcome challenges such as how to distinguish acts of cybercrime from, for instance, cyber incidents that are part of a hybrid campaign with strategic objectives and designed to target another country's national security. Such considerations have made cyber a mainstream topic in various international organisations. New international cyber defence cooperation initiatives are emerging, bringing together likeminded cyber defense communities, characterized by Deibert and Rohozinski as, "a slow but steady internationalization of critical infrastructure protection initiatives, policy coordination, and legislation" (Deibert and Rohozinski 2010a, p. 20).

The National Level

(a) Embracing digitization and ICT can be considered a vital economic interest of any state. According to the WEF's Global Information Technology Report 2013, and according to the study by Booz & Company for the report, a 10% increase in a country's digitization rate could have gained +0.75% of GDP per capita and −1.02% in unemployment (World Economic Forum 2013, p. vii). As digitization contributes significantly to a nation's GDP, there are tangible economic benefits to (a) investing into further digitalization and (b) safeguarding the existing digital ecosystem. Embracing digitalization as a national policy will likely require a shift in both a political mindset as well as in the allocation of national resources.

(b) Reliance on digital platforms such as e-government can be a matter of a critical rational choice for smaller nations, for example, to save on human resources and avoid a heavy bureaucracy. As an example, Estonia is believed to save one week of working time per citizen per year, by having made it mandatory for all to have a government-issued ID card, and thus, it enabled the transformation to digitally signed documents nation-wide as a result. For bigger nations, undertaking such task might be more challenging for various reasons such as a traditionally strong bureaucracy, the complexity of an existing national legislation and the seemingly smaller criticality of the same choice that faced the smaller countries.

(c) As discussed above, embracing digitalization and ICT has also led to cyber defense becoming a requirement in terms of state functions. On the one hand, the EU's NIS Directive will make it mandatory for EU member states to *inter alia* develop dedicated national structures (Competent National Authorities, Computer Security Incident Response Teams) (Lord 2017). On the other, along with digitization states would need to provide education and training to their societies to enable the introduction and use of new ICT products and services. Compared to earlier times, this challenge is one with a moving target, driven by the ever-increasing speed of developments in technology.

The Subnational/Sectoral Level

As an example of a subnational sector, for the purposes of this article the focus is on the defence sector. For the Western militaries, not dealing with cyber security has not been an option for decades already. The development of high-tech platforms for warfare on land, in the air and at sea (even not to mention space) and upgrading of existing weapon systems by way of digitalization can be said to have had a pioneering role for digitalization in general. Nowadays, defence experts say that attention to cyber details such as the integrity of information in the systems, which have been tampered with, needs to be enhanced to ask questions like "What if battle plans from a military commander were hacked without anyone's knowledge and troops were ordered to fire against friendly forces?" (Emmott 2017).

In today's wars, with the exception of a few hand-held weapons (like the conventional rifle), almost all weaponry is to some degree digital, from night-vision devices to ballistic missiles. In addition to conventional forces, cyber operations ["the employment of cyber capabilities to achieve objectives in or through cyberspace" (Schmitt 2017, p. 564)] as a stand-alone function in militaries around the world are also a reality. From a generic perspective, dealing with cyber issues is thus critical for the defense sector, at least in (but not limited to) the following areas:

1. Weapon systems
2. Command and Control (C2) systems
3. Different types of operational activities like intelligence collection, military operations, exercises etc.

The challenge of the matrix of matrices that nations need to cope with is visualized in Fig. 2.1. (The cloud motive hints that the listed items are only a few of the many cyber security issues that relate either to the *levels* or the *layers*, or both.)

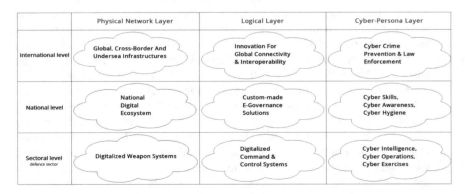

Fig. 2.1 Cyber security matrix of matrices. *Own source*, Alatalu 2018

Cyberwar and National Security

According to Buzan, national security means the "pursuit of freedom from threat and the ability of states and societies to maintain their independent identity and their functional integrity against forces of change, which they see as hostile" (Buzan 1991, p. 432). Traditionally, against an hostile "force of change"—threats of or the use of force or an armed attack by another state—the national armed/defence forces have been the designated response capability. When considering cyber threats as threats to national security, however, what might be the appropriate tool or capability a nation can use to deal with them? What will be the appropriate role of the military? When will isolated cyberattacks appear to be organised and appropriate to be called "cyber war"?

These questions once again became relevant in 2017 when two global outbreaks of malware attacks, known as WannaCry and NotPetya, caused damages on an unprecedented economic scale—and in unprecedented areas. WannaCry and NotPetya, later attributed respectively to North Korea and Russia, showed that a nation state can release a cyber attack on others that spreads globally. Even more importantly, they may well be willing to target areas that are critical to the society such as hospitals, IT systems, global logistics hubs or the banking system—but to date not necessarily always identified as potential targets.

2017 also marked the 25th anniversary of coining of the term "cyberwar" in an essay by Arquilla and Ronfeldt (1993, pp. 141–165). Among other things, referring to the success of the Mongol hordes in conquering an empire based on their dominance of battlefield information, Arquilla and Ronfeldt argued, "information is becoming a strategic resource that may prove as valuable and influential in the post-industrial era as capital and labour have been in the industrial age". They claimed that "the information revolution will cause shifts both in how societies may come into conflict, and how their armed forces may wage war" (Arquilla and Ronfeldt 2017, pp. 26–27).

Today, there are multiple definitions of "cyberwar" available. To highlight three of them:

- According to Arquilla and Ronfeldt, "Cyberwar refers to conducting, and preparing to conduct, military operations according to information-related principles. It means disrupting if not destroying the information and communications systems […] on which an adversary relies in order to 'know' itself: who it is, where it is, what it can do when, why it is fighting, which threats to counter first, etc. It means trying to know all about an adversary while keeping it from knowing much about oneself. It means turning the 'balance of information and knowledge' in one's favour, especially if the balance of forces is not. It means using knowledge so that less capital and labour may have to be expended" (Arquilla and Ronfeldt 2017, p. 30).
- Another, more straightforward, definition originally by Richard A. Clarke defines "cyber warfare" as "actions by a nation-state to penetrate another nation's computers or networks for the purposes of causing damage or disruption". As characterized by Paganini, this definition benefits from including two critical aspects, the

"nation-state commitment and the intent of the offensive that could be conducted with the purpose of causing damage or to spy on an enemy's networks" (Paganini 2012).

- A third definition offered by A.P. Liff defines "cyberwar" as a "state of conflict between two or more political actors characterized by the deliberate hostile and cost-inducing use of computer network attacks against an adversary's critical civilian or military infrastructure with coercive intent in order to extract political concessions, as a brute force measure against military or civilian networks in order to reduce the adversary's ability to defend itself or retaliate in kind or with conventional force, or against civilian and/or military targets in order to frame another actor for strategic purposes" (Liff 2012, p. 408).

The discussion on cyberattacks possibly targeting national critical infrastructure is not new. Already in his 2009 forecast, Clarke stated that "offensive cyberattacks have the potential to reach out from cyberspace into the physical dimension, causing giant electrical generators to shred themselves, trains to derail, high-tension power-transmission lines to burn, gas pipelines to explode, aircraft to crash, weapons to malfunction, funds to disappear and enemy units to walk into ambushes" (Clarke 2009, lk 31). In another reflection, Arquilla added that "striking, via cyberspace or by other means, at key information based control functions would have crippling effects on armies, fleets, and air forces—in a manner similar to the way in which classic blitzkrieg offensives early in World War II knocked out enemy communications hubs, creating much chaos" (Arquilla, Twenty Years of Cyberwar 2013, p. 80).

Wannacry and NotPetya were not the first cyberattacks with wide-reaching-impact. Ever since the 2007 cyberattacks against Estonia, there has been a trend of escalation. According to Brantley, cyberattacks that include destruction, denial or degradation of military or civilian communications platforms, have abound—such as the Mirai malware botnet attack of 2016, a form of which could be directed against, for example, IP addresses of the U.S. Federal Aviation Administration, emergency services or anything that is Internet-connected and has an IP address (Brantly 2018, p. 42).

On the other hand, developing offensive cyber capabilities has long since become mainstream among countries, regardless of whether they publicly declare doing it or not. An assessment voiced by NATO in 2010 said that "countries, such as China, have assembled within the people's liberation army up to 100,000 operators who work full time, as a full-time 9-to-5 job, in probing the systems of other countries and [...] about 100[1] countries in the world [...] are actively developing offensive, not defensive, but offensive—cyber capabilities" (Shea 2010). Among authoritarian states, Russia, China, Iran and North Korea are seen as the biggest cyber threats by the West (Pernik 2018).

It should then be fair to say that although not always considered elements of national military power, offensive cyber capabilities have become tools in nations' toolboxes for projecting national power. Building on the experiences of WannaCry

---

[1]Different sources offer even bigger numbers. For instance, the website of the Infosec Institute claims that at least 140 countries were developing cyber weapons.

and NotPetya (but also other targeted attacks on, for example, powergrids), it is essential that the complexity of the situation and the potential effects on national security are recognised as they are. This then assumes that the defence forces and their cyber capabilities might well have a call of duty when the national cyberspace becomes a target. In determining their exact role nevertheless, it is advisable that, as offered by Wallace, "the appropriate level of military involvement needs to be informed by both the dangers to national security and the alternatives available (including the risk of misemploying the military)" (Wallace 2013).

From Deterrence to Attribution

The realization of a threat and the need to actively engage one's defensive capabilities implies that one's deterrence—the "ex-ante dissuasion of adversaries through the threat of ex-post costs in response to potential adversary action" (Brantly 2018, p. 33) has failed. In the context of cyberspace, however, the credibility of one's deterrence posture is more nuanced than with conventional arms. While it is possible to assume that the cost-to-benefit ratio between an offensive weapon and a denial tool is 1000 to 1—as it was in the case of the U.S. approach to counter Russian S-300 and S-400 medium range missiles (Brantly 2018, pp. 35–36)—in cyberspace, the calculation is more difficult to make, as increasing investments are required to maintain the achieved status quo. The likelihood of the efficiency of a static denial tool decreases over time as it inevitably will be challenged by technological advances.

Also important is the question of the cost of failure of deterrence and of the ex-post response. Over time, deterrence by denial, becomes deterrence by threat of punishment, as the probable related crisis evolves with the adversary pushing the limits. Referring to Brantley, the matter-of-principle problem with resorting to punishment forms (other than cyberspace counter-operations)—such as the expulsion of Russian intelligence operatives and imposing sanctions in return for the Russian meddling in 2016 U.S. elections—however, would probably not undo the damage or offer satisfactory compensation. "The U.S. response imposed insignificant costs in comparison to the utility achieved" by Russia (Brantly 2018, p. 43).

When deterrence fails, the question of attribution rises. This is particularly important in the West, where *pacta sunt servanda*—agreements are to be kept, and any possible response needs to base on legally solid grounds. To quote a current NATO Deputy Assistant Secretary General's speech from 2010, "it often takes months, a very punctilious investigation for a whole range of countries and jurisdictions before you can be certain that you know where that attack came from. [...] cyber is, if you like, the conflict between 'the strong and the complacent' in Western societies and 'the weak, but motivated'" (Shea 2010). To characterize the complexity of attribution, Rid and Buchanan offer three general features of an attribution process:

– It is almost always too large and too complex for any single person to handle;

– It is likely to require a division of labor, with specialties and sub-specialties throughout;
– It proceeds incrementally on different levels:—immediate technical collection of evidence, follow-up investigations and analysis, and then legal proceedings and making a case against competing evidence in front of a decision authority (Rid and Buchanan 2015, p. 5).

It would thus appear evident that relying on deterrence alone in the context of cyberspace is not a comprehensive solution. Rather, a mix of denial tools (intrusion detection and prevention systems) in combination with a threat of punishment tools (expulsions, travel bans, sanctions, confiscation of assets, etc.) could establish the general ex-ante and ex-post limits of a cyber-based conflict. As for what remains between these two extremes, it is somewhat more unclear and remains a critical question of what solution states could and should have in-place in order to have assured delivery of cyber defense for their society and critical services.

An often-used term in this context is "resilience", defined by Wilner as the "ability to bounce back, to mitigate the effects of an attack, to recover quickly after getting hit" (Wilner, 2017, p. 310). According to Wilner, however, resilience falls under deterrence by denial (not between the two extremes), and compared to deterrence by punishment, it should dissuade the aggressor, because it reduces the benefits from the malicious activity, rather than adding to the costs.

The responses of a state to malicious cyber activities or even operations by another state are bound to (at least when it intends to adhere with international and national law) identifying whether the situation at hand is a cyberattack (an incident in or through cyberspace with an objective to degrade, deny or destroy) (Brantly 2018, p. 41), an act of cyber espionage (which, like conventional espionage has not been included as a violation of international law) or cybercrime (theft of data, intellectual property, etc.—indeed breaches but likely not crossing the threshold of an internationally wrongful act, bar use of force or armed attack).

Against the universal and complex background of the different nature, capacity, sources and targets of threats, the need to consider one's cyber deterrence posture and the *sine qua non* need for a solid attribution of the act-to-actor, the question to be asked is how best to achieve a comprehensive solution to deal with all of them in a coordinated manner. Is this an issue of national security and, as a result, should states pursue a military capabilities-based approach to dealing with the cyber challenge? The question is timely as, according to Smeets, similar to private enterprises, national governments look into organizational integration in search of a more effective provision of services. As a result, we are seeing a diverse institutional landscape emerge to deal with cyber security, "shaken up by the new cyber security challenges that countries face" (Smeets 2017, p. 26).

Different Roles of Cyber Commands

The establishment of the U.S. Cyber Command in 2010 signaled the start of a new era where cyber defense has become part of military command chains. In the past couple of years, many nations have announced following suit. Of NATO countries, at least 8 have established an "independent cyber command or service: France, Germany, Italy, the Netherlands, Norway, Spain, Turkey and the United States (Pernik 2018). This means that the emergence of civilian CERTs (Computer Emergency Response Teams) or institutions that safeguard the cyber security of networks—be them private (corporate CERTs) or public (national CERTs)—has been matched by a similar development for safeguarding military networks. Among the countries that have announced setting up new cyber organizations are Singapore, which also tops the International Telecommunications Union's Global Cybersecurity Index 2017 (International Telecommunications Union 2017), Estonia (ranked 5th globally and 1st in Europe) and Canada (ranked 9th globally).

The establishment of the Singapore Defence Cyber Organisation (DCO) was announced in March 2017. A unit under the country's Ministry of Defence, the 2400-strong DCO was stated to have four pillars: Cyber Security Division (operational response issues), Policy and Plans Directorate (capability development), Cyber Security Inspectorate (vulnerability assessments) and the Cyber Defence Group (security monitoring, incident response and audit, network security testing and training). The objective of the new institution, according to the Minister of Defence, was to prepare for an environment "where state-orchestrated cyber and information campaigns against another state are not only considered legitimate, but can be ongoing all the time" (NewsAsia 2017). In June 2017, he added that "Singapore has now found itself on someone's [target] list" while announcing that, in parallel to the efforts of the Ministry of Defence, the Singapore Armed Forces are to set up a new, 2000-strong C4 (Command, Control, Communications, Computers) Command (Leong 2017).

In June 2017, the Estonian government approved the establishment of the Estonian Defence Forces (EDF) Cyber Command. The objective of this 300-strong organization, which the Ministry of Defence sees as "an instrument that forms part of national defence toolbox [...] with an interdisciplinary informational and civil military dual use nature" is to complement Estonia's defense of its "territory, population, values, and way of life" (Kodar 2017). The EDF Cyber Command's mandate will also include "cyber offensive responses to an adversary's offensive attacks" (Tigner 2017).

In February 2018, the Canadian government announced the establishment of the Canadian Centre for Cyber Security, based on the existing Communications Security Establishment (the Canadian NSA). The objective of the new organization, to quote, is "by consolidating operational cyber expertise from across the federal government under one roof, the new Canadian Centre for Cyber Security will establish a single, unified Government of Canada source of unique expert advice, guidance, services and support on cyber security operational matters" (Pugliese 2018). In an earlier process, amendments were proposed to Canada's legislation to expand the mandate of the CSE to include "active cyber operations" to deal with "foreign groups, organizations, states and individuals who are involved in terrorist activity, are attempting to compromise national security, trying to disable key infrastructure, or spying on Canadians" (Scotti 2017).

As shown by the three recent examples, countries from around the world are pursuing a more active posture in cyber defense. There are, however, significant differences that stem from not only the different sizes of the mentioned states (and thus, the different sizes of their respective human resource pools). In light of the rather limited number of public announcements on the three cases, it is possible to generalize that they all have the same end—step up national efforts to deal with increasing threats. What differs are the ways and the means. While the latter will eventually come down to resources (both Singapore and Estonia make reference to also including volunteers in the national effort, where Estonia already has a solid record of accomplishment to show with its so-called Cyber Defence League), the former are different.

The Singaporean solution will be to have two organizations that appear to deal with a more comprehensive approach (from operational response to training), as well as with operationalizing cyber activities to be part of national defense efforts. The Estonian solution, too, appears to remain inside the defense forces, while being elevated to a service command status. The statements also refer to an approach that considers the importance of the respect for law—there will likely be an adjustment of the national legislation, to allow the new Cyber Command to be able to conduct cyber operations, including against an adversary on the offensive. The Canadian approach differs from the other two significantly for the fact that the new "one-stop-shop for defending federal networks and systems" (Boutilier 2018) will be a civilian organization, effectively an expansion of the mandate of the existing CSE (Canada's electronic spies). Compared to Singapore and Estonia, Canada also highlights the financial investment that comes with the new plan.

This diversity of tasks that the three countries have given to their new cyber commands, reflects the findings by Pernik who concluded that cyber commands' functions can include:

- looking after the cyber security of armed forces' data communication networks, information systems, infrastructure, weapons systems, etc.;
- procurement; and the recruitment and support of personnel and activities related to their career paths, education and training.
- the core task of planning, preparing for and conducting cyber operations in cooperation with other military structures, especially military intelligence and operations directorates (Pernik 2018).

Conclusions

Evidently, there is no one solution for a cyber command that fits all nations. What, then, should the national cyber defense institutions be able to deal with in real time and have the mindset for? To summarize the threats discussed earlier in this article, the likely challenge will continue to be the so-called malicious cyber activities—one or another form of cyberattacks directed against one or several sectors in one state that can be related to a stated or generally understood political objective of another state—that, however, take place in peacetime and remain below the threshold of an armed attack, where the sovereign right of self-defense and the use of countermeasures is perhaps not as evident compared to a conventional armed conflict.

It is, however, also understandable that even such a wide scope of tasks that the cyber commands are to fulfill would probably not cover the entire matrix of cyber security challenges that a nation needs to be able to handle. Therefore, whilst the role of cyber commands will continue to be essential in dealing with threats to national security, an equally prudent approach is also needed on the civilian side, both in the Governmental as well as in the private sector, especially in the critical infrastructure providers. In light of the limited resources, and especially human resource, it would appear feasible for a nation to consider a joint or closely coordinated civil-military approach to the extent possible.

Strategically, national security begins with knowing one's potential adversary—also in cyberspace. In the context of a global cyberspace and the speed of technological development, this too is a task with a moving target. To narrow the list down to the national priorities, a solid dose of common sense, coupled with efficient intelligence, is a prerequisite. To determine the intrusion potential and the rationale for it, the routine analysis should include different viewpoints, such as diplomatic, informational, military or economic. That implies that the institution responsible for cyber defense can have both techies and social scientists on its payroll for a very practical reason: "knowing an adversary's motivation and behaviour makes mitigating future breaches easier" (Rid and Buchanan 2015, p. 25).

Operationally, there needs to be a clear understanding of the mandate and "playground" for the institution. To use the U.S. Cyber Command as an example, "Cybercom's mission is, when ordered, to disrupt and destroy adversaries' networks. It is also to defend the nation against incoming threats to critical systems and to protect the military's computers from cyberattack. The NSA also has a defensive mission—to protect the government's classified networks—but is better known for its role in conducting electronic spying on overseas targets to gather intelligence on adversaries and foreign governments" (Nakashima 2016). Clearly, there are limits to what military (and civilian) structures can legally do, country by country. Therefore, it is unlikely that the Singaporean or the Estonian cyber commands in particular could replicate the NSA functions, for example. Nevertheless, there is merit in exploring the options of rotating staff between the civil and military institutions, providing equally sophisticated training, exercising joint teams in international cyber defence exercises, etc.

One should also consider what a former commander of U.S. Cyber Command has said about the tasks of his Command: "first, to protect U.S. and Allied freedom of action in cyberspace" and "second, when directed, to deny freedom of action in cyberspace for our adversaries" (Alexander 2011, p. 11). While there might be resource considerations that would limit the ambitions of smaller nations' cyber commands, the strategic objective for at least a NATO nation should not be any less.

On the tactical level, according to Smeets, sophisticated cyberattacks include four general stages: reconnaissance ("sniffing", "footprinting", "enumeration"), intrusion (either relying on user-level or administrator-level (root) credentials and rights), privilege escalation (exploitation of the vulnerability in a system or service or software package) and mission completion stage (denial of service, backdoor installation, data exfiltration, espionage, corruption, etc.) (Smeets 2017, p. 30). Evidently some of the attacks might happen quickly with very short time for each phase. To be able to reverse-engineer the damages, and as a result, a reverse engineering is required to undo the damages, in principle, a defender would need to have all the required skillsets available in-house.

Furthermore, to provide for adequate defenses on a broader, nation-wide scale so that the "cyber posture" of the given nation would actually deter and deny the adversary from intruding (refraining from reconnaissance might admittedly be unrealistic to achieve), the institution should be able to encourage, if not enforce, a similar skillset in, for example, national critical infrastructure institutions. For the other side of the deterrence coin (punishment) to work, the institution should also be able to field an offensive capability upon the adversary as soon as the situation requires—ideally for a nation, according to Smeets, "the 'cyber option' is always available as a potential (strategic) asset to use" (Smeets 2017, p. 31). A caveat to recognize here is that not all nations will be able to possess such capacity, and the 'kinetic option' cannot be ruled out either.

In addition to the self-generating need to continuously update one's cyber defence capabilities to match new threats, is there an option that could further enhance nations' home-grown cyber defense capacity? It will make sense to look for international solutions, in particular between like minded liberal democracies. For NATO members that are also member states of the European Union, the EU "Cyber Diplomacy Toolbox" might be of use on the strategical level. On the technical skills level, there are also now available new options for cooperation and information sharing. Upon the outbreak of WannaCry and NotPetya, the EU CSIRT network that stems from the NIS Directive was, for the first time, activated by the Estonian Presidency, to allow for a Union-wide coordination of the response to and mitigation of the effects of the outbreaks. By many accounts, this proved to be a success story of real-time operational cooperation across Europe. In addition, it will remain essential, including for political and legal reasons, that the functions explained above will not happen without due consultation with the National Competent Authorities that are the national Points of Contact for cyber security vis-à-vis the EU by definition.

References

Alexander, K. B. (2011). Building a New Command in Cyberspace. *Strategic Studies Quarterly, 5*(2), 3–12. Retrieved June 30, 2018 from http://www.airuniversity.af.mil/Portals/10/SSQ/documents/Volume-05_Issue-2/Alexander.pdf.

Arquilla, J. (2013, April 17). Twenty Years of Cyberwar. *Journal of Military Ethics, 12*(1), 80–87. https://doi.org/10.1080/15027570.2013.782632.

Arquilla, J., & Ronfeldt, D. (1993). Cyberwar is Coming!. *Comparative Strategy, 12*(2), 141–165. (Spring, Copyright 1993 Taylor & Francis, Inc.).

Arquilla, J., & Ronfeldt, D. (2017). *In Athena's camp. Preparing for conflict in the information age.* RAND Corporation. Retrieved June 26, 2018 from https://www.rand.org/pubs/monograph_reports/MR880.html.

Bartholomees, J. (2006). A survey of the theory of strategy. In J. B. Bartholomees, *U.S. army war college guide to national security policy and strategy.* U.S. Army War College. Retrieved from https://ssi.armywarcollege.edu/pdffiles/PUB708.pdf.

Boutilier, A. (2018, June 12). *Canada's spy agency expands its cyber security role.* Retrieved from The Star: https://www.thestar.com/news/canada/2018/06/12/canadas-spy-agency-expands-its-cyber-security-role.html.

Brantly, A. F. (2018). The cyber deterrence problem. In T. J. Minarik, (Ed.), *10th International Conference on Cyber Conflict CyCon X: Maximising Effects* (pp. 31–53).

Buzan, B. (1991). New patterns of global security in the twenty-first century. *International Affairs.*

Clarke, R. (2009, Nov-Dec). War from cyberspace. *The National Interest*, 31–36. Retrieved June 30, 2018 from http://nationalinterest.org/article/war-from-cyberspace-3278.

Collier, J. (2016). *Strategies of cyber crisis management: Lessons from the approaches of Estonia and the United Kingdom.* Oxford: University of Oxford DPIR. Retrieved June 30, 2018 from https://www.politics.ox.ac.uk/publications/strategies-of-cyber-crisis-management-lessons-from-the-approaches-of-estonia-and-the-united-kingdom.html.

Deibert, R. J., & Rohozinski, R. (2010a). Risking security: Policies and paradoxes of cyberspace security. *International Political Sociology, 4*, 15–32.

Deibert, R., & Rohozinski, R. (2010b). Control and subversion in Russian cyberspace. In R. Deibert, J. Palfrey, R. Rohozinski, J. Zittrain, R. Deibert, J. Palfrey, R. Rohozinski, & J. Zittrain (Eds.), *Access controlled: The shaping of power, rights and rule in cyberspace.* Cambridge, MA: MIT Press. Retrieved June 30, 2018 from https://pdfs.semanticscholar.org/48b5/50fe0dc602ea7e0a9d4f8f395d9ede34ae66.pdf.

Department of the Army HQ. (2017, April 11). *Field Manual (FM) 3–12. Cyberspace and Electronic Warfare Operations.* Retrieved June 30, 2018 from https://fas.org/irp/doddir/army/fm3-12.pdf.

Emmott, R. (2017, September 7). Critical alert: EU ministers test responses in cyber war game. *Reuters.* https://www.reuters.com/article/eu-defence-cyber/critical-alert-eu-ministers-test-responses-in-cyber-war-game-idUSL8N1LN56A.

European Commission. (2018, May 4). *Questions and answers: Directive on security of network and Information systems, the first EU-wide legislation on cybersecurity.* Retrieved from European Commission: europa.eu/rapid/press-release_MEMO-18-3651_en.pdf.

Friedman, T. L. (2005, April 3). It's a flat world, after all. *The New York Times Magazine.* Retrieved from https://www.nytimes.com/2005/04/03/magazine/its-a-flat-world-after-all.html.

Härmä, K., & Minarik, T. (2017, Sep 18). *European union equipping itself against cyber attacks with the help of cyber diplomacy toolbox.* Retrieved from NATO CCDCOE: https://ccdcoe.org/european-union-equipping-itself-against-cyber-attacks-help-cyber-diplomacy-toolbox.html.

Heidegger, M. (1977). The question concerning technology. Retrieved from http://www.psyp.org/question_concerning_technology.pdf.

Infosec Institute. (2012, October 5). *The rise of cyber weapons and relative impact on cyberspace.* Retrieved June 29, 2018 from Infosec Institute: https://resources.infosecinstitute.com/the-rise-of-cyber-weapons-and-relative-impact-on-cyberspace/.

International Telecommunications Union. (2017, July 6). *Global cybersecurity index 2017.* Retrieved from https://www.itu.int/en/ITU-D/Cybersecurity/Documents/Global%20Cybersecurity%20Index%202017%20Report%20version%202.pdf.

Klimburg, A. (2012). *National cyber security framework manual.* In A. Klimburg (Ed.), Tallinn: NATO CCDCOE.

Kodar, E. (2017). *NATO information assurance symposium'17. Mons,* Belgium. Retrieved June 30, 2018 from https://www.youtube.com/watch?v=PKC-nWRfez4.

Leong, J. (2017, June 30). *Singapore on target list for cyberattacks, new SAF command set up to handle threats: Ng Eng Hen.* Retrieved from Channel NewsAsia: https://www.channelnewsasia.com/news/singapore/singapore-on-target-list-for-cyberattacks-new-saf-command-set-up-8991770.

Liff, A. P. (2012, May). Cyberwar: A new 'absolute weapon'? the proliferation of cyberwarfare capabilities and interstate war. *Journal of Strategic Studies,* 401–428. https://doi.org/10.1080/01402390.2012.663252.

Lord, N. (2017, March 27). *What is the NIS directive? Definition, requirements, penalties, best practices for compliance and more.* Retrieved from Digital Guardian: https://digitalguardian.com/blog/what-nis-directive-definition-requirements-penalties-best-practices-compliance-and-more.

Nakashima, E. (2016, December 23). *Obama moves to split cyberwarfare command from the NSA.* Retrieved June 30, 2018 from The Washington Post: https://www.washingtonpost.com/world/national-security/obama-moves-to-split-cyberwarfare-command-from-the-nsa/2016/12/23/a7707fc4-c95b-11e6-8bee-54e800ef2a63_story.html?noredirect=on&utm_term=.e85bb0ea7a3f.

NATO. (2016, July 8). *Cyber defence pledge.* Retrieved from NATO: https://www.nato.int/cps/su/natohq/official_texts_133177.htm.

NewsAsia, C. (2017, March 3). *Singapore to set up new defence cyber organisation.* Retrieved from https://www.channelnewsasia.com/news/singapore/singapore-to-set-up-new-defence-cyber-organisation-8775266.

Paganini, P. (2012, October 5). *The rise of cyber weapons and relative impact on cyberspace.* Retrieved from Infosec Institute: https://resources.infosecinstitute.com/the-rise-of-cyber-weapons-and-relative-impact-on-cyberspace/.

Pernik, P. (2018, November 26). *Estonian cyber command: What is it for?* Retrieved from International Centre for Defence and Security: https://icds.ee/estonian-cyber-command-what-is-it-for/.

Pugliese, D. (2018, February 28). *Canada adds new cybersecurity center, hikes funding for electronic spy agency.* Retrieved from Defense News: https://www.defensenews.com/international/2018/02/28/canada-adds-new-cybersecurity-center-hikes-funding-for-electronic-spy-agency/.

Rid, T., & Buchanan, B. (2015). Attributing cyber attacks. *The Journal of Strategic Studies, 38*(1–2), 4–37. https://doi.org/10.1080/01402390.2014.977382.

Schmitt, M. N. (2017). *Tallinn manual 2.0 on the international law applicable to cyber operations.* Cambridge University Press.

Scotti, M. (2017, June 20). *Canadian security agency will soon be able to launch cyber attacks against terrorists.* Retrieved from Global News: https://globalnews.ca/news/3542168/canada-launch-cyber-attack-terrorism/.

Shea, J. (2010, February 2). *Lecture 6—Cyber attacks: hype or an increasing headache for open societies?* Brussels: NATO. Retrieved June 29, 2018, from https://www.nato.int/cps/en/natolive/opinions_84768.htm.

Smeets, M. (2017). Organisational integration of offensive cyber capabilities: A primer on the benefits and risks. In *2017 9th International Conference on Cyber Conflict. Defending the Core* (pp. 25–42). NATO CCDCOE.

Tigner, B. (2017, December 1). *Estonia to incorporate offensive capabilities into its future cyber command.* Retrieved from Jane's 360: http://www.janes.com/.

Wallace, I. (2013, December 16). *The military role in national cybersecurity governance.* Retrieved from Brookings: https://www.brookings.edu/opinions/the-military-role-in-national-cybersecurity-governance/.

Wilner, A. (2017). Cyber deterrence and critical-infrastructureprotection: Expectation, application, and limitation. *Comparative Strategy, 36*(4), 309–318. https://doi.org/10.1080/01495933.2017. 1361202.

World Economic Forum. (2013, March 30). *The global information technology report 2013.* Retrieved June 30, 2018 from World Economic Forum: http://www3.weforum.org/docs/WEF_ GITR_Report_2013.pdf.

World Economic Forum. (2018). *The global risks report 2018.* Retrieved from World Economic Forum: http://www3.weforum.org/docs/WEF_GRR18_Report.pdf.

Siim Alatalu joined the Estonian Information System Authority in 2019 and is in charge of implementing the EU CyberNet project. His previous career in cyber security inter alia includes being the Head of International Relations of the NATO Cooperative Cyber Defence Centre of Excellence (CCDCOE) since January 2015 where his primary role was to lead the development of the Centre's relations with its growing network of partners from government, military, academia and industry overseeing its enlargement to an international organisation with currently 25 member countries. In 2018 he joined the Centre's Strategy Branch, being in charge of cyber strategy and policy research and training related to NATO and the European Union, as well as providing subject matter expertise to the Centre's other flagship projects. This article is a result of his research conducted at the CCDCOE. His prior professional career includes several advisory and managerial positions at the Estonian Ministry of Defence and the Estonian Delegation to NATO. During the Estonian Presidency of the Council of the European Union in 2017, he also co-led the development of EU's cyber policy and strategy, being the Vice Chair of the Council's Horizontal Working Party for Cyber Issues. Siim Alatalu is a graduate of the Maxwell School of Syracuse University (Master of Arts in International Relations in 2006, as a Fulbright fellow), as well as of the Baltic Defence College (Higher Command Studies Course 2011) and the University of Tartu (B.A. in history in 2001). He is currently pursuing his Ph.D. at the Tallinn University of Technology.

Chapter 3
Digital Innovation Hubs and Their Position in the European, National and Regional Innovation Ecosystems

Maurits Butter, Govert Gijsbers, Arjen Goetheer and Kristina Karanikolova

Introduction

Digitization has been recognized as a key driver of competitiveness for the European industry (European Commission 2018a). Yet, companies experience difficulties in the uptake and incorporation of new technologies in their business operations. This is also the case in Europe. To address this issue and unlock the potential of new technologies, the Commission has launched a number of policies and strategies. Following the 2015 Digital Single Market Strategy (European Commission 2015), in 2016, the Commission launched the Digitizing European Industry initiative (DEI, see Fig. 3.1) aiming "to ensure that every industry in Europe, in whichever sector, wherever situated, and no matter of what size can fully benefit from digital innovations" (European Commission 2016).

Digital Innovation Hubs (DIHs, see Fig. 3.1) are one of the five main pillars of the DEI initiative and are seen as a one-stop-shop vehicle to help the digital transformation of businesses and support them in the adoption of new digital technologies (European Commission n.d.). More recently, the Commission has presented the first concept of the Digital Europe Program. The proposed Program aims to strengthen the European position in Artificial Intelligence, High Performance Computing, and

M. Butter · G. Gijsbers · A. Goetheer · K. Karanikolova (✉)
The Netherlands Organization for Applied Scientific Research (TNO), The Hague, The Netherlands
e-mail: kristina.karanikolova@tno.nl

M. Butter
e-mail: maurits.butter@tno.nl

G. Gijsbers
e-mail: gwgijsbers@gmail.com

A. Goetheer
e-mail: arjen.goetheer@tno.nl

© Springer Nature Switzerland AG 2020
D. Feldner (ed.), *Redesigning Organizations*,
https://doi.org/10.1007/978-3-030-27957-8_3

Cyber Security. Next to these content areas and the development of "advanced digital skills", the fifth focus area of the Program addresses "deployment of digitization". DIHs are seen as the core of this fifth area and as an instrumental tool for the other focus areas (European Commission 2018c).

Coined as a concept in 2016 by the Commission (see also Fig. 3.2), DIHs are seen as an important instrument to enhance collaboration between research and industry and in this way accelerate the digital transformation and help companies transform technology innovations into business practices. DIHs provide a regional one-stop-shop "where SMEs and mid-caps test the latest digital technologies and get training, financing advice, market intelligence and networking to improve their business"

Fig. 3.2 The background of
the DIH concept, based on
XS2I4MS project. *Own
Source* based on Gijsbers
et al. (2018)

The concept of a DIHs was largely developed as part of the **XS2I4MS project**. The XS2I4MS (Access to I4MS) project was a coordination and support action aiming to advance the European I4MS community.

One of the core activities of the project was the XS2I4MS **Mentoring programme** which supported 29 candidate DIH projects in the development of a feasibility study to establish DIHs across Europe. As part of this mentoring, the business models and business plans for DIHs were addressed.

(European Commission 2018b). One of the core benefits of DIHs is that they provide a useful instrument to share costly infrastructure and expertise. This is considered a crucial mechanism to reduce economic risks for SMEs, allowing them to adopt digital technologies (see de Heide and Butter 2016; Butter et al. 2015).

Nowadays, the idea of sharing of equipment/expertise and supporting the transformation of technology into business is seen as an important policy instrument not only by many Member States, but also outside Europe. Within Europe, some interesting examples include: the Netherlands (Fieldlabs), UK (Catapult Centres), Germany (Mittelstand 4.0-Kompetenzzentren) and Spain (Digital Innovation Hubs), etc. Outside of Europe, the USA Innovation Institutes and Chinese Manufacturing Innovation Centers represent similar concepts. In 2016–2017, a consortium led by TNO carried out a mapping exercise to identify existing DIHs in Europe. Published by the Joint Research Center, a database with more than 400 self-registered innovation hubs across Europe is now available. Yet, it can be expected that the list of DIHs will be expanded in the coming years, also including different "stages of evolution" (Butter and Karanikolova 2018; Goetheer and Butter 2017). The DIHs Catalogue clearly shows a vibrant but scattered European landscape that is still very much under development on regional and national levels.

Although initially promoted by the European Commission to target digital transformation in the manufacturing sector (notably in SMEs), the DIH concept is now being adopted in other industries as well: for example, in agriculture, health and chemicals, etc. (Gijsbers et al. 2018). More broadly, the European Commission has embraced the concept of sharing infrastructures and expertise to support transformation outside of the realm of digitization via similar mechanisms with different names (e.g., Open Innovation Testbeds, Technology infrastructures, KETs Technology Centres, etc.). Following support for individual DIHs in previous years, the European Commission is now focusing on the development of a pan-European network of DIHs. This is expected to bring about an exchange of knowledge, "complementarily in competences and infrastructure as well as economic aspects of a network collaboration" (DIH Working Group 1, May 2018).

As presented above, the conceptual landscape on DIHs is still very much under development and is often not fully clear to many policy-makers, industry players and research organizations. To help them navigate this landscape, the following sections will discuss the concept of DIHs and their position within the European, national and regional innovation ecosystems. Section 3.2 will set the context in which the DIH concept was coined by discussing the process of digitization and the argumentation of its needs and legitimization. In Sect. 3.3, we will define the concept of DIHs, focusing on the services they usually offer. In the final section, we will look into recent developments. This chapter will also map the multi-layered system in which DIHs operate and the challenges for cooperation.

Great Opportunities from Digitization but Unrealized Potential and a Need for Policy Support

Opportunities and Challenges in Digitization

Digitization and the Fourth Industrial Revolution are expected to add more than €110 billion of annual revenue to the European industry by 2020 (European Commission 2018a). The benefits of digitization to companies have been widely proclaimed. Digital leaders across industries are given as examples, showing higher productivity, faster revenue growth and faster improvements in profit margins compared to less-digitized companies (Manuika 2017). New technologies—such as Artificial Intelligence (AI), machine learning and IoT—are expected to bring even more new opportunities to boost the productivity of companies and bring benefits to society at large.

And yet, digitization has proven a challenge for many companies. According to PwC, close to two-thirds of the globally surveyed companies have not yet started, or have just initiated, their digital transformation. European companies are also affected: research shows that only 5% of the studied manufacturers in EMEA become "digital champions" compared to 19% in Asia (Geissbauer et al. 2016). In Europe, more than 90% of SMEs and 60% of large enterprises feel that they are lagging behind in digital transformation (European Commission 2018b). What is more, the digital adoption and maturity differs per country and sector, thus leading to an uneven landscape (Bughin et al. 2016). This uneven progress is only expected to deepen as frontrunners in the digital adoption start applying AI solutions, while others are still uncertain about what AI can do for them and how they can integrate it in their business (Bughin et al. 2017).

Digitization is especially challenging for SMEs. Some of the main barriers to digitization are that companies find it difficult to assess what digitization would mean for their enterprise, which technologies will yield the best results and when they should start investing, as well as how these changes will be funded (Stephen et al. 2018, p 4, p 15). Digitization often requires the introduction of new, complex and often costly solutions. What is more, digital transformation is often accompanied by organizational and operational changes, along with adapting or developing new business models, and might require development of new skills and expertise (see also Fig. 3.3, KFW 2017). Therefore, the associated risks of (digital) innovations increase the costs and uncertainty for companies. This discourages companies from investing at the stages of pilot production and demonstration, since they estimate that the uncertainty of the innovation process is high and there is risk that the required investment would not be recovered (de Heide and Butter 2016).

Fig. 3.3 Summary of barriers identified by KFW 2017 Business Survey. *Own Source* based on KFW (2017)

A Business Survey (KFW 2017) found that enterprises in Germany consider the biggest barriers to digitization to be:

- adapting organizational structure and work organization (33%)
- requirements of data privacy and data security (31%)
- lack of IT expertise (28%)
- followed by inadequate quality of internet connection, "difficulties in the conversion of existing IT systems and uncertainty over future technologies and standards" (between 26% and 28%)

Costs and Uncertainties as Drivers for Sharing

In 2015, the Multi-KETs Pilot Lines project concluded that this economic risk is at the core of the so-called "valley of death" (Butter et al. 2015, see Fig. 3.4). One of the main recommendations from the project was that shared facilities for pilot production should be supported, but that such initiatives should expand their activities beyond technological services. Business support services (e.g. market articulation, access to finance) were seen as crucial to reduce the uncertainty associated with the market potential of innovative products. This reflected the growing awareness that technological barriers were not the only challenge that companies face in order to cross both the technological and commercialization "valley of death" (Butter

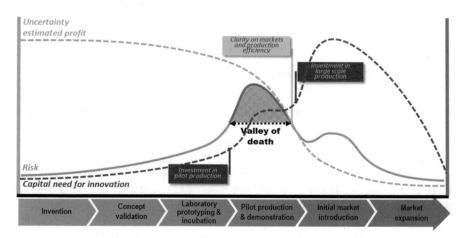

Fig. 3.4 The valley of death. *Source* Butter et al. (2015), as part of the work in the Multi-KETs Pilot Lines project

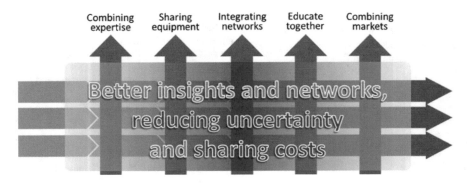

Fig. 3.5 Sharing equipment and expertise will reduce economic risks. *Source* Butter et al. (2015)

et al. 2015; Jenkins and Mansur 2011). Especially for SMEs the economic risk of engaging in innovative activities was found to be at its highest during pilot production, due to the fact that reduction of uncertainty is out of sync with the increase of investments needed to generate information on the performance and behavior of the future products and to get a better understanding of the market (Butter et al. 2015; Jenkins and Mansur 2011) (Fig. 3.4).

"Shared facilities for pilot production" were therefore seen not only as a way to reduce costs of equipment, but also as a mechanism to set the foundation of a network where expertise can be pooled to increase the overall understanding of the technological, business and economic opportunities and risks associated with innovations (see Fig. 3.5). The study, as well as experience from the EU-GREAT! and XS2I4MS projects, showed that it was crucial to combine technology services, ecosystem development and business-oriented services in helping companies translate research into business opportunities. This conclusion has formed the basis of the concept that we now call Digital Innovation Hubs.

Innovation Hubs More Than Traditional Technology Transfer

Traditionally, innovative technology infrastructures are provided by Research and Technology Organizations (RTOs), universities and some private initiatives. Organizations like Fraunhofer, TNO, VTT, as well as technical universities own state-of-the-art equipment that is used for (applied) research and technology development. Sharing this equipment is key as such infrastructures are seen as "public goods" where governmental support is legitimized. Companies make use of this equipment because of reduction of costs while educational institutes use them for training students.

DIHs however address more than technological challenges. Where technology infrastructures mainly focus on the *technological* valley of death, the DIHs are connecting with business expertise to support also crossing the *commercial* valley of death (Jenkins and Mansur 2011).

Fig. 3.6 Innovation—the intersection of ecosystem, technology, and business mindsets. *Source* TNO (2018)

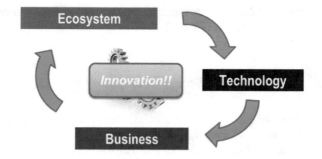

Especially, many SMEs need business support to transform innovative technologies into actual business. One of the key barriers mentioned during the discussions within the Multi-KETs Pilot lines project was that the mindsets of scientists, technological engineers, business developers, marketing/sales and entrepreneurs are fundamentally different. DIHs aim at structurally enabling the convergence of these mindsets (see also Fig. 3.6). DIHs are about creating a structural multi-actor relay game to get the "technology baton" into the hands of the entrepreneur.

What Are DIHs?

Tracing DIHs Throughout the History of Innovation Theory

DIHs provide a solution to one of the key issues in innovation research—how to ensure that new technologies and innovations reach all actors, and how to support small companies in the adoption of these innovations. As such, DIHs fit in a long tradition of innovation theory and practice. The notion of Digital Innovation Hubs builds on the idea of technology transfer, public-private partnerships and open innovation (Bozeman 2000; Chesbrough 2003; OECD 2004; Van der Zee et al. 2016). The core idea behind these schools of thought is that the increasing speed of technological development and the need to respond to more complex, multi-disciplinary problems require a much wider range of knowledge and technologies than can be found in in-house silos. Cooperation and interaction among different actors in both research and application is, therefore, key and can help actors access expertise, technology and markets.

Building on this idea of collaboration, DIHs are by their very nature multi-actor, public-private partnerships (PPPs) in which companies, (public) research institutions, intermediary organizations, and government agencies cooperate to support the transformation of technologies into business and involve all relevant ecosystem actors to support continuous innovation (see Fig. 3.6). Governments see such PPPs as a tool to organize and increase the efficiency of governmental support for RD&I, while for companies, collaboration reduces the uncertainty of R&D by providing access to tacit knowledge and research infrastructure found in universities, RTOs and centers of excellence (Cervantes 1998).

A Multi-actor, Ecosystem Approach

The concept of Digital Innovation Hubs is still developing and continuously evolving. DIHs take a variety of forms and structures depending on their focus and their local situation. The core idea, in line with the notion of PPPs, is that a range of actors work together, offering a variety of services to stimulate and support companies in their uptake of digital technologies. Based on that, a stylized picture of what a DIH is can be painted (see Fig. 3.7).

The customers or end-users of a DIH are companies. Although the mission of a DIH is usually to support SMEs and start-ups, often large enterprises create a long-term foundation of activities that allow the DIH to support SME and start-ups in a sustainable manner. Each of these end-users may have a different interest in the DIHs (Gijsbers et al. 2018)—large companies might be interested in lowering cost of infrastructure, collaborative research and access to talent and networks, while SMEs and start-ups might need help developing their business plans and access to costly testing infrastructure they do not have.

A core characteristic of a DIH, from an organizational point of view, is that it is a multi-actor, ecosystem-oriented network of organizations which together can support all facets of the innovation and industrial adaptation of a (digital) technology. The organization of DIHs is often (semi-) open, to allow new partners to be added and to address the needs of new customers (Butter 2016).

Fig. 3.7 Stylized depiction of digital innovation Hubs. *Source* TNO (2018), as part of the work for the XS2I4MS project

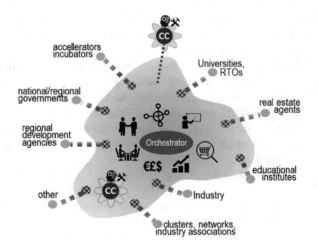

DIHs are usually built around one or more Competence Centers (CCs), that form the technological core of the DIH. CCs are often established as part of, or managed by, Research and Technology Organizations (RTOs), universities, or corporate labs. These CCs usually provide access to technological infrastructure and the relevant expertise on technical issues. But DIHs also aim to be one-stop-shops for companies and assist them with different types of services (Butter 2016). Therefore, DIHs usually need to connect to different stakeholders and involve other actors: incubators/accelerators to support new start-ups, development agencies to get access to finance, educational institutions to co-develop highly skilled personnel and training programmes, governments to co-fund and connect to policy, real estate agents to provide housing and intermediary organization to facilitate brokerage. All these actors work together to establish a vibrant and dynamic ecosystem (see Fig. 3.7).

Defining DIHs by What They Do

During the first steps of developing the concept of DIH in projects such as XS2I4MS (see Fig. 3.8) and the DIHs Catalogue, there was much discussion about the precise definition of the concept. Yet, providing such a definition proved complex and without much added value. Rather, it was concluded that looking at the services provided would create a strong and operational view on what a DIH is (see Goetheer and Butter 2017; Butter and Karanikolova 2018).

Much of the added value of DIHs is connected to their ability to act as a one-stop-shop and offer companies a variety of services. Although one DIH significantly differs from another, previous research has identified some key activities that could be offered by DIHs. Based on the practical experience of the XS2I4MS project,

	Service	Activities
Ecosystem	Community building	Scouting, brokerage, awareness creation, dissemination, ecosystem building
	Strategy development	Market intelligence, market assessments, roadmapping, technology watch
	Ecosystem learning	Workshops, seminars to share knowledge and experience
	Representation, promotion	Representing interests during meetings & conferences, organizing (country) visits, roadshows
Technology	Strategic RDI	Joint, pre-competitive R&D, secondment from companies
	Contract research	Specific R&D, technology concept development, proof of concept
	Technical support on scale-up	Concept validation, prototyping, small series production
	Provision of technology infrastructure	Renting equipment, low rate production, platform technology infrastructure, Lab facilities
	Testing and validation	Certification, product demonstration, product qualification
Business	Incubator/accelerator support	VoiceOfCustomer, market assessment, business development, legal, IPR, location, sales strategy
	Access to finance	Financial engineering, connection to funding sources, investment plans
	Skills and education	Courses, bilateral mentoring, workshops, technological infrastructure for education, secondment
	Project development	Identification of opportunities, creating consortia, development of proposals
	Offering housing	Office space and space for experimentaton and pilot manufacturing

Fig. 3.8 Services and activities offered by DIHs. *Source* TNO as part of the work for the XS2I4MS project

the services can be broadly clustered into three groups (Gijsbers et al. 2018): (1) Ecosystem services; (2) Technology services and (3) Business services. Core of the DIHs are the technological services, as these often provide the infrastructure and expertise to offer customers insight in the RD&I related issues. However, the outcomes of the services are without any value if not implemented in the actual business of the customer. This adoption process requires insights from non-technical perspective and offering business related services: on questions related to the market potential, the economic feasibility, access to (different) funding sources, potential need to introduce changes to the organizational structure, organization of the value chain, attracting skilled employees, etc. All these issues are crucial to a successful implementation.

The third group of services is focused on creating a dynamic ecosystem, needed to develop a sustainable inflow of new partners and customers. Activities are focused on building a sound and active community, but also on creating a strategic view on digitization. Part of these activities are also related to creating a learning mechanism between organizations in the ecosystem. An important aspect of a DIH is that its ecosystem is often regional. The core of the DIH customers are regional SMEs and the financial basis of the DIHs is often regional. With the development and establishment of a DIH, national, European and global customers can also be attracted and serviced to ensure sustainability. Yet, the business and ecosystem services are often highly influenced by the regional characteristics.

The Evolving Collaboration Among DIHs: Mapping the Emerging Landscape

As shown, the DIH concept can be seen as the next evolutionary step from the technology focused CCs to an entity that also includes functions supporting business creation. But a second evolutionary step is emerging in the last years—a coordination of DIHs in regional networks.

At the same time, on the European level coordination and innovation-oriented projects are specifically focused on supporting the cross-regional and cross-border collaboration (research, as well as innovation). Where many of the EC programs in the recent past aimed at supporting the initiation of DIHs, a shift is now seen towards European networks in which DIHs and EC programs collaborate. This European coordination is considered more and more important to optimally use the benefits from DIHs throughout Europe.

Fig. 3.9 Illustration of a
regional DIH network.
Source TNO (2018) as part
of the work for the
DIHNET.EU project

Regional DIH Networks Emerging as Cross-sectoral and Multi-technology Hubs

In many regions, a scattered growth of number of DIHs is seen. With the increase in the number of DIHs across Europe, some regions are starting to coordinate activities by different actors to increase efficiency and effectiveness (i.e. economies of scale) (Butter and Karanikolova 2018). These efforts aim to develop a multi-domain one-stop-shop for the regional innovation ecosystem (see Fig. 3.9). Examples of such coordination networks have already emerged—for instance, in the Basque country and in the Netherlands, where there is an initiative to coordinate the existing 33 field labs into regional networks (see, for example, Smart Industry website (smartindustry.nl) and SMITZH—Digital Innovation Hub Smart Manufacturing supporting companies in the province of Zuid-Holland, Innovation Quarter 2017).

CCs and DIHs often focus on a specific technology such as AI, robotics, or big data or on a sector such as manufacturing or agriculture. By coordinating activities, a regional network can transform into a single doorway for companies, providing links to different DIHs/CCs, different technologies and capabilities. In this way, they optimize the operations of the existing DIHs and CCs.

The development of such regional networks has just started, and different structures and initiatives are emerging. The emergence of regional networks might, however, result in shifting the function of a one-stop-shop for businesses to the regional network as well as transferring some of the DIHs' services to these regional structures. Otherwise, this additional layer will only bring more complexity in the system and create tension between sectoral hubs (on robotics, AI, etc.) and regional networks.

Supporting European Initiatives Shifting Toward Pan-European Collaboration

While the focus of the regional and national levels is on developing individual DIHs, the focus of the European Commission has shifted toward the concept of "European added value" and linking existing national and regional initiatives in a pan-European network (see Fig. 3.10). The creation of these networks can stimulate cooperation

Fig. 3.10 Illustration of a
pan-EU DIH network.
Source TNO (2018), as part
of the work for the
DIHNET.EU project

among different regions, thus offering to SMEs pan-European access to specific
innovations and infrastructures. This cooperation among regions and regional DIH
networks requires support and organization. One of the most important challenges
involves identifying the added value of collaboration for each of the participating
entities—from CCs, to DIHs and regional networks.

There are examples of initiatives on the EU level that attempt to promote collaboration. These include already established networks—such as EIT's Knowledge and
Innovation Communities, as well as H2020 programs like I4MS (ICT Innovation for
Manufacturing SMEs) and SAE (Smart Anything Everywhere) and many more. Yet,
more is needed to support this development and really make such a pan-EU network
effective.

Evolving into a Complex, Multi-layered Ecosystem

Looking at the many initiatives, one can conclude that there is a dynamic ecosystem
in which DIHs, CCs, regional networks and support structures are operating inside
and across sectors, technologies and borders (see Fig. 3.11). Different layers in the
landscape seem to be emerging. Each of these layers has its own objective and
priorities, but they also co-exist and cooperate at different levels. Therefore, creating
the right incentives for the different entities to cooperate is of key importance.

This, however, brings its own challenges: what is the benefit of collaboration for
each of the players in this ecosystem?; how can collaboration among the different
players (DIHs, CCs, regional networks, regional and European clusters and initiatives, etc) be coordinated to support optimal use of resources and provide access
to infrastructure and expertise beyond the capacity of the individual players?; what
business models should these players employ individually and to support such collaboration?; what role does specialization (regional, on technologies, sectors, customer focus) play in this collaborative ecosystem? Answers to these questions are

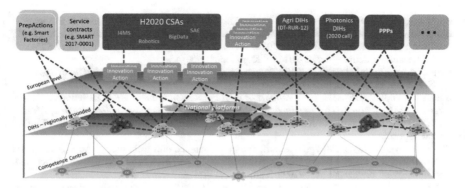

Fig. 3.11 Different layers in the landscape of DIHs. *Source* TNO (2018), as part of the work for the DIHNET.EU project

still evolving, but there are already initiatives looking into these challenges (e.g. the DIHNET.EU project and the third phase of the I4MS initiative).

Conclusions

The notion of DIHs has developed over time to now encompass a one-stop-shop entity that can provide companies with technical and business support in the uptake of digital technologies. DIHs are developing into multi-stakeholder entities offering businesses support via a number of technologies, business and ecosystem services. The coordination poses challenges related to the development of a business model, organizing the collaboration and sharing of resources.

The proliferation of DIH initiatives has, however, also resulted in a number of new developments where we see the formation of regional DIH networks, national platforms and a drive to develop a pan-European network. This creates a complex, multi-layered innovation ecosystem that is still very much under development. The coordination of activities and responsibilities in these layers poses its own challenges, including establishing the added value for each organization.

Bibliography

Reports and Studies

Bozeman, B. (2000). Technology transfer and public policy: A review of research and theory. *Research Policy*, 29(4–5), 627–655.
Bughin, J., et al. (2016), *Digital Europe: Pushing the Frontier, capturing the benefits*. McKinsey & Company.

Bughin, J., et al. (2017). *Artificial intelligence; The next digital frontier*, McKinsey & Company.

Butter, M. (2016). *Defining and demarcating Digital Innovation Hubs*. Presentation at the XS2I4MS DIH Summer School on Sep 23, 2016. Available at: https://i4ms.eu/documents/XS2I4MS-SummerSchool-defining-DIHs-2016-0920.pptx.pdf

Butter, M., & Karanikolova, K. (2018). *Support to development of a basque digital innovation hub*. TNO report, Project reference code: 931101.

Butter, M., et al. (2015). *Assessing support of pilot production in multi-KETs activities: D7 Final report of the multi-KETs Pilot lines project, including the tentative policy roadmap*. Multi-KETs Pilot Lines project. Available at: http://www.mkpl.eu/fileadmin/site/final/mKETs_D7_final_report.pdf.

Cervantes, M. (1998). Public/private partnerships in science and technology: An overview. *STI Review* No. 23, Special Issue, OECD. Available at: https://www.oecd-ilibrary.org/science-and-technology/sti-review_sti_rev-v1998-2-en.

Chesbrough, H. (2003). *Open innovation: The new imperative for creating and profiting from technology*. Boston: Harvard Business School Press.

Digital Innovation Hubs Working Group 1. (May 2018). Report from the 3rd Working Group Meeting on Networking among DIHs, role of Member States and regions for the development of DIHs, Catalogue of DIHs. *DG CONNECT*. Available at: https://ec.europa.eu/digital-single-market/en/news/report-third-meeting-working-group-digital-innovation-hubs.

de Heide, M., & Butter, M. (2016). *Deliverable 5.3 report assessment match/mismatch and issues with combined funding*. EU-GREAT!, Ref. Ares (2016) 6998311.

European Commission. (2018a). *Digitising European Industry: Progress so far, 2 Years after the Launch*. Brochure. Luxembourg: Publications Office of the European Union, 2018. ISBN 978-92-79-80325-3. Available at: https://ec.europa.eu/digital-single-market/en/news/digitising-european-industry-2-years-brochure.

European Commission. (2018b). *Digital single market: The 'digitising European industry' initiative*. Factsheet. Available at https://ec.europa.eu/digital-single-market/en/news/digitising-european-industry-initiative-nutshell.

Geissbauer, R., et al. (2016). *Global digital operations study 2018 digital champions: How industry leaders build integrated operations ecosystems to deliver end-to-end customer solutions*. PwC. Available at: https://www.strategyand.pwc.com/media/file/Global-Digital-Operations-Study_Digital-Champions.pdf.

Gijsbers, G., et al (2018). *Deliverable 6.3. Final report: Cross-case report analyzing the results from the digital Innovation Hub feasibility study projects*. XS2 I4MS project.

Goetheer, A., & Butter, M. (2017). *Digital innovation Hubs catalogue final report*, TNO 2017 R11340, TNO in cooperation with Civitta, CPI, D'Appolonia, EDATER, European Dynamics, Jennifer Harper, LIST, Technology Centre CAS, SmartIS City, Tecnalia, University of Zagreb, VDI and VTT.

Innovation Quarter. (2017). *SMITZH: Nieuw loket voor smart manufacturing*. Available here: https://www.innovationquarter.nl/nieuws/smitzh-nieuw-loket-smart-manufacturing/.

Jenkins, J., & Mansur, S. (2011). *Bridging the clean energy valleys of death*. Breakthrough Institute. Available at: https://thebreakthrough.org/blog/Valleys_of_Death.pdf.

KFW (2017). *Business survey 2017: Most businesses plan to go digital in the next two years*. Press release. Available at: https://www.kfw.de/KfW-Group/Newsroom/Latest-News/Pressemitteilungen-Details_418304.html.

Manuika, J. (2017), *10 imperatives for Europe in the age of AI and automation*, McKinsey Global Institute.

OECD. (2004). *Public-private partnerships for research and innovation: An evaluation of the Dutch experience*. Paris: OECD.

Smart Specialisation Platform (n.d.). *Digital innovation hubs catalogue*. Accessed on October 05, 2018. Available at: http://s3platform.jrc.ec.europa.eu/digital-innovation-hubs-catalogue.

Stephen, L., et al. (2018). *Embracing a digital future: How manufacturers can unlock the trans-formative benefits of digital supply networks.* Deloitte. Available at: https://www2.deloitte.com/insights/us/en/focus/industry-4-0/digital-supply-network-transformation-study.htm.
Van der Zee, F., Goetheer, A., & Gijsbers, G. (2016). *De staat van Nederland innovatieland 2016.* The Hague: TNO.

Legislation

European Commission. (2015). Communication from the Commission to the European Parliament, the Council, the European Economic and Social Committee and the Committee of the Regions: A Digital Single Market Strategy for Europe", COM (2015) 192 final. Available at: https://eur-lex.europa.eu/legal-content/EN/TXT/?uri=COM%3A2015%3A192%3AFIN.
European Commission. (2016). Communication from the Commission to the European Parliament, the Council, the European Economic and Social Committee and the Committee of the Regions: Digitising European Industry Reaping the full benefits of a Digital Single Market. COM (2016) 180 final, available at: https://eur-lex.europa.eu/legal-content/EN/TXT/?uri=CELEX:52016DC0180.
European Commission. (2018c). *Proposal for a Regulation of the European parliament and of the council establishing the digital Europe programme for the period 2021-2027.* COM (2018) 434 final, 2018/0227(COD). Available at: https://eur-lex.europa.eu/legal-content/EN/TXT/?uri=COM%3A2018%3A434%3AFIN.
European Commission. (n.d.). *Pillars of the Digitising European Industry initiative.* Accessed on November 23, 2018. Available at: https://ec.europa.eu/digital-single-market/en/pillars-digitising-european-industry-initiative.

Maurits Butter is an expert in industrial innovation policy. He graduated at the faculty of chemical engineering at the Delft University of Technology and started his career as an environmental consultant with Dutch consultancy firm. In 1994–1998, he worked for the Netherlands' Ministry of Housing, Spatial Planning and Environment (VROM). In 1998 he joined TNO as senior advisor innovation policy. He is involved in the interface between technological innovation and policy, focusing on industrial innovation policy. He was the research leader for the TNO research program on "Renewal of Innovation" and now is expert in connecting research to industrial innovation and related policy. He supports organizations that offer shared facilities to SMEs to transform research into business and advises regional/national governments and the European Commission in their development of industrial innovation policy. His present focus is on understanding the valley of death, especially on the development and financing of public private partnerships that support SMEs to cross the valley of death by technological, business and community building services. He has been mentoring both individual Digital Innovation Hubs and regional authorities on business models, finance, governance and development of business plans. He is part of the Regional Advisory Board of the EC funded I4MS Go and trainer for the TNO internal Orchestrating Innovation program.

Govert Gijsbers specializes in research and innovation policy, with a focus on foresight studies, monitoring, evaluation and impact assessment, applied in projects to specific sectors and technologies. His recent activities include projects on innovation in manufacturing ('smart industry'), biotechnology, and agri-food. Another research area relates to the impact of technology on employment, labor markets, skills and jobs. He has led a major European project on digital innovation for manufacturing SMEs (I4MS). Other recent projects include a project on the impact of robotics, smart skills for smart industry, an economic analysis of the Dutch biotechnology sector, a study on technology and employment, an analysis of Key Enabling Technologies (KETs) in Horizon 2020, and a project on the impact of the financial crisis on the job creation potential of SMEs. He is an experienced trainer in both EU project (mentoring I4MS) and in TNOs corporate program Orchestrating Innovation. Govert joined TNO in 2004 and worked as senior researcher and advisor at the TNO Strategy and Policy Group until July 2019. Before that he worked for the Consultative Group for International Agricultural Research (CGIAR), the Netherlands Ministry of Foreign Affairs and the United Nations. His work for the CGIAR in Asia formed the basis for his dissertation on agricultural innovation systems in Asia.

Arjen Goetheer works as Innovation Policy and Strategy Researcher at TNO's Strategy & Policy Department, where he carries out projects in the broad field of science, technology and innovation policy. His main areas of interest are innovation policy measures, start-ups, venturing and new business concepts, foresight and trend analysis, evolution of innovation networks, dynamics of eco systems, industry-university interaction and decision-making processes. Arjen's methodological expertise lies in the area of developing conceptual frameworks, quantitative and qualitative foresight and trend analysis, impact assessment and monitoring and evaluation, and he is familiar with a range of research methods, varying from workshops and qualitative case studies, to indicator development, quantitative statistical analyses and benchmark studies. Before joining TNO Arjen worked as Trend and Innovation Analyst in Switzerland, where he analyzed global and local trends in innovation, technology, start-ups, new business concepts and consumerism, carried out quantitative and qualitative trend, innovation and market analyses for clients and advised on strategic business opportunities and innovation activities. As Innovation Officer of a Swiss start-up company he formulated and implemented the company's open innovation strategies and launched big data innovation projects and initiated strategic partnerships with international companies and universities.

Kristina Karanikolova, LL.M works as a Researcher at TNO's Strategy and Policy Department since August 2016. Kristina is involved in a number of European projects where she carries research on topics related to innovation policy. Kristina has recently contributed to projects related to Digital Innovation Hubs (DIHs), including XS2I4MS (focusing on mentoring projects on the development and establishment of DIHs), the Digital Innovation Hubs Catalogue, and providing training to individual DIHs. Kristina is also participating in the DIHNET.eu project which focuses on smart specialization and DIHs. Other recent project includes a study on the barriers, drivers and possible steps towards circular economy and a study to monitor the business and regulatory environment affecting the collaborative economy. Prior to joining TNO, Kristina was a Blue Book trainee at the Automotive and Mobility Industries Unit of DG GROWTH. Kristina holds two Master's degrees from the University of Amsterdam—in European competition law and regulation and in European and International labor law.

Chapter 4
Transatlantic Privacies—Lessons from the NSA-Affair

Russell A. Miller

Introduction

One of the most significant—and overlooked—lessons of the NSA-Affair, which Edward Snowden triggered with his massive disclosures about American intelligence operations, is that a vast chasm exists between American and German perspectives on privacy. This has been a favorite theme of well-known comparative law scholars such as Yang (1966), Walsh (1976), Barnett (1999), Bignami (2007), Lachmayer (2014), and Krotoszynski (2014). In fact, the different American and German reactions to Edward Snowden's leaks demonstrate that there is hardly another issue about which transatlantic attitudes diverge so sharply. Americans do not understand Germans' outrage over the collection of seemingly meaningless and mostly innocent information that, when deployed creatively, has pragmatic value for promoting security and commercial innovation. At the same time, Germans do not understand Americans' seeming indifference toward the profound personal privacy implicated by access to highly-revealing telecommunications and Internet data. The so-called "NSA-Affair"—as it is referred to in Germany—once again proves that there are "significant privacy conflicts between the United States and the countries of Western Europe—conflicts that reflect unmistakable differences in sensibilities about what ought to be kept private" (Whitman 2004).

Transatlantic disagreement over the social, political, and legal meaning of privacy calls into question the widespread conviction that privacy is a shared and fundamental Western value, not to mention the view that privacy is a universal norm. That is a confounding conclusion for any discussion about managing our digital and data-centric future.

This chapter is an extensively edited and revised version of the author's contributions to the book *Privacy and Power: A Transatlantic Dialogue in the Shadow of the NSA-Affair* (Miller ed., 2017).

R. A. Miller (✉)
Washington and Lee University School of Law, Lexington, USA
e-mail: MillerRA@wlu.edu

© Springer Nature Switzerland AG 2020
D. Feldner (ed.), *Redesigning Organizations*,
https://doi.org/10.1007/978-3-030-27957-8_4

The critical insight I intend to advance with my contribution to this expansive collection grappling with the issues of data privacy and digital security is this: Privacy, especially when expressed as a norm within a domestic legal framework, necessarily reflects a society's culture (Altman 1977; Richter and Albrecht 2013). Our different legal notions of privacy are rooted in different histories, different social forces, different political traditions and institutions, different legal cultures, and different economic conditions and orientations. On these terms, there is no privacy. There are only privacies (Nissenbaum 2009).

First, I will document the dramatically different reactions to the NSA-Affair in the United States and Germany. This substantiates my fundamental claim that America and Germany have different views about privacy. Second, I will describe some of the social and political factors that unavoidably influence the two countries' very different legal understandings of privacy.

Different Reactions to the NSA-Affair

I have used the phrase "NSA-Affair" to refer to the political and legal turmoil—both domestic and international—loosed by Edward Snowden's disclosures. And that is precisely how Germans view the NSA operations exposed by Snowden: the NSA's activities were scandalous, unethical and illegal. But the developments swirling around Snowden's revelations are not seen in singularly appalling terms by most Americans. It is telling, for example, that Snowden's revelations have not widely earned the label "NSA-gate" or "Snowden-gate" in the popular American coverage of the story. That would have been in keeping with the tiresome American practice of borrowing the suffix "-gate" from the Nixon-era "Watergate scandal" to create a catchy label for every contemporary controversy worthy of Americans' attention. But Americans apparently view Snowden's revelations as less problematic than under-inflated footballs ("Deflategate") and "wardrobe malfunctions" during the Super Bowl halftime show ("Nipplegate").

Besides the very different ways in which Americans and Germans speak about the NSA-Affair, other anecdotes point to the radically different responses to Snowden's revelations in the two countries. Germans have sought to recognize Snowden as an advocate for freedom, bestowing honorary degrees or other awards of distinction on him, or by naming plazas and streets after the former NSA contractor. The Academic Senate of the Free University of Berlin granted Snowden an "honorary membership" in appreciation for his "exceptional commitment to transparency, justice and freedom" (ASTA FU 2014). Just blocks from Dresden's marvelously restored baroque Frauenkirche, a private landowner has named a plaza in Dresden's Neustadt district "Edward Snowden Platz" (Noack 2015). There have been few such gestures of veneration in the United States, where Snowden still faces a federal criminal indictment that could result in a lengthy prison sentence—if the American authorities can get their hands on him. Perhaps worse than the government's strong condemnation, it seems that the American public quickly lost interest in Snowden. One commentator wondered if Snowden's revelations have grown stale or have "proven to be inaccessible or not titillating enough for the American public" (Chandler 2015).

There can be little doubt about Americans' and Germans' dramatically different responses to Snowden and the NSA intelligence-gathering operations he disclosed. Germans are inclined to see Snowden as a hero who cast light on highly intrusive and unnecessary surveillance programs. Americans are inclined to see Snowden as a well-intentioned criminal who jeopardized valuable anti-terrorism programs. A large majority of Germans (61%) approved of Snowden's actions, even if they were illegal. Sixty percent see him as a "hero" and not as a "criminal" (Spiegel 2013). PEW Research (2013), on the other hand, registered an increase in the percentage of Americans who believe the government should pursue a criminal case against Snowden (Motel 2014). On the basis of a survey conducted in the days immediately following the media's initial extensive coverage of Snowden's disclosures, PEW reported that a narrow majority of Americans (56%) found the NSA's intelligence-gathering operations to be acceptable (Cohen 2013). In June 2013—at the height of the sensational coverage of Snowden's leaks—a majority of Americans (53%) believed that the NSA's programs helped prevent terrorist attacks (PEW 2013).

American and German differences with respect to personal information privacy and intelligence-gathering—and the resulting different reactions to Snowden's revelations—are not just reflected in labels and anecdotes. Social science research and survey data confirm the differences.

Research that draws on the characteristics of national culture described by G. Hofstede (1980, 1991) assigns the United States and Germany to different (albeit adjacent) clusters of national culture, identified respectively as the "Anglo" and the "Germanic Europe" cultural groups (CCL 2014). Building from these claims, many authors in the area of Information Science argue that they have "identified a relationship between national culture and attitude to information privacy" (Cockcroft 2007). Concerns about personal information privacy are stronger, the research suggests, in societies characterized by higher levels of power equality, higher levels of communitarianism, and higher levels of uncertainty avoidance (Bellman et al. 2004). In one study, Germans were found to be twice as likely as Americans to be concerned about personal information privacy (IBM 1999). Social and Information scientists seem willing to attribute this result to a German national culture that is—at least relative to America—more egalitarian, more communitarian and more averse to uncertainty.

The 2014 "Privacy Index" produced by the German Internet and technology consultancy EMC (see also Rosenbush 2014), and a parallel survey produced by the Boston Consulting Group (Rose et al. 2014), substantiate the claim that Americans and Germans have different expectations with respect to personal information privacy. The former report found, for example, that Germans are much less willing than Americans—by almost 20 percentage points—to trade some privacy for greater convenience (EMC 2014). Underscoring their general aversion to trading privacy for convenience—even in the commercial or consumer context—EMC's "Privacy Index" reported that Germans were more likely than Americans to believe that the law should prohibit businesses from buying and selling data without an individual's consent. While 92% of Germans thought that businesses should be legally barred from selling consumer information without consent, only 88% of Americans felt the same way. The latter report shows that, across a broad range of categories, Germans are significantly more likely than Americans to consider data to be "moderately" or "extremely" private (Rose et al. 2014), including: social network information (14% higher); information about media usage and preferences (10% higher); dialed-phone-number history (9% higher); exact location data (6% higher); and surfing history (5% higher).

The significant American and German differences regarding personal information privacy are also evident in the work of scholars and commentators.

German privacy scholars, for example, are inclined to see technology almost exclusively as an ominous threat. They devote large parts of their work to documenting the new and ever-deeper ways technology is intruding upon our privacy. In 2009, Peter Schaar, the former Federal Commissioner for Data Protection and Information Freedom (*Bundesbeauftragter für Datenschutz und Informationsfreiheit*) published a representative manifesto entitled *Das Ende der Privatsphäre* (*The End of Privacy*). His alarm taps into Germans' awareness of the fact that IBM punch card technology was used in the Nazis' 1938 census, which helped the *Reich* develop the demographic profiles it needed to implement the Holocaust.

Computing technology, Schaar warns, can lead to an all-encompassing surveillance state of the kind Orwell imagined. The first fifty pages of Schaar's book constitute a careful accounting of the many ways in which Orwell's vision is now being realized. Schaar concludes, for example, that the Internet "has a shadowy side". He warns against the state's collection of data about our normal activities and the grave risks for data protection that result from our deepening "*Vernetzung*" (increasing use of the Internet). The most threatening possibility, Schaar notes, comes from the role technology is coming to play in the health care sector, including digital and networked records-keeping and data-driven or biometric research and treatment. Christine Hohmann-Dennhardt, in a 2006 essay published during her service as a Justice at the Federal Constitutional Court, also lamented the way in which technology seems to have rendered privacy an "antiquated description of an idyllic condition that belongs to the past". The year before he joined the Federal Constitutional Court as the reporting justice for matters concerned with, inter alia, personal liberty and data-protection, Wolfgang Hoffmann-Reim wrote about the "new risks" resulting from "new technologies", a development he compared to an arms race (1998). In 2014, Spiro Simitis, one of Germany's best-known experts in the field of data protection, took a similar approach, expressing particular concern about the ways in which technology is helping businesses track—perhaps even manipulate—consumers' shopping activities (Simitis 2005). If they do so at all, these German scholars only reluctantly acknowledge the ways that the same technologies have improved our lives.

The general skepticism towards technology in German privacy scholarship is accompanied by a contrasting pastoral, quasi-spiritual conceptualization of privacy. Wolfgang Schmale and Marie-Theres Tinnefeld represent the most extreme version of this posture. In their 2014 book, *Privatheit im digitalen Zeitalter* (*Privacy in the Digital-Age*), they draw on the *Bible*'s "Garden of Eden" as a metaphor for privacy, because it points to the deeply-rooted cultural significance we place on the need for a protected retreat in which we can think and compose ourselves in full acceptance of nature and our bodies. The tangible garden, Schmale and Tinnefeld believe, should allow us to understand the abstract notion of data protection in more concrete terms. Sadly, they miss the chance to point out that it is in no small part the technology of a company called "Apple" that has chased us from privacy's paradise—or that the Garden of Eden may have been the most comprehensively surveilled place in human history (thanks to God's worrisome monitoring of the actions taken by Adam and Eve). In any case, Schmale and Tinnefeld see the European Union, with its culture and tradition of rights, as the "paradise" in which privacy can be restored. Hohmann-Dennhardt, taking a more secular turn, compared privacy to Rousseau's garden, in which one lives in simple harmony with nature. Simitis also understands data protection as part of an effort to fashion a utopian paradise. Schaar sees something of the sacred in privacy. He approvingly quotes Philippe Quéua, the former Director of the UNESCO's Division on Information and Society, who called privacy the "foundation of human dignity and the sacred nature of the human person". It is this quasi-spiritual approach to privacy that helps make sense of Schmale's and Tinnefeld's appeal for data and information aestheticism.

Daniel Solove, America's leading privacy scholar, sees things differently. First, Solove takes a more balanced approach to technology. He acknowledges that technology raises concerns about privacy. But Solove leaves space for an alternative view of these developments by acknowledging that "not everyone is concerned". He is less willing than the German privacy scholars to see technology in exclusively menacing terms. On the one hand, he characterizes many of the problems facing privacy as traditional or historical concerns, including risks to communications privacy (going back to the eras of letters, telegraphs, and telephones), risks resulting from information collection and surveillance (going back to ancient Jewish law and the original "peeping Tom" in the middle-ages), and risks resulting from information processing and aggregation (going back to the accelerating use of computers in the 1960s). These are old problems that are not exclusively linked to advances in technology. Nor is technology, for Solove, exclusively a threat. He is able to acknowledge the benefits of modern technology, even in areas (such as consumer data aggregation) that Schaar vilifies. "Identification is connecting information to individuals", Solove explains. While accepting that "identification" creates special problems, Solove recognizes that it also provides many benefits. Solove obviously cares a great deal about privacy. But he does not succumb to German scholars' Neo-Luddism.

Solove's most significant contribution to the theory of privacy is precisely his rejection of the broad and abstract approach adopted by German privacy scholars. Solove proposes a pragmatic, context-specific understanding of privacy. His is a "pluralistic" and not a "unitary" theory. Most conceptions of privacy suffer, Solove explains, because they are too broad. This is true of Louis Brandeis' and Samuel Warren's famous conclusion that privacy is the "right to be let alone". It is true of the notion that privacy involves a right to limit others access to the self, which

Solove sees as "too broad and vague". It is also true of the idea that privacy involves the right to control one's personal information. This approach is too broad, Solove explains, "because there is a significant amount of information identifiable to us that we do not deem as private". According to Solove, each of these general theories of privacy (and others I have not mentioned here) suffers from being "too vague" or "too broad". To solve this problem—which plagues the German approach to privacy— Solove proposes treating "privacy" as an "umbrella term that refers to a wide and disparate group of related things". Those "things", Solove urges, must be assessed pragmatically in their specific contexts. He quotes Serge Gutwirth, who observed that "Privacy ... is defined by its context and only obtains its true meaning within social relationships". With this admonition in mind, Solove proposes differentiated concepts of privacy for distinct circumstances, including private relations in the family, privacy relating to one's body and sex, privacy associated with the home and privacy connected with communications.

America and Germany have different cultural expectations of personal information privacy. The question remains: how are these cultural differences reflected in the two countries' legal regimes for privacy?

Different Transatlantic Privacies

The different conceptions of privacy in America and Germany are shaped by and reflected in discordant regulatory regimes for the protection of privacy, especially in the context of the state's surveillance and intelligence-gathering activities. Our different notions of privacy are the consequence of different histories, different social and cultural forces, different political traditions and institutions, different legal cultures and different constitutional regimes. I will highlight only a few of these differentiating factors, including different American and German histories regarding privacy and intelligence-gathering; the two countries' different political cultures, which lead policy makers in the two systems to strike different balances with respect to the protection of privacy and the threat posed by terrorism; and the ways in which their different constitutional regimes operationalize different legal conceptions of privacy.

Different Histories

A common explanation for Americans' and Germans' different responses to the NSA-Affair is that their reactions reflect the disparate experiences they have made with respect to terrorism and their countries' use of personal surveillance. On both points, America and Germany have very different histories.

On the one hand, while the American government has long had an excessive interest in collecting information about its citizens, Americans have not had to confront brutal and invidious totalitarian dictatorships, such as those that used personal information to terrorize all Germans between 1933 and 1945 and East Germans between 1949 and 1990. On the other hand, the contemporary American acceptance of government intelligence gathering reflects the still-recent trauma of the 11 September 2001 terrorist attacks in the United States. Germany has its own history with terrorism. And Germany is a target of the current brand of Islamist terrorism (Deutsche Welle 2015; VICE 2016). Yet, the terror of the German Autumn is now several generations old, and the country—unlike its European neighbors in Spain, England, and France—has so far avoided large-scale Islamist terror attacks. The experience Germans had (and have been socialized to remember in subsequent generations) with Nazi and East German authoritarian surveillance and control helps to explain why German law places such a high priority on personal information privacy as a fundamental liberty protection (Gujer 2010). Germans have deep and profound historical reasons to prioritize privacy and no recent terrorist trauma that would suggest the need to sacrifice privacy in the name of security.

America's spies, domestic and foreign, have not been angels. But comparisons with the Gestapo and Stasi are fallacious. The FBI has played a role in curtailing personal freedoms. The CIA has killed and sown the seeds of bloody discord around the world. But it cannot be said that the American intelligence community was a central cog in one of history's largest and most gruesome genocides, or that it implemented one of history's most thorough, invasive and sinister regimes of surveillance and social control. It is an unfortunate fate, but those are distinctly German histories. The intrusions on privacy with which the American public has been confronted—including the programs revealed by Snowden—are a pale reflection of the domestic terror German governments have (relatively recently) inflicted on their citizens with the help of secret, state-sanctioned surveillance and intelligence gathering.

But it is not only the different quality (or quantity) of intelligence abuses that distinguishes the American and German histories. The consequences of the abuses, once exposed, also differ in significant ways. Americans have come to understand that intelligence abuses inevitably come to light and can be met with democratic responses inside the state's institutions and structures. This is the enduring lesson of the Church Committee (Miller 2008). It has also been true in the post-9/11 era. The scandal involving the Terrorist Surveillance Program prompted President George W. Bush to discontinue the NSA initiative and to place future surveillance programs under the authority of the Foreign Intelligence Surveillance Act and the Foreign Intelligence Surveillance Court (NYT 2007). Snowden's revelations have generated significant reform, including President Obama's Policy Directive 28 and the USA Freedom Act (The White House 2014). By contrast, the only outcome of the extreme intelligence abuses Germans endured in the 20th century (under the Nazis and in East Germany) was the complete dissolution of the respective states. External forces were needed in both cases—with respect to the German Reich and the German Democratic Republic—to overcome the political cultures that had fostered and facilitated massive surveillance regimes. Unlike the Americans, the Germans have not experienced the corrective possibility of an existing democratic system confronting their worst intelligence abuses.

Different Political Cultures

A country's response (legal or otherwise) to the threat of terrorism is affected by many factors. It is the most straightforward republican calculation, but one factor is the degree to which the political class is required to be attuned and accountable to popular sentiments, such as fear of terrorism. The American and German political systems calibrate this dynamic differently. According to typologies originally mapped by comparative political scientists such as Lijphart (1999), American politics are seen as more majoritarian while German politics are seen as more consensual (Dickovick and Eastwood 2013). Democracies classified as majoritarian are characterized by high levels of subsystem autonomy and intense competition for majoritarian support among elites (Lijphart 1969). Consensual democracies are characterized by limited subsystem autonomy and deliberate efforts on the part of elites to take actions that counteract the potentially destabilizing impulses of shifting majorities. Confirming Germany's classification as a consensual democracy, Ralf Dahrendorf famously described German politics as "government by elite cartel" (1967). Elsewhere, Lijphart has used the concepts "mass political culture" and "elite political culture" to describe these distinct democratic approaches (1971).

A number of features in the two systems confirm these labels. America is a heterogeneous society with strong subsystem autonomy. Politics in the United States harnesses these forces through multi-level and nearly constant competition in the formation of governing majorities and for the framing of policy. The majoritarian and accountability elements of the systems are institutionally secured through biennial, direct elections for Congress and the (seeming) direct election of the president (U.S. Const. art. II, § 2–3; Dahl 2003). The autonomy of subsystems can be seen in the relative lack of party discipline and the mélange of civil society advocates, activists and lobbyists (Beutler 2014). Germany is a more homogenous society with weaker subsystem autonomy. Elites in Germany have seized on these factors to fashion and maintain a governing consensus. In its most benign form, this has served as a curative to the highly-fractious and unstable politics of the Weimar era (Schwarz 2010). Germany's consensus politics are facilitated by a number of structures, including the so-called *Parteienstaat* (which almost exclusively privileges the traditional political parties in the democratic process); proportional, party-based election of half the parliamentarians; and the proportional-parliamentary election of the chancellor (Kommers and Miller 2012). Grand coalitions featuring the largest center-right parties (CDU-CSU) and the largest center-left party (SPD) are a prominent example of Germany's consensus politics (Lijphart 1969). Three of the last four governments have been formed through grand coalitions of this type.

The distinct political cultures, and the institutions that reinforce them, produce different conditions with respect to the control and oversight of intelligence services—and, by extension, the two societies' understanding of privacy. The strict separation of powers in America's Madisonian system, for example, permits Congress to play a significant role in overseeing the executive's intelligence-gathering operations. This can be reinforced by frequent partisan splits between the presidency and the Congressional majority. Again, this was the lesson of the Church Committee. Especially in the wake of the 11 September 2001 terrorist attacks, an inter-branch and bipartisan security consensus formed in America that undermined the possibility that checks and balances would be an adequate brake on the government's intelligence-gathering activities. Still, Congress found extremely rare common ground to enact the USA Freedom Act in 2015, a move that one commentator described as a signal "that the days when Congress gave maximal deference to the executive branch might finally be over" (Lemieux 2015). The success of this reform was also a product of America's strong subsystem autonomy, which helps to explain the emergence and political success of the "Tea Party" movement (Beutler 2014). The Tea Party movement has, in part, been animated by libertarian concerns about government overreach, including on issues of intelligence gathering and security (Clement 2013). The 2012 Senate Intelligence Committee's historic "Study of the Central Intelligence Agency's Detention and Interrogation Program" may be a more inspiring example of the possibility in the American system for inter-branch oversight. The *Bundestag's* intermingled relationship with the chancellor and her cabinet—typical of the parliamentary model—leaves the German parliament with a smaller role in controlling the executive's intelligence-gathering function. The success of the *Bundestag*'s Investigative Committee on the NSA-Affair, for example, will largely depend on the engagement of the parliamentary opposition, which held only four of the Committee's sixteen seats. With these structural differences in mind, it is not surprising that the American judiciary has shown more restraint than the German judiciary in its review of privacy and intelligence-gathering cases.

Different Constitutional Laws of Privacy

An examination of the two countries' constitutional systems reveals dramatic differences with respect to privacy. These regimes are distinguished, in part, by their different constitutional texts and a resulting, very different jurisprudence of privacy.

If constitutional text is the beginning of constitutional analysis, then American and German constitutional law start from very different places with respect to the issue of personal information privacy. It is an old trick, for example, to note that the U.S. Constitution never uses the term "privacy" while the German constitution does. Article 10 of the Basic Law provides that "the *privacy* of correspondence, posts and telecommunications shall be inviolable". More than its mere invocation of the term "privacy", Article 10 is significant because it establishes a concrete constitutional protection for the exact activities involved in the NSA-Affair. With its modern outlook, Article 10 also seems to better anticipate the contemporary forms of electronic communication—such as email and smartphone usage—that are central to reimagining privacy for our digital age (Fetzer and Yoo 2013). Naturally, America's 18th century text is more awkwardly suited to that project.

Of course, constitutional law is not bound to the narrowest construction of the charter's text. Slightly broader readings of both constitutions reveal a number of liberty protections that serve the same interests as those we imagine to be involved in privacy (de Vries 2013). Without using the term "privacy", the Fourth Amendment to the U.S. Constitution nevertheless protects Americans from unreasonable searches and seizures "in their persons, houses, papers, and effects". This has been extended to include some of the forms of communication covered by Article 10 of the German Basic Law (Fetzer and Yoo 2013). Both constitutions also protect against government intrusions into the home (U.S. Const. amends. III, IV, Basic Law art. 13).

The Basic Law, however, prominently includes text that identifies and protects liberty interests and values that can be more easily read to be constituent elements of privacy. Articles 2 and 1 of the Basic Law, for example, are clear and very expressive commitments to personal freedom and human dignity. The human condition to which these protections aspire—including the relationship to state power—obviously involves an inviolable intimate sphere (Kommers and Miller 2012). America's Due Process Clause has been put to similar use, but without the same clarity and expressive force (Stephens 2015). Similarly, the Eighth Amendment to the U.S. Constitution, which prohibits "cruel and unusual punishments", asserts a dignified (privacy-respecting) image of the human condition, but does so chiefly in the limited circumstances of the state's penal function.

On the basis of their distinct constitutional texts and traditions, the American and German courts have developed dissimilar jurisprudences on the issues of personal information privacy. German law resorts to a general concept of privacy derived from Articles 2 and 1 of the Basic Law. American law recognizes discrete privacy interests to which it extends distinct legal protections. The privacy interests implicated by the NSA-Affair, for example, are chiefly a concern of the Fourth Amendment to the U.S. Constitution.

A number of recent cases decided by the Constitutional Court have recognized a right to personal information privacy in surveillance or data-collection scenarios on the basis of the Court's pioneering jurisprudence that conceived a right to "informational self-determination". The right was first articulated in the 1983 *Census Act Case* (65 BVerfGE 1, 1982). The Constitutional Court demanded that the parliament amend the federal census statute to ensure that there would be no abuses in the collection, storage, use and transfer of the personal data gathered during the census. The Constitutional Court demonstrated remarkable foresight—with respect to technology, data collection and the potential for the chilling effects of surveillance—when articulating the basis for the new right.

The Constitutional Court derived the right to informational self-determination from the general personality and dignity protections secured by Articles 2 and 1 of the Basic Law. That constitutional doctrine has provided a foundation for a general concept of privacy that finds relevance in a number of settings. The easy application of this general privacy interest in circumstances as various as transsexual rights and information privacy is a consequence of the German legal culture's preference for abstract concepts.

The right to informational self-determination has taken on increasing relevance as Germany pursued its own counter-terrorism measures in the wake of the 11 September 2001 attacks in the United States. Perhaps the most dramatic example of the Constitutional Court's promotion of privacy in the context of those policies was the *Online-Durchsuchungen Entscheidung (Online Computer Surveillance Case)*, which was decided in 2008 (120 BVerfGE 274). The Court issued a landmark ruling in defense of the "right to the confidentiality and integrity of information technological systems". The Constitutional Court derived this right from the general personality protection secured by Articles 2 and 1 of the Basic Law. The case involved challenges to a state law that empowered intelligence officials to conduct surveillance and collect data by covertly infiltrating computer systems through the Internet. The decision extended the Basic Law's privacy protection to personal computers. The Constitutional Court explained that "today's personal computers can be used for a wide variety of purposes, some for the comprehensive collection and storage of highly personal information… corresponding to the enormous rise in the importance of personal computers for the development of the human personality". The right to informational self-determination, said the Constitutional Court, protects individuals against the disclosure of personal data unless surveillance and data collection is necessary to avoid a "concrete danger" to human life or the security of the state. The Constitutional Court noted, "the fundamental right to the integrity and confidentiality of information technology systems is to be applied… if the empowerment to encroach covers systems that, alone or in their technical networking, contain personal data of the person concerned to such a degree that access to the systems facilitates insight into significant parts of the life of a person or indeed provides a revealing picture of his or her personality". The Constitutional Court concluded that general exploratory online searches based on mere suspicion of some remote danger, however serious, is constitutionally impermissible.

American courts, if they are willing to engage with the issues raised by the NSA-Affair, have largely been concerned with the discrete privacy protection provided by the Fourth Amendment to the U.S. Constitution. The broader notion of privacy, anchored in the due process clauses of the Fifth and Fourteenth Amendments, has not played a role.

The Fourth Amendment was implemented in response to the British practice of issuing general search warrants that lacked probable cause (Pittman 1954). And, to the degree that it secures protection of the individual against the overwhelming power of the state, the Fourth Amendment also is a reflection of the founding precepts of American democracy (Newman 2007). In its seminal decision in *Katz v. U.S.*, the Supreme Court rejected the traditional jurisprudence that had aligned the Fourth Amendment's privacy protection with notions of property and trespass (389 U.S. 347, 1967). The Court in *Katz* emphatically declared that "the Fourth Amendment protects people, not places". The substance of this protection consists in the requirement that government searches may be performed only when authorized by a detailed and specific warrant that has been issued by a neutral and detached magistrate on the basis of sworn evidence demonstrating probable cause (McInnis 2009). The Court has, however, identified a number of exceptions to the Fourth Amendment's warrant requirement, permitting searches that are otherwise "reasonable". Some contend that these exceptions have swallowed the rule, leaving the Fourth Amendment a hollow form that no longer provides meaningful privacy protection (Starkey 2012).

A threshold question is what constitutes a "search" for Fourth Amendment purposes. Far more than the substantive scope of Fourth Amendment protection, this preliminary issue has complicated the application of the Fourth Amendment to intelligence-gathering cases. After *Katz*, the occurrence of a "search" no longer depended on evidence that the state had made a physical intrusion into a private space. Instead, the Court found an intrusion into Katz's *personal sphere of privacy*. In *Katz*, a wiretap had been placed on the outside of a glass pay phone box permitting law enforcement officers to listen to Katz's phone conversation. Although no physical intrusion into the pay phone box had taken place, the Court reasoned that Katz had a subjective expectation that "the words he utters into the mouthpiece will not be broadcast to the world" and that society would accept Katz's expectation as reasonable. This is now the standard for determining whether a "search" has taken place, without which the substantive protections of the Fourth Amendment will not apply: (1) a person "has exhibited an actual (subjective) expectation of privacy"; and (2) society is prepared to recognize that this expectation is (objectively) reasonable.

The Supreme Court applied this standard in *Smith v. Maryland* and found that a Fourth Amendment search had not occurred (442 U.S. 735, 1979). This is relevant because the circumstances of the *Smith* case might be seen as closely analogous to those involved in the NSA-Affair. In *Smith*, law enforcement officers collected evidence of the suspect's telephone contacts by installing a "pen register" on his telephone line at the telephone company's offices. An electronic device, the pen register records only the numbers called from a particular telephone line. The content of phone calls is not documented. The Court concluded that neither of the elements necessary for a Fourth Amendment *search* existed in the case.

First, Smith did not have a subjective expectation in the privacy of the telephone numbers he dialed because "people in general [do not] entertain any actual expectation of privacy in the numbers they dial". The Court reasoned that telephone users know that the phone company registers the numbers they dial and keeps permanent records of that information for billing purposes.

Second, the Court found that a subjective expectation of privacy with respect to the phone numbers one dials—as unlikely as that expectation would be—cannot be regarded as reasonable. Society appreciates, the Court explained, that electronic equipment is used extensively to track and catalogue the telephone numbers called from any particular phone. At the very least, the Court concluded, this is common (and commonly known), because it is necessary for the telephone company to keep billing records. The Court ruled that, in dialing the telephone numbers, Smith held that information out to third parties (at least the telephone company). Exposing information in such an indiscriminate way, which stripped it of any subjective or objective expectation of privacy, meant that the government's collection of the telephone numbers involved only the acquisition of non-private information. On the basis of the third-party doctrine, the Supreme Court ruled that a search had not occurred and that the Fourth Amendment had no applicability to the case whatsoever.

Judge William Pauley of the Southern District of the New York Federal District Court drew on the obvious parallels between the facts in the *Smith* case and the NSA's bulk telephony metadata collection program when he dismissed a Fourth Amendment challenge to the NSA's surveillance measures in December 2013. Citing *Smith*, Judge Pauley ruled that phone users had no reasonable expectation of privacy that would give them Fourth Amendment rights, especially with respect to information they voluntarily provide to third parties, such as telephone companies (*American Civil Liberties Union v. James Clapper*). In 2015, on appeal from Judge Pauley's order, the United States Court of Appeals for the Second Circuit found that the NSA's bulk telephony metadata collection program exceeded the surveillance authority established by the relevant statutory provisions. But the Appeals Court refused to rule on the constitutional issues in the case, even as it expressed grave misgivings about the continuing adequacy of the *Smith* case and the third party doctrine for ensuring privacy under present technological conditions.

In at least two other lawsuits filed in response to the NSA-Affair (*Smith v. Obama* and *U.S. v. Muhtorov*), the first-instance courts found that the *Smith* precedent and the third-party doctrine precluded a Fourth Amendment challenge to the NSA's data-collection programs.

The high courts in both countires are increasingly aware of the challenge modern telecommunications technology poses to their respective privacy traditions. The German jurisprudence seems better adapted to the new circumstances. The German jurisprudence, dating back to the *Census Act Case*, has been conscious of the distinct privacy harm that could result from the accumulation of personal information data. In its recent cases, the Constitutional Court has sought to strengthen constitutional privacy protection in response to the sweeping personal portrait our ever-more extensive use of technology makes it possible for the state to develop from mere telecommunications metadata. This approach is in line with what is referred to as the "mosaic" theory of privacy, which seeks to account for intrusive conduct as a "collective whole", rather than as isolated or sequential incidents (Kerr 2012).

The American jurisprudence is just beginning to struggle with the dramatic challenge contemporary telecommunications technology poses for privacy. If the courts will hear the cases at all, then so far they have hewn to the traditional sequential approach to enforcing the Fourth Amendment. The clearest move in the direction of the mosaic approach occurred in the *Carpenter v. United States* case decided by the Supreme Court in 2018. Carpenter challenged the government's use of cell-site location information (CSLI) as evidence of his proximity to a number of robberies. CSLI is created as users' cell phones constantly scan their vicinity for the most effective available cell tower. This record produces an increasingly precise record of a cell phone's geographic location. Cell phone network providers routinely document this information. The Supreme Court ruled that the use of the CSLI records should be distinguished from the dialed phone numbers involved in *Smith*, and that their use in Carpenter's trial constituted an illegal "search" under the Fourth Amendment. Chief Justice Roberts wrote the opinion for the Court. First, the Court nodded towards the mosaic theory of privacy by noting that CSLI records represent a different quantity and quality of information, which is comprehensive, encyclopedic and effortlessly compiled. This information, Justice Roberts explained, touches on the concern the Supreme Court has shown for the privacy owed to "the whole of a person's physical movements". Justice Roberts warned that CSLI represents near perfect surveillance of a person's location—going back almost five years. The risk of excluding CSLI from the protection of the Fourth Amendment, the Court insisted, results in part from the pervasive and insistent role of cell phones in modern life. Second, the Court distinguished the CSLI from the dialed phone numbers in *Smith* by noting that the information involved is not voluntarily shared in any common understanding of that term. Justice Roberts explained that cell phones automatically and constantly produce CSLI, even when the phone's user is not actively employing one of the phone's applications.

For all their differences with respect to the constitutional protection of privacy, the Supreme Court's decision in the *Carpenter* case suggests that American and German jurisprudence might be converging in this discrete but profound way. It would be wrong, however, to rush from this conclusion to any claim of harmonization or a hoped-for universalization of privacy rights. The Supreme Court insisted in *Carpenter* that its decision was narrow and does not disturb the broad and general application of the third-party doctrine articulated in the *Smith* case.

History, politics and law confirm and explain the different notions of privacy prevalent in America and Germany.

Conclusion

The different responses to the NSA-Affair in America and Germany are the product of the two countries' different notions of privacy. Those distinctions are embodied in—and foster—very different legal regimes for the protection of privacy.

The American approach shows greater confidence in the political process for striking the balance between privacy and security. When the courts become involved, they enforce a specialized constitutional privacy right that has been calibrated to respond to the state's surveillance and intelligence-gathering activities. This jurisprudence, so far, has only cautiously embraced the mosaic approach to privacy, which seems to better account for the comprehensive and intimate uses to which we put technology today. The German approach emphasizes the judicial enforcement of a broad and general concept of privacy. In its sensitivity to technology's ubiquity and the deeply revealing portraits that can be developed through the accumulation of a vast amount of discrete data, the German jurisprudence has been a pioneer of the mosaic approach to privacy.

Most profoundly, operating in their unique socio-legal contexts, the two constitutional privacy regimes offer very different visions of personhood. On the one hand, the German Constitutional Court has imagined and enforced a substantive and objective vision of personhood that includes a protected private and intimate sphere. The state is obliged to help realize this vision. On the other hand, the American courts have reinforced individuals' freedom of action, including the autonomy to dispose of one's privacy. This is an autonomous and subjective vision of personhood.

The challenge posed by the NSA-Affair—a challenge that underlies all our discussions about privacy in our increasingly digitalized and data-centric future—is not to envision and enforce a harmonized approach to privacy, but to come to accept with William Shakespeare that "a rose by any other name would smell as sweet". Our best chance for acting productively to ensure privacy is to appreciate and respect our different notions of privacy.

References

Articles

Altman, I. (1977). *Privacy regulation: Culturally universal or culturally specific? Journal of Social Issues, 33*(66).

Barnett, S. R. (1999). *The right to one's own image: Publicity and privacy rights in the United States and Spain. American Journal of Comparative Law, 47*, 555.

Bignami, F. (2007). *European versus American Liberty: A comparative privacy analysis of antiterrorism data mining. Boston College Law Review, 48*, 609.

Chandler, A. (2015, April 6). What it takes to make people care about NSA surveillance. *The Atlantic*. http://www.theatlantic.com/national/archive/2015/04/naked-selfies-and-the-nsa/389778/.

Dickovick, J. T., & Eastwood, J. (2013). *Comparative Politics, 452*.

Fetzer, T., & Yoo, C. S. (2013). New technologies and constitutional law. In M. Tushnet, et al. (Eds.), *Routledge handbook of constitutional law* (Vol. 485, pp. 490–492).

Hoffmann-Riem, W. (1998). Informationelle Selbstbestimmung in der Informationsgesellschaft—Auf dem Wege zu einem neuen Konzept des Datenschutzes. *Archiv des öffentlichen Rechts, 123*, 513, 517–518.

Hohmann-Dennhardt, C. (2006). Freiräume—Zum Schutz der Privatheit. *Neue Juristische Wochenschrift, 59*, 545, 546.

Kerr, O. S. (2012). The Mosaic theory of the fourth amendment. *Michigan Law Review, 111*, 311, 316–317, 320.

Krotoszynski, R. J., Jr. (2014–2015). Reconciling privacy and speech in the era of big data: A comparative legal analysis. *William & Mary Law Review, 56*, 1279.

Lachmayer, K. (2014). The challenge to privacy from ever increasing state surveillance: A comparative perspective. *UNSW Law Journal, 37*, 748.

Lijphart, A. (1969). Consociational democracy. *World Politics, 21*, 207.

Lijphart, A. (1971). Comparative politics and the comparative method. *American Political Science Review, 65*, 682, p. 213.

Pittman, R. C. (1954, May). The supremacy of the judiciary: A study of preconstitutional history. *ABAJ, 40*, 389, 391 (quoting Justice Horace Gray Jr.).

Starkey, B. S. (2012). A failure of the fourth amendment & equal protection's promise: How the equal protection clause can change discriminatory stop and frisk policies. *Michigan Journal of Race & Law, 18*, 131.

Stephens, O. H., et al. (2015). II *American Constitutional Law* (6th ed., 406).

Walsh, B. (1976–1977). The judicial power and the protection of the right of privacy. *Dublin University Law Journal, 1*, 3.

Whitman, J. Q. (2004). The two western cultures of privacy: Dignity Versus liberty. *Yale Law Journal, 113,* 1151, 1155.
Yang, T. L. (1966). Privacy: A comparative study of english and american law. *International and Comparative Law Quarterly, 15,* 175.

Books

Dahl, R. A. (2003). *How democratic is the American Constitution?* (73).
Dahrendorf, R. (1967). *Society and democracy in Germany* (276).
de Vries, K. (2013). Privacy, due process, and the computational turn. In M. Hildebrandt & K. de Vries (Eds.), *Privacy, due process and the computational turn: The philosophy of law meets the philosophy of technology* (9, 17–19).
Gujer, E. (2010). Germany: The long and winding road. In G. J. Schmitt (Ed.), *Safety, liberty, and islamist terrorism: American and European approaches to domestic counterterrorism* (62, 63).
Hofstede, G. (1980). *Cultural consequences: International differences in work related values.*
Hofstede, G. (1991). *Cultures and organizations: Software of the mind.*
Kommers, D. P., & Miller, R. A. (2012). *The constitutional jurisprudence of the Federal Republic of Germany* (3rd ed., 269, at 405).
Lijphart, A. (1999). *Patterns of democracy.*
McInnis, T. (2009). *The evolution of the fourth amendment* (51–56, 75).
Newman, B. A. (2007). *Against that "Powerful Engine of Despotism": The fourth amendment and general warrants at the founding and today.*
Miller, R. A. (Ed.). (2008). *U.S. National Security, Intelligence and Democracy: From the Church Committee to the War on Terror.*
Nissenbaum, H. (2009). *Privacy in context.*
Schaar, P. (2009). *Das Ende der Privatsphäre* (pp. 38, 40, 42, 49, 54, 75–78).
Schmale, W., & Tinnefeld, M.-T. (2014). *Privatheit im digitalen Zeitalter* (11, pp. 18, 86, 244).
Schwarz, H. P. (2010). Woran scheitern deutsche Bundeskanzler? In C. Hillgruber & C. Waldhoff (Eds.), 60 *Jahre Bonner Grundgesetz – Eine geglückte Verfassung?* 29.
Simitis, S. (Ed.). (2014). *Bundesdatenschutzgesetz—Kommentar* (8th ed.).
Simitis, S. (2005). Datenschutz—eine notwendige Utopie. In R. M. Kiesow, et al. (Eds.), *Summa— Dieter Simon zum 70. Geburtstag* (pp. 511, 527).
Solove, D. J. (2008). *Understanding privacy* (4, 5, 9, 16, 18, 20, 25, 29, 45, 47, 50, 61–62, 107–108, 118, 122–123).

Judgements and Laws

American Civil Liberties Union (ACLU) v. James Clapper, No. 13-3994 (S.D. New York December 28, 2013), 959 F.Supp.2d 724.
American Civil Liberties Union v. James Clapper, 785 F.3d 787 (2d. Cir. 2015).
Carpenter v. United States, No. 16-402, 585 U.S. ___ (2018).
Katz v. United States, 389 U.S. 347 (1967).
Smith v. Maryland, 442 U.S. 735 (1979).
Smith v. Obama, 24 F.Supp.3d 1005 (2014).
United States v. Muhtorov, No. 1:12-cr-00033-JLK (D. Colo.).
Census Act Case, 65 BVerfGE 1 (1982).
Online Computer Surveillance Case, 120 BVerfGE 274 (2008).

Constitution of the United States, https://www.senate.gov/civics/constitution_item/constitution.htm.

German Basic Law (C. Tomuschat, D. P. Currie, & D. P. Kommers, Trans.). https://www.gesetze-im-internet.de/englisch_gg/englisch_gg.pdf.

Media and Art

AStA FU. (2014, June 23). *AStA FU Informs Edward Snowden about his Honorary Membership and Demands Asylum for Edward Snowden*, AStA FU. https://www.astafu.de/content/asta-fu-informs-edward-snowden-about-his-honorary-membership-and-demands-asylum-edward-snowd.

Beutler, B. (2014, May 21). The American right, not the 'Tea Party,' is the GOP's big liability. *New Republic*. https://newrepublic.com/article/117845/gop-primary-victories-aside-republicans-still-have-gop-base-problem.

Clement, S. (2013, July 26). Tea party privacy concerns skyrocket, poll finds. *Washington Post*. https://www.washingtonpost.com/news/the-fix/wp/2013/07/26/tea-party-privacy-concerns-skyrocket-poll-finds/.

Cohen, J. (June 10, 2013). Most Americans back NSA tracking phone records, prioritize probes over privacy. *Washington Post*. https://www.washingtonpost.com/politics/most-americans-support-nsa-tracking-phone-records-prioritize-investigations-over-privacy/2013/06/10/51e721d6-d204-11e2-9f1a-1a7cdee20287_story.html.

Deutsche Welle. (2015, February 28). *Police in German city warn of islamist terrorism threat*. http://www.dw.com/en/police-in-german-city-warn-of-islamist-terrorism-threat/a-18286410.

Lemieux, S. (June 8, 2015). How congress learned to stop bowing to president Obama on national security. *The Week*. http://theweek.com/articles/558953/how-congress-learned-stop-bowing-president-obama-national-security.

Noack, R. (2015, June 23). The global cult of Edward Snowden keeps growing. *Washington Post*. https://www.washingtonpost.com/news/worldviews/wp/2015/06/23/the-global-cult-of-edward-snowden-keeps-growing/.

Richter, S., & Albrecht, J. P. (2013, October 30, 7:19 AM). NSA spying on Europe reflects the transatlantic culture gap. *The Guardian*. http://www.theguardian.com/commentisfree/2013/oct/30/nsa-spying-europe-transatlantic-culture-gap.

Rosenbush, S. (June 12, 2014, 5:16 PM). EMC's new index shows public is deeply conflicted over privacy. *Wall Street Journal*. http://blogs.wsj.com/cio/2014/06/12/emcs-new-index-shows-public-is-deeply-conflicted-over-privacy/.

Spiegel Online. (2013, November 8). Spying fallout: German trust in United States plummets. http://www.spiegel.de/international/germany/nsa-spying-fallout-majority-of-germans-mistrust-united-states-a-932492.html.

The New York Times, (2007, January 17). Letter from Alberto R. Gonzalez, Attorney General of the United States, to Senator Patrick Leahy and Senator Arlen Specter. http://graphics8.nytimes.com/packages/pdf/politics/20060117gonzales_Letter.pdf.

The White House, Press Release, Office of the Press Sec'y, Presidential Policy Directive/PPD-28, at 5. (2014, January 17). Uniting and Strengthening America by fulfilling rights and ending eavesdropping, dragnet-collection and online monitoring act [USA FREEDOM Act], Pub. L. No. 114-23, 129 Stat. 268 (2015).

VICE. (Jan. 1, 2016, 10:49 AM). Germany still on alert after tip about possible Islamic State terror attack. https://news.vice.com/article/germany-still-on-alert-after-tip-about-possible-islamic-state-terror-attack.

Studies and Reports

Bellman, S., et al. (2004). international differences in information privacy concerns: A global survey of consumers. *The Information Society, 20*, 313, 315, 321.

Center for Creative Leadership. (2014). Leader effectiveness and culture: The GLOBE study. http://www.ccl.org/leadership/pdf/assessments/GlobeStudy.pdf.

Cockcroft, S. (2007). Culture, law and information privacy. In *Proceedings of European and Mediterranean Conference on Information Systems* 2007 (EMCIS2007), June 24–26, 2007, Polytechnic University of Valencia, Spain. http://www.academia.edu/7688223/CULTURE_LAW_AND_INFORMATION_PRIVACY.

EMC, The EMC Privacy Index Global & In-Depth Country Results. (2014). http://www.emc.com/collateral/brochure/privacy-index-global-in-depth-results.pdf.

IBM. (1999, October). *IBM multi-national privacy survey consumer report 14*. IBMftp://www6.software.ibm.com/software/security/privacy_survey_oct991.pdf.

Motel, S. (2014, April 15). NSA coverage wins Pulitzer, but Americans remain divided on Snowden leaks. Pew Research Center. http://www.pewresearch.org/fact-tank/2014/04/15/nsa-coverage-wins-pulitzer-but-americans-remain-divided-on-snowden-leaks/.

Pew Research Center. (2013, June 17). Public split over impact of NSA leak, but most want Snowden prosecuted. http://www.people-press.org/2013/06/17/public-split-over-impact-of-nsa-leak-but-most-want-snowden-prosecuted.

Rose, J., et al. (2014, February 19). Data Privacy by the Numbers, *bcg.perspectives*. https://www.bcgperspectives.com/content/Slideshow/information_technology_strategy_digital_economy_data_privacy_by_the_numbers/#ad-image-0.

Senate Select Committee on Intelligence—Committee's Study of the Central Intelligence Agency's Detention and Interrogation Program. (December 13, 2012). Available at https://web.archive.org/web/20141209165504/http://www.intelligence.senate.gov/study2014/sscistudy1.pdf.

Professor Russell Miller is the J.B. Stombock Professor of Law at Washington and Lee University. His teaching and research focus on public law subjects and comparative law theory and methods. Professor Miller, an expert in German law and legal culture, is the author/editor of a number of books. He has been recognized for his work on German law and transatlantic affairs. In 2013 Professor Miller was named a KoRSE Fellow (http://www.sicherheitundgesellschaft.uni-freiburg.de/) at the University of Freiburg. Professor Miller was a 2009/2010 Fulbright Senior Research Fellow in residence at the Max Planck Institute for Comparative Public Law and Public International Law (http://www.mpil.de/ww/en/pub/news.cfm) in Heidelberg, Germany. In appreciation for his work on behalf of the Robert Bosch Fellowship and German-American relations, Professor Miller was recognized as the Bosch Alumni of the Year in 2012. Professor Miller has contributed to or been quoted or cited in a number of global media sources, including The *Los Angeles Times*, *the Frankfurter Allgemeine Zeitung*, *Reuters*, and *Der Spiegel*.

Chapter 5
Regulating the Internet—Necessary Evil or Squandered Opportunity?

Ruth Barber

Introduction

The birth of the Internet heralded a potential new world of freedom for trade and expression, and free from governmental interference, a fundamentally U.S. libertarian worldview. As the Internet moved from a U.S. based online community of technical enthusiasts to become a global communication network, attempts were made to regulate cyberspace by competing commercial, security and moral interests. Governments attempted to assert control through technical and regulatory measures. Control measures imposed include geo-blocking, ISP regulation and hardware control (China), the repeal of Net Neutrality rules (USA), the Network Enforcement Act (Germany) and the General Data Protection Regulation (EU).

However, the global and networked nature of the Internet makes it inherently resistant to geographical control, and regulatory measures imposed tend to result in jurisdictional overreach. How then should the Internet be regulated, assuming that regulation is now deemed necessary? Cyberspace is increasingly dominated by a few transnational platform providers. Is the regulation of the Internet now effectively controlled by these organizations, whose economic power and political influence exceeds the power of many nation states? If so, has the Internet become the independent jurisdiction that its early creators dreamed it should be?

R. Barber (✉)
London, Great Britain
e-mail: ruth.barber@ruthbarberconsulting.co.uk

Berlin, Germany

© Springer Nature Switzerland AG 2020
D. Feldner (ed.), *Redesigning Organizations*,
https://doi.org/10.1007/978-3-030-27957-8_5

79

The Information Sharing Revolution

The development of the Internet has enabled an information sharing revolution. Ordinary citizens can now bypass the publishing houses that previously controlled information sharing. Citizens have the independent power to share information quickly and cheaply with millions of other people throughout the globe.

The information sharing phenomena took off via social media platforms such as Facebook, YouTube and Twitter. These sites are designed to facilitate information sharing to maximize user engagement. They do not charge a fee for their use but are funded by advertising revenue. Maximum user engagement means maximum advertising revenue. The more information the sites have about their users, the better the advertising can be targeted. Cambridge Analytica, a political advertising company, harvested millions of Facebook users' data through a third-party application in order to better target political advertising (Cadwalladr and Graham-Harrison 2018).

Social networking sites claim to be information sharing platforms rather than publishers, and therefore exempt from defamation laws (Jarvis 2018; Kiss and Arthur 2013; House of Lords 2018). The European eCommerce Directive 2000/31/EC Article 12 provides a liability exemption for online information hosts providing they have no prior knowledge of unlawful activity and act quickly to remove the offending material when notified. This provision is mirrored in the U.S. in section 230 of the Communication Decency Act.

Social networking platforms claim to exercise no editorial control over content published save via their community guidelines (Facebook 2018), which filter for obscene and violent content. The platforms do control what material appears in a user's news feed by use of an algorithm, periodically tweaked (Peters 2018). Facebook has admitted to experimenting on groups of users by deliberately manipulating the material appearing in their news feeds. The use of this algorithm weakens Facebook's assertion that it is a mere information conduit.

A 2018 court case has eroded Facebook's claim to be a mere information sharing platform. Martin Lewis, a British champion for consumer rights, sued Facebook when his image was appropriated without consent for advertising on Facebook. He argued that since he does not engage in Facebook advertising, it should be a simple matter for Facebook to remove every advert with his image without the need for him to report the advertisements. However, advertising is not third-party content, and the court held Facebook jointly liable (FT 2018).

As Facebook gets increasingly hit with lawsuits, it appears to now be willing to define itself as a publisher in order to take advantage of the first amendment of the US Constitution protection for publishers, which protects free speech. It still seems unfair to hold social media platforms, who are not editors, liable for user postings. It may be that historical terms such as "publisher" are no longer adequate to describe the modern media landscape and the scope of legal liability (Levin 2018).

Social Networks and Psychology

The social network business model has a number of unfortunate social effects. Profit is made from users responding to and sharing information, and human nature tends to respond automatically to information that surprises, shocks or offends. Since no control on social media is made for truth, profitable "fake news" proliferates. Much of this fake news is generated by individuals with no agenda other than to make money (Ball 2017).

Nefarious actors have also realized the potential for spreading political propaganda. Russia established troll farms for the purpose of spreading disinformation on social media sites and generating social unrest in the West (Green 2018). The development of "deep fake", an AI driven application, allows for digital impersonation and the production of convincing videos of people of doing and saying things they never did (Chesney and Citron 2018).

Rather than presenting a range of views that might challenge or inform, social networking sites give us more of what we like. This has the effect of reinforcing and entrenching existing views. It is alleged that the use of "fake news" had an influence on both the Brexit Vote and the U.S. presidential election in 2016 (Fiadh 2017).

When challenged, the response of social networking sites has been reluctant and ineffectual (Hill 2018). Their business models rely on the maximum amount of user engagement, both in terms of time and number of participants.

States that were previously keen to court the social networking companies with favorable tax rates have become increasingly hostile (Cadwalladr 2018). Users are also becoming increasingly disillusioned and are leaving the platforms and sharing less personal information (Locklear 2018).

Concern about user disengagement as a result of privacy concerns has resulted in greater efforts by social networking sites to demonstrate privacy protections. However, a business model that is predicated on harvesting data for profit is arguably always vulnerable to abuse. A subscription model would remove the reliance on advertising revenue, but it would also discourage many users, reducing market share.

A Borderless Internet?

Governments cannot stop electronic communications from coming across their borders, even if they wanted to do so. Nor can they credibly claim a right to regulate the Net based on supposed local harms caused by activities that originate outside their borders and that travel electronically to many different nations. One nation's legal institutions should not monopolize rule-making for the entire Net. Even so, established authorities will likely to continue to claim that that they must analyze and regulate the new online phenomena in terms of physical locations. After all, they argue, people engaged in online communications still inhabit the material world, and local legal authorities must have authority to remedy the problems created in the physical world by those acting on the Net. (Johnson and Post 1996)

The prophesy of Johnson and Post is illustrated in the cases of LICRA and UEJF v. Yahoo! and Microsoft Corp v. United States. Both attempt to exert extraterritorial jurisdiction to the Internet. In LICRA and UEJF v. Yahoo!, attempts are made to regulate a U.S. website visible in France. In the Microsoft case, the question relates to a U.S. company storing information "in the cloud" but using a Dublin-based server.

LICRA and UEJF v. Yahoo! Inc

French users of the Yahoo! online auction site were able to view and purchase Nazi memorabilia. The offering of Nazi memorabilia for sale is prohibited in France. The League against racism and antisemitism and the Union of French Jewish Students brought a case against Yahoo! Inc. in France. The French court found a sufficient nexus to establish jurisdiction, since the items were viewable by French citizens and this was known to Yahoo! Inc., since they targeted French users with French advertising. Yahoo! Inc. had a French subsidiary, Société Yahoo! France.

The French court found in favor of the applicants and ordered Yahoo! to block Nazi memorabilia from French citizens. Yahoo! resisted the order and attempted to argue that selective blocking of French users was impossible; experts disputed the claim and argued that 90% blocking was possible. The court required Yahoo! to impose the blocking or be subject to fines of 15,244 Euros per day.

Yahoo! then filed a case against the applicants in a U.S. court (the District Court for the Northern District of California), arguing that the French decision was not binding upon them in the U.S. The applicants argued that any finding of the U.S. court was not binding on them in France.

The U.S. court stated, "A basic function of a sovereign state is to determine by law what forms of speech and conduct are acceptable within its borders". The issue was "whether it is consistent with the Constitution and laws of the United States for another nation to regulate speech by a U.S. resident on the basis that such speech can be accessed by Internet users in that nation" (LICRA v. Yahoo!). Yahoo!'s application was granted. LICRA appealed and the decision was reversed on the basis that the District Court did not have jurisdiction over LICRA.

Judge William Fletcher stated in a judgement on 12 January 2006:

> Yahoo! is necessarily arguing that it has a First Amendment right to violate French criminal law and to facilitate the violation of French criminal law by others. […] the extent — indeed the very existence — of such an extraterritorial right under the First Amendment is uncertain.

Following the judgement, Yahoo! elected to remove all Nazi memorabilia from its site.

Ultimately, Yahoo! Inc. was obliged to comply with the French judgement as a result of its economic interests in France. Businesses that have no economic interests in a country cannot be regulated so easily by legal means.

Microsoft Corp. v. United States

In 2013, Microsoft challenged a warrant issued under section 2703 of the Stored Communications Act (SCA) by the U.S. federal government to turn over the emails of an account that was stored in Ireland (Microsoft v. U.S.). It argued that the Act could not be used to compel American companies to produce data stored in servers outside the U.S. The judge at first instance found against Microsoft on the basis that the Act was not subject to territorial restrictions. The Irish government filed a brief in the proceedings, arguing that the decision violated the European Data Protection Directive and Ireland's data privacy laws. Ireland argued that the emails could only be disclosed on request to the Irish Government. Microsoft won on appeal, with the court holding that legislation only has national effects unless it is clearly expressed to the contrary. The warrant also did not specify whether the owner of the emails was a U.S. citizen or resident. The Department of Justice appealed to the Supreme Court.

While the Supreme Court hearing was pending, Congress passed the Clarifying Lawful Overseas Use of Data Act (CLOUD Act), which amended the SCA to specifically include cloud storage of U.S. providers, regardless of where the servers may be located. This act was supported by both the U.S. Government and Microsoft. The Supreme Court hearing was then moot, and the appeal court hearing vacated.

China

China realized the limited options for exercising legal control over U.S. providers of information technology that have no significant assets in the country, and instead developed indigenous versions of Google and Facebook products, which come preloaded on smartphones sold in China. The Chinese government has a monopoly on all national internet connections via the Ministry of Information Industry (MII). The number of ISPs is restricted, and each provider must be approved by the MII. ISPs are required to block websites named by the Public Security Bureau. In addition, the authorities monitor and control all traffic going through China's primary gateways to the global Internet (Chew 2018).

Compared to the U.S.—with its emphasis on freedom of expression embodied in the First Amendment—the Chinese approach is arguably restrictive and repressive. However, the right to freedom of expression is upheld in the U.S. at the expense of personal privacy, which in the U.S. has only minimal legal protection.

Germany

The Snowden revelations of 2013 revealed the extent of U.S. online intelligence operations that routinely harvest the data of users of U.S. based information technology services (Macaskill and Dance 2013). A backlash against U.S. information services occurred in Germany, when it was discovered that U.S. intelligence services had eavesdropped on the private telephone calls of Angela Merkel. Germans, who have a particular cultural sensitivity to personal data harvesting due to data abuses undertaken by the Nazi regime (Freude and Freude 2016), increasingly moved to onshore their data and transfer to national information service providers (Spiegel 2013).

Germany, long a champion of European privacy rights, also pushed for the enactment of the General Data Protection Regulation, which provides for strong privacy protections for European Citizens. The regulation also recognized the global effects of the web and the need to be able to enforce against businesses with neither a legal or a physical presence within the EU jurisdiction. Article 3(2) allows the application of the EU Regulation to non-EU based data processors that process the data of individuals in the EU in two situations: first, if they are offering goods and services in the EU; and second, if they are monitoring of the behavior of people in the EU.

"The Regulation amounts to a unilateral expansion of the application of European law to non-EU businesses. No one could deny that this expansion is justified by the borderless domain of the Internet, which in response requires also a borderless application of the law. In a way, there is no doubt that effective data protection on the Internet does not get along with a domestic scope of application. Nonetheless, the EU dares to go much further than any other state on this aspect, and with the highest level of standards in the world" (Azzi 2018). Effective enforcement is still largely reliant upon the companies having enough economic or political nexus with the EU.

Having seen the problems that arose from alleged online electoral interference in the UK referendum and the U.S. election and fighting battles against the rise of right-wing extremism that flourished on social networking sites, Germany pushed through the controversial Network Enforcement Act on 1 October 2017. This Act placed a legal obligation on large social networking platforms to remove content that offended specific provisions of the German Criminal Code or be subject to swinging fines. The Act was criticized by free speech champions also for its extraterritorial effect (Jash 2017). Complaints can only be made under the Act "in Germany", but targeted material may come from any jurisdiction.

United States of America

On 14 December 2017, the U.S. Federal Communications Commission voted to repeal Net Neutrality. Net Neutrality is the principle that internet service providers treat all data on the internet equally and do not discriminate or charge differently by user, content, website, platform, application, type of attached equipment or method of communication (Gilroy Gilroy 2011). The basis for the decision was to increase competition (Selyukh and Greene 2017), however, critics argue that the repeal of Net Neutrality permits government censorship (Skorup 2016) as it permits the preferential online delivery of, for example, government approved news channels. Twenty-two U.S. states are appealing against the ruling and a hearing is due in February 2019.

Legal Measures

Civil legal measures are only effective in so far as the regulated entity has assets in the jurisdiction of the complainant, or to the extent that judgements in one jurisdiction are enforceable in another. Countries with close political and trading arrangements may choose to honor judgements.

Online criminal enforcement is only possible between sympathetic jurisdictions with similar criminal law standards. A website removed in one country under a notice and take down procedure, can simply reappear hosted on a server in a more libertarian or sympathetic jurisdiction.

Some alt-right groups purged from Facebook and Twitter have joined the Russian Facebook clone VKontakte (Zavadksi 2017), which has greater tolerance for white supremacist views. Jihadi groups joined the encrypted social networking site Telegram. Some pornographic and holocaust denial websites taken down from European servers have found a home in the U.S.

Technical Measures

Geo-blocking—i.e., restricting Internet access based on a user's supposed location—can be thwarted by the use of a proxy server. Surveillance of online activity can be thwarted by the use of a VPN (Virtual Private Network), which encrypts internet traffic. Platforms and electronic service providers are however becoming increasingly aware of these techniques and are developing responses to combat them.

Search engine results can be manipulated to produce no results in certain jurisdictions in response to regulated search terms. This raises questions of censorship and is likely to deter only passive or naïve searchers.

Internet traffic can be filtered on the basis of restricted key words or site blacklists. This method frequently results in incorrect blocking or overblocking. This is a method used by the industry-led Internet Watch Foundation, which has successfully reduced the amount of child pornography hosted in the UK, although this material has frequently reappeared in another jurisdiction.

The global nature of the internet means that jurisdictional efforts to regulate online content and behavior are likely to deter only the most passive or naïve. This may help to prevent influencing from foreign political propaganda but will do little to prevent the determined and technologically savvy from accessing information.

The Dark Net

The Dark Net is the part of the Internet not open to public view and only accessible using the Tor browser, which permits anonymized browsing. It contains websites and file locations that are not indexed by conventional search engines and are, therefore, hard to find. Criminals and extremists can avoid monitoring by using a series of "redirects": links that must be followed by invited users to reach certain sites. If content is not indexed by search engines, it is not possible for regulators to tweak search results to hide it. If its location is not known, filtering with reference to blacklists will not work, as the material will not make it onto these lists in the first place. If authorities cannot locate content, they cannot attempt to remove it (Stevens 2009).

The anonymity afforded to users of the Dark Web provides safe internet access for criminals but also for investigative journalists and users blocked from accessing information by repressive regimes.

The Dark Web continues to act as a marketplace for the exchange of illegal goods and services, notwithstanding the shutting down of its infamous Silk Road marketplace and the imprisonment of its founder, Ross Ulbricht, in the U.S. in 2013. The investigation and action by the FBI were intended to send a message that even the Dark Web was not outside the control of U.S. law enforcement.

Payment for goods on the site was made by the untraceable cryptocurrency Bitcoin. Ulbricht created the marketplace to function without government oversight but found it difficult to verify anonymous transactions, since the anonymity afforded to buyers and sellers prevented relationships of trust from being established. Ironically, scammers complained about being scammed, since anonymity precluded accountability. Ulbricht started increasing oversight. He added measures to ensure trustworthiness with implementation of an automated escrow payment system and automated trader review system similar to the features of the Amazon legal online trading platform. As the site became increasingly profitable, Ulbricht suffered threats from the traders. In the absence of state mechanisms for enforcement, he allegedly sought to hire thugs to enforce his business (Farrell 2015).

In a letter to the judge before his sentencing, Ulbricht stated that his actions via Silk Road were committed through libertarian idealism and that "Silk Road was supposed to be about giving people the freedom to make their own choices" (Snyder 2015). Unfortunately, giving people the freedom to make their own choices meant they made choices in their own interest to his detriment. His experiment in creating an online libertarian utopia was ultimately a failure. In the words of Robert Lee Hale "There is government whenever one person or group can tell another what to do, and when those others have to obey or suffer a penalty" (Samuels 1992).

Accountability is a necessary feature of a functioning online community. The question is whether this accountability can only be provided by nation state based systems.

The primarily illegal nature of the goods on the Silk Road site meant that traders and purchasers were uniquely vulnerable. But what if Ulbricht had been selling lawful goods? Such a business model exists and is wildly successful.

Amazon.com

Amazon is a U.S. based global electronic trading platform, operating with national subsidiary websites throughout the world. Traders register with the site and supply goods to customers who log in with a password and email. Amazon does not verify the traders beyond basic identifying details, but a trust system is in operation where customers can rate the traders. The company offers its own escrow system and takes a fee from the seller for every transaction. Anyone in the world can sell to anyone in the world on the site (save in countries where is the service is blocked). The national subsidiary sites cater to national markets, but there is no block on, say, a UK user buying from the U.S. site, although national restrictions may block the purchase of certain goods.

It is no accident that the trading model of Amazon and Silk Road are almost identical, since they have evolved to make the most efficient use of the architecture of the Internet. Amazon recognizes national jurisdictions through its subsidiaries but ultimately still provides a global trading platform. It ensures accountability deliverable though its own dispute resolution system.

As national jurisdictions struggle to regulate the global nature of the Internet, it seems that transnational global service providers are filling the gap. We are seeing the rise of "Corporation as Courthouse" (Van Loo 2016).

Just as Amazon can now exercise jurisdiction over online trade disputes, Facebook can exercise jurisdiction over online expression. After many years of holding out, Facebook has increased the strictness and enforcement of its "community guidelines" in the face of national regulation. These standards are public and of global application. Facebook even offers its own appeals process (Newton 2018).

The loss of territorial sovereignty is being replaced with functional sovereignty (Pasquale 2017). It may be more efficient, but as the power of the platforms grows relative to national power, how are the platforms themselves to be held accountable? Are we entering a neo-feudal arrangement where the power to obtain "justice" is not based on the rule of law but on an individual's economic leverage on the platform? Or have the platforms already become so powerful and ubiquitous that they are effectively a digital public space, wielding the ultimate sanction of digital social exclusion against recalcitrants?

Conclusion

> Many of the jurisdictional and substantive quandaries raised by border-crossing electronic communications could be resolved by one simple principle: conceiving of Cyberspace as a distinct place for purposes of legal analysis … What procedures are best suited to the often-unique characteristics of this new place and the expectations of those who are engaged in various activities there? What mechanisms exist or need to be developed to determine the content of those rules and mechanisms by which they can be enforced? Answers to those questions will permit the development of rules better suited to the new phenomena in question, more likely to be made by those who understand and participate in those phenomena, and more likely to be enforced by means that the new global communications media make available and effective. (Johnson and Post 1996)

Libertarians have discovered that the borderless nature of the Internet is the perfect architecture for the market, rather than nation states (Boushey 2017). This model is expanding across the globe via the rise of multinational service providers and fits the model of regulating cyberspace as a separate entity, proposed by Johnson and Post. China has resisted, but only through the use of pervasive technological blocking techniques. Does the rise of the Net mean that future regulation of citizens is now polarized between market forces or absolute state control? Or can democracy evolve to regulate the Internet (Schlechtman 2018)?

References

Articles

Azzi, A. (2018). The challenges faced by the extraterritorial scope of the general data protection regulation. *JIPITEC 9*, 126 para 30.

Johnson, D. R., & Post, D. G. (1996). Law and borders—The rise of law in cyberspace. *Stanford Law Review, 48*, 1367, 1378–1379, 1390–1391.

Pasquale, F. (2017, December 6). From territorial to functional sovereignty: The case of Amazon. *Law and Political Economy*. https://lpeblog.org/2017/12/06/from-territorial-to-functional-sovereignty-the-case-of-amazon/.

Samuels, W. J. (1992). *Essays in the history of Heterodox Political Economy* (p. 184). Macmillian.

Stevens, T. (2009, April). *RUSI Journal, 154*(2), 28–33.

Van Loo, R. (2016). The corporation as courthouse. *Yale Journal on Reg, 33*.

Books

Ball, J. (2017). *Post-truth: How bullshit conquered the world.*

Judgements

"LICRA et UEJF v. Yahoo! Inc and Yahoo! France" Tribunal de Grande Instance de Paris, 22 May 2000.
United States v. Microsoft Corporation, Supreme Court of the United States 584 U.S.__ 2018.

Laws

Cloud Act, H.R. 4943—Clarifying Lawful Overseas Use of Data Act, 23 March 2018. https://nsarchive.gwu.edu/news/cybervault/2018-04-02/hr-4943-clarifying-lawful-overseas-use-data-act-cloud-actMedia.
Network Enforcement Act. https://germanlawarchive.iuscomp.org/?p=1245.

Media

Boushey, H. (2017, August 15). How the radical right played the long game and won. *New York Times*. https://www.nytimes.com/2017/08/15/books/review/democracy-in-chains-nancy-maclean.html.
Cadwalladr, C. (2018, November 24). Parliament seizes cache of Facebooks internal papers. *The Guardian*. https://www.theguardian.com/technology/2018/nov/24/mps-seize-cache-facebook-internal-papers.
Cadwalladr, C., & Graham-Harrison, E. (2018, March 17). How Cambridge Analytica turned Facebook 'likes' into a lucrative political tool. *The Guardian*. https://www.theguardian.com/technology/2018/mar/17/facebook-cambridge-analytica-kogan-data-algorithm.
Farrell, H. (2015, February 20). Dark Leviathan. *Aeon Essays*. https://aeon.co/essays/why-the-hidden-Internet-can-t-be-a-libertarian-paradise.
Fiadh, M. (2017, December 18). The dangers of echo chambers, complacency, and fake news. *The Irish Times*. https://www.irishtimes.com/student-hub/the-dangers-of-echo-chambers-complacency-and-fake-news-1.3331458.
Green, J. J. (2018, September 17). Tale of a troll: Inside the 'Internet Research Agency' in Russia. *Washington's Top News*. https://wtop.com/j-j-green-national/2018/09/tale-of-a-troll-inside-the-Internet-research-agency-in-russia/.
Hill, A. (2018, March 19). Facebook's floundering response to scandal is part of the problem. *The Financial Times*. https://www.ft.com/content/42491a80-2b5b-11e8-a34a-7e7563b0b0f4.
Jarvis, J. (2018, August 10). Platforms are not publishers. *The Atlantic*. https://www.theatlantic.com/ideas/archive/2018/08/the-messy-democratizing-beauty-of-the-Internet/567194/.
Kiss, J., & Arthur, C. (2013, July 29). Publishers or platforms? Media giants may be forced to choose. *The Guardian*. https://www.theguardian.com/technology/2013/jul/29/twitter-urged-responsible-online-abuse.
Levin, S. (2018, July 2). Is Facebook a publisher? In public it says no, but in court it says yes. *The Guardian*. https://www.theguardian.com/technology/2018/jul/02/facebook-mark-zuckerberg-platform-publisher-lawsuit.
Macaskill, E., & Dance, G. (2013, November 1). NSA FILES: DECODED. *The Guardian*. https://www.theguardian.com/world/interactive/2013/nov/01/snowden-nsa-files-surveillance-revelations-decoded#section/1.

Newton, C. (2018, April 24). Facebook makes its community guidelines public and introduces an appeals process. *The Verge*. https://www.theverge.com/2018/4/24/17270910/facebook-community-guidelines-appeals-process.

Selyukh, A. & Greene, D. (2017, May 5). FCC chief makes case for tackling net neutrality violations 'After The Fact'. *National Public Radio*. https://www.npr.org/sections/alltechconsidered/2017/05/05/526916610/fcc-chief-net-neutrality-rules-treating-Internet-as-utility-stifle-growth?t=1543780680159.

Spiegel Online. (2013, August 26). *German email services report surge in Demand*. http://www.spiegel.de/international/germany/growing-demand-for-german-email-providers-after-nsa-scandal-a-918651.html.

Schlechtman, J. (2018, April 13). *The internet is killing democracy, whowhatwhy.org*. https://whowhatwhy.org/2018/04/13/the-Internet-is-killing-democracy/.

Skorup, B. (2016, June 20). Net neutrality is government censorship. *National Review*. https://www.nationalreview.com/2016/06/net-neutrality-government-control/.

Snyder, B. (2015, May 27). Silk Road mastermind pleads for light sentence. *Fortune*. http://fortune.com/2015/05/27/silk-road-sentencing/.

"In online advertising, Facebook is a publisher," *The Financial Times*, April 23, 2018. https://www.ft.com/content/9206f5f2-46f8-11e8-8ee8-cae73aab7ccb.

Zavadski, K. (2017, November 3). American Alt-Right Leaves Facebook for Russian Site VKontakte. *The Daily Beast*. https://www.thedailybeast.com/american-alt-right-leaves-facebook-for-russian-site-vkontakte.

Studies and Guidance

Congressional Research Service, & Gilroy, A. A. (2011, March 11). *Access to broadband networks: The net neutrality debate* (Report) (pg. 1). DIANE Publishing. ISBN: 978-1437984545.

Freude, A., & Freude, T. (2016, October 1). *Echoes of history: Understanding German data protection, Bertelsmann Foundation*. https://www.bfna.org/research/echos-of-history-understanding-german-data-protection/.

House of Lords Library Briefing (2018, January 8). *Social media and online platforms as publishers*. https://researchbriefings.parliament.uk/ResearchBriefing/Summary/LLN-2018-0003.

Web References

Amazon Buyer Dispute Program. https://pay.amazon.com/us/help/201751580.

Chesney, R., & Citron, D. (2018, February 21). Deep fakes: A looming crisis for national security, democracy and privacy? LAWFARE Blog.

Chew, W. C. (2018, May 1). How it works: Great firewall of China. https://medium.com/@chewweichun/how-it-works-great-firewall-of-china-c0ef16454475.

Facebook Community Standards. (2018). https://www.facebook.com/communitystandards/.

Peters, B. (2018, February 8). The new Facebook algorithm: Secrets behind how it works and what you can do to succeed. Buffer Social Blog. https://blog.bufferapp.com/facebook-algorithm.

Locklear, M. (2018, September 5). Facebook users are changing their social habits amid privacy concerns. *Engadget*. https://www.engadget.com/2018/09/05/facebook-changing-social-habits-privacy-concerns/?guccounter=1.

Jash, S. (October 27, 2017). *Outsourcing censorship, attacking civil liberties: Germany's NetzDG*. https://www.hertie-school.org/the-governance-post/2017/10/outsourcing-censorship-attacking-civil-liberties-germanys-netzdg/.
Yahoo! Inc v LICRA and UEJF No. 01-17424 United States Court of Appeals for the 9th Circuit 443 433 F.3d 1199 (9th Cir. 2006)

Ruth Barber is currently completing a Master's Degree in IT and Communications Law with Queen Mary University London and researching the regulation of social media. She completed a law degree in 1995, was admitted to the Bar of England and Wales in 1996 and became a Solicitor (England and Wales) in 1998. She gained Higher Rights of Audience in 2005. She was admitted to the UK Attorney General's list of Counsel in 2007 and was shortlisted for Solicitor Advocate of the Year in 2010. She specializes in regulatory criminal law and works as an international legal consultant. Ruth is the co-author of the Confiscation Law Handbook 2011, published by Bloomsbury Professional.

Chapter 6
Data Ownership

Winfried Bullinger and Sophie Terker

Introduction

Is there such a thing as ownership of data? The keywords "data ownership" are frequently used in connection with digital business models. The commercial value of data is undisputed: they are, so to speak, the blood in the cardiovascular system of the digital body. Companies generate data from the traces of the users of their digital products; these data are of high value if they are systematically analyzed. For example, they reveal the interests, purchasing power and foreseeable behavior of their users.

The collection and analyzation of data are not only relevant for the targeted placement of advertisements for individual internet users. It can help monitor of the behavior of a person as a whole. In addition to the analysis of economic activities, conclusions can be drawn about the user's political views, health, physical activity and other habits. As a matter of fact, the customer in the digital world often pays with his data for seemingly free services. Based on the premise that data have a high economic value, the question of the legal protection of data directly follows. Can a company that generated data legally defend itself if others access and use the data for their own purposes? Can rights to data volumes be the subject of transactions and exclusive license agreements? This article deals with the protection of data from a legal point of view in the sense of an owner's rights to data. The starting point is absolute intellectual property rights that may be applicable to data, as well as legal provisions that only protect data reflexively. To anticipate it: There is no abstract ownership of data per se, neither literally nor figuratively [*Volkszählungsurteil* (1983) Federal Constitutional Court, reference number: 1 BvR 209, 269, 362, 420, 440, 484/83].

W. Bullinger (✉) · S. Terker
CMS Hasche Sigle, Berlin, Germany
e-mail: claudia.berger@cms-hs.com

S. Terker
e-mail: s.terker@web.de

© Springer Nature Switzerland AG 2020
D. Feldner (ed.), *Redesigning Organizations*,
https://doi.org/10.1007/978-3-030-27957-8_6

The protection of data in the sense of an ownership-like right is instead composed of an amalgam of several different rights. These rights are the focus of this article.

To introduce the topic, the term "data" should first be defined.

Depending on the legal context, data are defined differently; in the area of criminal law, a different definition is necessary than in other areas of law due to the specific purpose of this field of law. For the purpose of this article, which deals mainly with civil law, data can be defined as any type of information, regardless of whether it is communicated orally or in writing (Specht 2016).

When asked about the legal protection of data, most people will first think of "data protection (law)". However, data protection law is a special area of law; the term is not to be confused with the generic term "legal protection of data". As far as Germany is concerned, data protection regulations can be found in the (European) General Data Protection Regulation (GDPR), the German Federal Data Protection Act ("FDPA" *Bundesdatenschutzgesetz, BDSG*) and in the data protection laws of the federal states.

It should be noted that these regulations and acts only concern the protection of personal data. Since the GDPR came into force on 25 May 2018, Art. 4(1) GDPR contains a uniform EU-wide legal definition of "personal data": Personal data means any information relating to an identified or identifiable natural person ("data subject"); an identifiable person is a natural person who can be identified directly or indirectly, in particular by reference to an identifier such as a name, an identification number or to one or more factors specific to the physical, genetic or cultural identity.

EU legislature attributes the power of disposal over personal data to the data subject; any use of such data by other persons requires justification, for example, the consent of the data subject (Art. 6 GDPR). However, the data subject does not "own" his data in the traditional understanding of the word "ownership": Goods and real estate belonging to one person can be transferred to another person. The power of disposal over personal data, on the other hand, cannot be transferred to a third person (See point (b) of Art. 5(1) GDPR, according to which data may only be "collected for specified, explicit and legitimate purposes" and the right to erasure ('right to be forgotten'), Art. 17 GDPR). The intention behind the European data protection law is to grant each individual the greatest possible control over his own data (including the possibility of revoking the above-mentioned consent, see also Recital 7(2) of the GDPR.).

However, if a set of data cannot be classified as personal data, the GDPR is not applicable. Non-personal data is therefore only protected in accordance with other legal provisions. What this protection looks like and under which conditions it is effective will be discussed below.

Copyrights for Linguistic Works

Requirements for Protection of Data as a Linguistic Work

All data—personal as well as non-personal data—may be protected as a linguistic work under s. 2(1) No. 1 German Copyright Act ("GCA", *Urheberrechtsgesetz, UrhG*) if they can be communicated in written or oral form.

In the case of personal data, copyright protection takes place in addition to protection under the GDPR.

The central concepts of copyright law are the "work" and the "author". According to s. 2(2) GCA, a work in the sense of copyright law is any personal intellectual creation; the author is the person who created the work. He is entitled to all rights to the work as described in the GCA (Klass 2018).

The linguistic work is a subcategory of the work. This category includes classical linguistic works such as novels, poems, plays, etc. However, the protection of a work result as a work does not presuppose that it is an artistic creation. This follows from s. 1 GCA, which stipulates that works of "literature, science and art" may be protected. Art and literature are only mentioned as exemplary genres of works; consequently, the protection of merely rational creations—such as data, for example—is not excluded from copyright law (Bullinger and Czychowski 2011).

Nevertheless, not all data fall under s. 2 GCA; rather, the general requirements for "works" must be fulfilled. A linguistic work is a personal intellectual creation whose content is expressed through language; the type of fixation or transmission is irrelevant (i.e., oral statements as well as written records are meant) (*Unlizenzierte Nutzung von Interviewfrage* (2012) District Court Hamburg, reference number: 308 O 388/12). Nor should the term "language" be interpreted too narrowly; data that is communicated with mathematical or other symbols can also be protected as a work of language. Programming languages are therefore also languages in the sense of the GCA (Horns 2001).

Finally, the data in question must be a personal intellectual creation. According to jurisprudence, a linguistic design shaped by individual creativity leads to copyright protection just as much as an original way of selecting or presenting the content (Kitz 2007). Therefore, only the person who created the data is the "author"—not the person to whom the data refer or the person who merely formulated data according to given instructions.

But when precisely can data be classified as "personal intellectual creations"? The only thing that is for certain is that the mere stringing together of information, for example, in registration forms (surname and first name, address, telephone number, etc.), cannot be a copyright work. Nor is the mere reproduction of facts a linguistic work, since facts stem from the outside world, not from the creative mind of a certain person. Something that already exists cannot be considered anyone's "creation" (Graef 2012).

According to constant jurisprudence, there are no excessive demands to be made on originality (Hoeren and Herring 2011). However, one must not lose sight of the fact that copyright law is primarily intended to protect artistic and scientific achievements; unlike U.S. law, copyright law in Germany has a considerable personal rights component. In most cases, however, data is merely a representation of the factual. For the purpose of fast and simple communication, this reproduction is usually limited to the essential and is therefore presented in as compressed a form as possible. Due to the brevity of the formulation of data and the limited room for creativity in the design, data will rarely fulfill the requirements for a "personal intellectual creation".

Due to their diversity, it cannot be categorically negated that data might, in some cases, be protected under copyright law as linguistic works; on the basis of the above considerations, however, it is fair to assume that copyright will only apply in absolutely exceptional cases.

Scope of Protection

But what does it mean for the author of data if they enjoy protection as a linguistic work? The relevant legal provisions are s. 12 and 15 GCA. These sections seem to name only the (positive) powers of the author at first glance: He may publish, reproduce, distribute, exhibit, and publicly reproduce his work (in an incorporeal manner)—for example, by a presentation or by uploading it to a website. According to the law, these rights are "exclusive" (cf. s. 15(1) GCA). This means that publication and exploitation are prohibited for all other persons. This is the (intellectual) property right of the author to "his" data. If, for example, someone copies these data without being entitled to do so by law or on the basis of a permission of the author (referred to as a "license" in copyright law), the author can demand from him, among other things, elimination of the disturbance, omission of the infringing act for the future and compensation for the damage incurred.

Copyrights for Databases

Not only data can be subject to copyright; databases may also be protected under German copyright law if they can be classified as a database work.

Requirements for Protection as a Database Work

The database is defined in s. 4(2)(a) GCA as a compilation whose elements are arranged systematically or methodically and are individually accessible by electronic means or in some other way. Compilations are collections of works, data or other independent elements that are a personal intellectual creation due to the selection or arrangement of the elements (s. 4(1) GCA). The principles that apply to linguistic works (see above) also apply to database works: standardized ways of selecting, arranging and presenting data cannot be classified as personal creations, nor can collections found in the outside world (Schricker 1996). Either the selection or the arrangement of the data in the database must have a certain originality. The prerequisite is that there is some scope for an individual arrangement of the data. Jurisprudence has so far negated this for alphabetically ordered telephone directories (*Tele-Info-CD* (1999) Federal Supreme Court, reference number: I ZR 199/96), a collection of biographical data without conceptual design (*Hubert-Fichte-Biographie* (1996) Higher Regional Court Hamburg, reference number: 3 W 53/96), and chronologically arranged television program overviews (*Rundfunkprogramme* (1933) Reichsgericht, reference number: I 250/32): In these databases, the data are arranged in a sequence that is determined by mere logic.

The necessary scope for design exists, for example, in medical or pharmaceutical databases or in a list of poem titles containing "The 1100 most important poems of German literature from 1730 to 1900", since the "importance" of a poem can be measured on the basis of various different assessment criteria (prominence of the poem, prominence of the author, frequency of publication in anthologies, etc.; see *Gedichttitelliste I* (2007) Federal Supreme Court, reference number: I ZR 130/04).

The author of the database is the person who created the concept for the selection and/or arrangement of the data; whoever implements this concept (i.e., created the database) is irrelevant (*Gedichttitelliste I* (2007) Federal Supreme Court). If the head of a research group instructs his employees to create a collection of data according to criteria he has specified, he is the author of the database; the employees are not to be regarded as (co-)authors.

Scope of Protection

As far as the author's powers and prohibition rights are concerned, what has been said about data as linguistic works applies in principle (see B.II above).

It is worth noting that the rights of the author are not only infringed upon when an unauthorized third party exploits the database work in its entirety; it is rather sufficient if the third party exploits a part of the database, provided that the exploited part enjoys legal protection as a database work in its own right (*Newton-Bilder* (2011) Higher Regional Court Cologne, reference number: 6 U 118/11). In the case of a list of poem titles comprising 1100 titles, the German Federal Court of Justice affirmed that the copying of 856 titles constitutes a violation of copyright (*Gedichttitelliste I* (2007) Federal Supreme Court).

Limitations in Licensing Database Works

In principle, the author of a work is authorized to allow the exploitation of the same to other persons by granting them a license. This generally also applies to database works; however, a restriction applies insofar as personal data within the meaning of Art. 4(1) GDPR are affected. Since the GDPR applies to them as well as the GCA, the provisions of the GDPR regarding the permissibility of data processing must be observed whenever personal data are affected. If the person whose personal data is made available or passed on to others has not given his consent to such use of his data, licensing is only legal if one of the justifications in Art. 6 GDPR intervenes.

Fair Use, Data Mining

Another problem in connection with the (commercial) use of data has only recently—on 1 March 2018—been regulated by German legislature: so-called data mining. S. 60d GCA, newly inserted into the GCA for this purpose, is difficult to understand due to its complicated wording. Data mining is the software-supported evaluation of large amounts of data (Spindler 2016). A prerequisite for such an evaluation is the mass reproduction and structuring of large amounts of text or data. Specifically designed software determines statistical frequencies or correlations in the content structured in this way and, thus, enables scientific analysis and evaluation. However, if—as will often be the case—entire copyrighted works or essential parts of databases were to be analyzed, this was previously not permissible under German copyright law. Licensing models were necessary.

In essence, the new regulation is a barrier to the right of reproduction and to the right to make a work publicly available which applies if a work is used for the purpose of scientific research. The user may only pursue non-commercial purposes (s. 60d(1)(b) GCA). If this is the case, there are no further restrictions. Anyone who wants to carry out automated scientific research can refer to s. 60d GCA. He may then also have the necessary actions carried out by third parties—for example, employees of a library with the necessary qualifications who are participants in a research project.

As far as data mining is concerned, German law differentiates between the source material and the so-called corpus. The former is made up of a large number of works that are to be evaluated automatically, the content of which is therefore the subject of research; which types of works are evaluated is not relevant.

Works with texts, data, images, sounds or audiovisual data can be evaluated. However, the researcher may only evaluate works to which he has already gained access legally; s. 60d is not intended to create a right of access to protected original material (de la Durantaye 2017).

The corpus, on the other hand, is the result of the reproduction of the source material. A uniform corpus is a prerequisite for text and data mining, since only appropriately prepared texts and data can be evaluated with the corresponding procedures.

Specifically, s. 60d GCA allows the normalization, structuring and categorization of the source material. For example, normalizing texts involves adapting different file formats; structuring involves converting them into machine-readable text; categorizing involves dividing them into sentences and individual words and other categories and creating a corresponding database (Spindler 2016).

The described use of copyrighted works is only legalized by s. 60d GCA as long as the repsective research project is still in progress. After completion of the research work, the corpus and other reproductions of the original material must be deleted and all public access must be terminated (s. 60d(3)(a) GCA).

The use is to be remunerated according to s. 60 h GCA. While data mining is legal without the permission of the author of the original material, it is not free of charge. With regard to the amount of the remuneration, s. 60 h(1) GCA stipulates that it must be "appropriate". S. 60 h(3) GCA regulates how to determine which remuneration is appropriate: the actual use and its scope are decisive.

Protection of Data as Trade Secrets

Data may also receive special legal protection if they are trade secrets. Under previous German law, in-house know-how that could not be patented or classified as a work or a trademark was protected only to a very limited extent against unauthorized access and disclosure. The betrayal of trade secrets was punishable under s. 17 Act Against Unfair Competition ("UCA", *Gesetz gegen den unlauteren Wettbewerb, UWG*), and the injured party could claim compensation for the damage incurred from the party who disclosed a trade secret (McGuire 2016). If the injuring party only disclosed the trade secret to individual competitors, the keeper of the trade secret could also demand that the latter refrain from exploiting the trade secret commercially (e.g., by reproducing a machine). However, if the injuring party revealed the secret to the general public—for example, by publishing it in a journal or uploading it to a website (so-called leaking)—the know-how lost its quality as a "secret", the consequence being that everyone was henceforth free to exploit it commercially (McGuire 2016). This was the critical deficiency of trade secrets compared to traditional intellectual property: The patentee who discovered that a third party was making unauthorized use of the subject matter of his patent could always demand injunctive relief from the infringer. However, if, for example, the (non-patentable) recipe for a soft drink were made available to the public, the producer of the drink could only demand compensation for the damage caused, and only from the person who distributed the recipe; he could not prevent other companies from producing soft drinks according to the same recipe. This resulted in two considerable economic disadvantages. Firstly, the product lost its unique selling point; the production of similar products by competitors could result in considerable financial losses, which might even endanger the existence of the company. Secondly, the injuring party rarely had the financial means to settle the entire damage incurred—in essence, the claim for damages was virtually void.

European legislature has now acknowledged that even know-how that is not protected by existing intellectual property rights can often only be acquired at considerable financial and time expense, and that the previously existing protection of the owner was inadequate against this backdrop. Directive (EU) 2016/943 (Trade Secrets Directive) of 8 June 2016 is intended to remedy this situation.

As an EU directive, the Trade Secrets Directive has no direct effect in the EU member states; it had to be transposed into national law by 9 June 2018. The German legislature missed this deadline; the German Trade Secret Act ("TSA", *Gesetz zur Schutz von Geschäftsgeheimnisse, GeschGehG*) entered into force on April 26, 2019. The TSA is strictly in line with the provisions of the EU directive; the main principles are explained below.

According to s. 2(1) TSA, a trade secret is information which is not, either in its entirety or in its precise arrangement and composition, generally known among or readily accessible to persons within the circles which normally deal with the kind of information in question and is therefore of economic value and is the subject of appropriate confidentiality measures by its legal owner. This definition includes, for example, data of (potential) customers of the company.

The three central actions whose legality the Act is intended to regulate are the acquisition, use and disclosure of trade secrets. Unfortunately, these terms are not defined in the Trade Secrets Directive or in the TSA; the following definitions are proposed in legal literature:

– Acquisition means the obtaining of information that constitutes the trade secret,
– Use means the use of trade secrets for a specific, not necessarily economic purpose,
– Disclosure means the disclosure of business secrets to third parties or to the public (Köhler and Bornkamm 2010).

S. 4 of the TSA stipulates when such actions are illegal. While s. 4(1) TSA regulates the unlawfulness of the *acquisition* of a trade secret, s. 4(2) TSA specifies when a trade secret may not be used or disclosed.

S. 4(3) TSA prohibits the acquisition, use or disclosure of a trade secret by anyone who acquired the information from a third person and knew or should have known that the respective information is a trade secret and that it has been used or disclosed unlawfully. This means that even if the trade secret has been made accessible to the public, it may not be commercially exploited if the offender had knowledge or negligent ignorance of the unlawful disclosure. Thus, the owner of the secret is granted stronger protection than under previous German law (see above).

In addition to the prohibitions, the TSA also provides grounds for justification of infringements. S. 3(1) stipulates that the acquisition of a trade secret is not illegal if someone learns about it through so-called reverse engineering or product tests, provided that the product itself has been made publicly available, or that it is in the legitimate possession of the investigating person without any restriction. Such restriction may result, for example, from a contractual confidentiality clause.

The TSA also provides for a justification for so-called whistleblowers. According to s. 5 TSA, the disclosure of a trade secret is not illegal if its disclosure is necessary to protect a legitimate interest. According to the (non-exhaustive) list in s. 5 TSA, legitimate interests include the uncovering of an illegal act or other misconduct. This legal provision is meant to encourage and protect whistleblowers—people (often employees) who uncover grievances in the interest of the general public. It is intended to encourage employees to disclose unethical behavior and illegal actions of their employers to the public by prohibiting employers from imposing sanctions on whistleblowers (Directive (EU) 2016/943, recital 20).

If an act fulfills an offense according to s. 4 TSA and no justification applies, the infringer is subject to the legal consequences according to s. 6 et seq. TSA. This includes the obligation to remedy and refrain from the infringing act, destruction, surrender, recall, removal and withdrawal from the market of the infringing product, information about the infringing products and compensation for the damage incurred. However, according to s. 10 TSA, the obligation to pay damages only applies to the infringer who acted intentionally or negligently.

Protection of Data Under Unfair Competition Law

German Unfair Competition law has no regulations tailored to data issues, so the interpretation of generally applicable regulations is essential for the protection of data under unfair competition law. In this area, many legal questions have not been conclusively clarified either in case law or legal literature, so there is considerable legal uncertainty. Due to the diversity of possible constellations, a comprehensive presentation of the current state of opinion is not possible. Unfair competition law only protects data which are relevant to competition. This relevance particularly exists if the data itself is offered as a product or is part of a product, or if it is for the manufacture of a product.

If the data themselves are the product or part of a product, they may be protected "as a product" according to points (a) and (b) of s. 4(1) UCA against unfair use; if data are only required for production, they may nevertheless enjoy protection "in products" according to point (c) of s. 4(3) UCA (Becker 2017).

Post-contractual Restraint

The protection of customer data also becomes relevant when employees leave a company—for example, due to termination of employment or the resignation of a managing director or a shareholder. Employees, managing directors or shareholders who leave the company often take up a similar occupation with a competitor or set up their own business in the same industry. The temptation to take customer data from the previous employer and make it available to the new employer or to use it to build up a customer base of one's own is great. Whoever takes customer data with him when leaving a company does not only violate the GDPR if he does so without consent of the customers; if he manages to entice customers away from his old employer, he may cause him considerable damage. In order to prevent departing employees from harming the former employer in this way, legislature and courts have developed principles for the admissibility of such a post-contractual non-compete obligation.

A post-contractual restraint is only legally valid if it serves the protection of a legitimate interest of the company and it must not unreasonably complicate the exercise of the profession and economic activity of the departing employee in terms of place, time and subject matter (Laskawy 2012).

The post-contractual restraint must, therefore, be confined to a certain geographical area (Laskawy 2012). How far the circle of prohibition is to be drawn depends on how large a catchment area the old company obtains its customers from. A doctor's office with no particular specialization will have a smaller catchment area than the office of a highly specialized sports physician; accordingly, the latter may impose a more extensive post-contractual restraint on ex-employees than the owner of the non-specialized doctor's office.

The temporal scope of application of the non-compete obligation should be limited to two years. For employees, this already results from s. 74a(1) German Commercial Code; for non-employees, this limitation has been established in case law (*Unverbindlichkeit der Wettbewerbsabrede* (1997) State Labor Court Düsseldorf, reference number: 3 Sa 1644/96).

The "subject matter" of the post-contractual restraint, i.e. the specification of the prohibited professional activity (Laskawy 2012), must be determined in consideration of all relevant circumstances of the individual case. For example, a departing managing director cannot generally be prohibited from working as a managing director; a non-compete obligation must be limited to certain types of companies or a certain industry.

Conclusion

If a set of data falls under the definition of personal data in the sense of Art. 4(1) GDPR, it enjoys the protection of the General Data Protection Regulation. However, if a set of data cannot be classified as personal data, protection can only be considered in accordance with other statutory provisions. Due specifically to the broad definition of "data", it is often difficult to assess the legal situation—not only for the legal layman. Moreover, the question of to whom data "belongs" can seldom be answered by referring to a single person; most data are integrated into a complex system of legal interdependencies. This system, in turn, is subject to the restrictions of various legal norms. Unfair competition law in particular "does not permit precise allocation of use (like legal intellectual property law), but it is in a position to prevent the all too direct exploitation of someone else's achievements with great consideration to the particularities of the individual case" (Becker 2017). To avoid an infringing use of data, it is therefore always necessary to determine as carefully as possible which person is entitled to which rights to the data in question.

References

Becker, M. (2017). Lauterkeitsrechtlicher Leistungsschutz für Daten. In *Gewerblicher Rechtsschutz und Urheberrecht (GRUR)* (346–355).

Bullinger, W., & Czychowski, C. (2011). Digitale Inhalte: Werk und/oder Software?—Ein Gedankenspiel am Beispiel von Computerspielen. In *GRUR* (19–26).

de la Durantaye, K. (2017). Neues Urheberrecht für Bildung und Wissenschaft – eine kritische Würdigung des Gesetzentwurfs. In *GRUR* (558–567).

Graef, R. O. (2012). Die fiktive Figur im Urheberrecht. *Zeitschrift für Urheber- und Medienrecht (ZUM)*, 108–117.

Horns, A. H. (2001). Anmerkungen zu begrifflichen Fragen des Softwareschutzes. In GRUR (1–16).

Hoeren, T., & Herring, E.-M. (2011). Urheberrechtsverletzung durch WikiLeaks? Meinungs-, Informations- und Pressefreiheit vs. Urheberinteressen. *MultiMedia und Recht (MMR)*, 143–148.

Kitz, V. (2007). Die Herrschaft über Inhalt und Idee beim Sprachwerk - Anmerkung zu LG München I, GRUR-RR 2007, 226 - Eine Freundin für Pumuckl. In *Gewerblicher Rechtsschutz und Urheberrecht – Rechtsprechungsreport (GRUR-RR)* (217–218).

Klass, N. (2018). Kunst und (Urheber-)Recht – Einleitung zu dem gleichnamigen Symposium des Instituts für Urheber- und Medienrecht am 13.4.2018 in München. *ZUM 2018, 481–484.*

Köhler, H., & Bornkamm, J. (2010). *Gesetz gegen den unlauteren Wettbewerb*. Munich: Beck.

Laskawy, D. H. (2012). Die Tücken des nachvertraglichen Wettbewerbsverbots im Arbeitsrecht – Stets geliebt und doch verkannt! *Neue Zeitschrift für Arbeitsrecht (NZA)*, 1011–1018.

McGuire, M.-R. (2016). Der Schutz von Know-how im System des Immaterialgüterrechts. In GRUR (1000–1008).

Schricker, G. (1996). Der Urheberrechtsschutz von Werbeschöpfungen, Werbeideen, Werbekonzeptionen und Werbekampagnen. In GRUR (815–826).

Specht, L. (2016). Ausschließlichkeitsrechte an Daten – Notwendigkeit, Schutzumfang, Alternativen. In *Computer und Recht (CR)* (288–299).

Spindler, G. (2016). Text und Data Mining – urheber- und datenschutzrechtliche Fragen. In *GRUR* (1112–1120).

Professor Dr. Winfried Bullinger is one of Germany's leading copyright lawyers. He advises his clients on all aspects of intellectual property law, including media law and know-how protection and also represents them in court where he is regularly involved in landmark cases. Winfried's clients include numerous large German and international technology and media companies as well as service providers. Winfried joined CMS in 1998 and has been a partner since 2002. He is head of the copyright group of the firm's IP team. Since 1998, he has been teaching copyright and media law at Berlin's Humboldt University as well as at Brandenburg University of Technology Cottbus-Senftenberg where he is also a honorary professor. Winfried has authored numerous academic publications and papers on copyright and media law and is the co-editor of the Praxiskommentar Urheberrecht (commentary on the German copyright law), published by C. H. Beck Verlag. Besides his outstanding legal work, he is also recognized photographer, published several books and exhibits his work on a regular basis.

Sophie Terker is a research associate at CMS Hasche Sigle EEIG. Before graduating from high school with the highest possible score in 2012, she won the first prize at the "Bundeswettbewerb Fremdsprachen" (German Competition for Foreign Languages) in 2010. Like all winners of this competition, she was awarded a scholarship from the German Academic Scholarship Foundation, Germany's oldest scholarship foundation. From 2012 to 2017, Sophie studied law at Berlin's Humboldt University, specializing in copyright, competition and media law. After graduating from law school, she joined the IP team at CMS Hasche Sigle.

Chapter 7
Redesigning Data Protection

Frederick Richter

Introduction

When it comes to redesigning a consolidated system of judicial perceptions and well-known rules like those of continental data protection, first of all, it would be necessary to justify and to provide reasons for this intention. And as any modification of a law brings more than negligible expenditures with it, a need for justification and reasons applies to larger reforms as well as to minor changes.

So, do recent rules for protecting people's private information and personal rights need to go through a complete transformation, or is it only a matter of fine adjustment? Are there any severe shortcomings? Below, some seemingly small points will be lifted out of the broader discussion. These points need some more light shed on them in order to make them more visible in recent debates regarding the use of personal data. Quite often in these debates, well-known assumptions and theorems are not scrutinized sufficiently. For a German Lawyer, it's even harder to do so, because our common attitude is to deem our law as nearly perfect and worth being exported and spread globally.

This is exactly what we did in the area of personal rights. We codified the world's first data protection law in 1970 in the German federal state of Hesse. With our Federal Data Protection Act of 1977, we gave strong guidance for the EU's 1995 Data Protection Directive. Finally, in 2012, it was mainly German experts in the European Commission and Parliament who wrote the draft of the General Data Protection Regulation—very likely always keeping in mind the German law on data protection, which had been time-tested for nearly four decades.

F. Richter (✉)
German Foundation for Data Protection, Leipzig, Germany
e-mail: richter@stiftungdatenschutz.org

© Springer Nature Switzerland AG 2020
D. Feldner (ed.), *Redesigning Organizations*,
https://doi.org/10.1007/978-3-030-27957-8_7

The Conventional Idea

Such knowledgeable tradition could be seen as a perfect starting point to build upon when it comes to creating future-proof rules on handling personal data. But in some respect, there is a danger of seeing the well-known way of regulation as the only way. Sometimes it seems that German and European lawmakers have become blind to some already notorious shortcomings of continental data protection law.

In the 1970s, when the seeds of recent European data protection laws were sewn in Germany, it was conceived as a right of defense against actions of the state—a state that was increasingly able to store and process large and structured amounts of data. One such state action in the following decade was the planned census, which was feared to incommensurately investigate German citizens.

The German idea for how to comprehensively protect human beings from dangers that might occur to their privacy and personal rights was nearly a perfect one. We simply put the right to decide what other people—and most of all, organizations like public authorities and private companies—know about the affected person into the hands of that very person. The inventor of this idea called it "informational self-determination" (Steinmüller 1971). In 1983, this legal concept figure was elevated to the rank of a fundamental right (Bundesverfassungsgericht 1983). Since then, the German state has been called upon to ensure that a citizen is always able to know *who* knows *what* about him, *when* and *on what occasions*.

Some believe that, even then, this objective was merely idealistic. In any case, in order to ensure such comprehensive knowledge of the individual's data use in the long term, it would be necessary to limit the handling of data to a level that is still manageable, or at least graspable, for the individual person. Technology continued to develop over the decades, while the core principle of data protection law stayed the same—regardless of the amount of data that the individual should be able to decide about for himself. The amount of data has increased massively. And it keeps rising, every month. It continues to rise with each new method of recording, with every single sensor being added the myriads of devices connected in the "Internet of Things", as well as with each reduction in costs for storage space.

Although the wording "informational self-determination" has not made it into either the German or the European data protection law, its basic idea of perfect personal protection by total data control still prevails: A natural person shall have the widest possible control over the data concerning her or him at any time. Any use of such personal data is initially forbidden by data protection law until either the person concerned permits the use or another legal permission takes effect.

There Is No Such Thing as an Informed Consent

If a person has legalized the processing of their data by consent, then it should only be natural that this consent is given on the basis of knowledge and understanding.

If the person does not know and understand what shall happen to her or his data, then free self-determination can no longer be seriously assumed.

If the means and ways of using, evaluating and linking personal data are becoming increasingly complex, then the law must respond to this. It must ensure that the information base on which consumers make decisions regarding the handling of their data remains broad. With further technological development, the information base must become all the better in order to continue ensuring self-determination. What does the recent law have to offer in this context? It simply tries to cope with the growing challenge of informing people as much as possible by offering specifications for the linguistic design of privacy policies and for the design of information pages for digital services. This answer is not sufficient.

The General Data Protection Regulation, adopted in 2016, defines consent as a "freely given, specific, informed and unambiguous indication of the data subject's wishes" in a statement or a clear affirmative action (Article 4 (11) GDPR). If declarations of consent are pre-formulated by a company, they should use "clear and plain language" (Recital 42 GDPR). At first glance, these requirements seem very plausible and, moreover, easy to implement for the data-using economy. But what does compliance with the law mean in concrete terms at this point? Using the much sought-after "simple language", pre-formulated explanations of consent will become even longer than before, because frankly it takes many more words to explain a legally and technically difficult issue in simple terms. To the same extent that a privacy policy grows, the willingness of users to read it drops. It is therefore possible that comprehensibility increases, while the level of true information on the user's side decreases. It is not enough for the legislator to rely on the existence of an ideal user—someone with a lot of time in today's fast moving daily routines and with the disposition to read lengthy terms and conditions of digital services.

A Solution on One Page?

There are a few ways to address the problem. One approach is to try to display the most important parts of the content of a privacy policy on only one page, so that as few users as possible are discouraged from reading simply by the sheer amount of information to be read. These one-pagers, promoted by the German Ministry of Justice and for Consumer Protection, are merely an additional source of information; they are not meant to replace any formal data protection declaration, because they summarize and inevitably simplify in a way (BMJV 2016). It is assumed that users will not read the full declaration after having read the one-pager. But also the proponents of the one pager concept would have to rely on users to read the page. And this is the crucial and sore point: in the vast majority of cases no data protection declaration gets read at all. With regard to the effectiveness of this promising approach , it has been scientifically found that, among consumers, the feeling of being informed does not increase with the summary being on one page (ConPolicy 2018). Further efforts will therefore be needed to approach a truly informed consent.

Transparency and Control

Knowledge and awareness are therefore indispensable for the data subject—and citizens cannot acquire sovereignty with regard to their own data without an overview and certain control. How many people know to whom and where and when they have given which consent for the use of their data? How many of these permissions to use personal information were not revoked, even though they were no longer in the interest of the data subject—just because the person no longer knew that they had given this consent? They need overviews and controls—via some digital dashboard or other technical solutions. Perhaps it is not primarily a task of data protection law to foster such innovative tools. But lawmakers and politicians are called upon to be more open minded toward news ways of dealing with the widespread problem of the uninformed customer.

The German Foundation for Data Protection took this unsatisfactory finding—too little openness concerning new solutions among the data privacy community—as an occasion for a project dealing with potential technology-supported improvements to the situation of consent. Some questions formed the starting point for the investigations: To what extent could technical consent assistants and consent platforms ensure the strengthening of rights to information, the automation of the consent process, the clarity and intelligibility of consent and the transparency of data processing purposes? What solutions—both internationally and in Germany—already exist, and where is further research necessary and worth promoting? We compared a number of very different practical experiments and theoretical approaches from the realm of the "Personal Information Management Systems—PIMS" (Horn et al. 2017).

The evaluation of technical possibilities and solutions in the field of PIMS shows that many approaches could create the preconditions for the legal processing of the relevant data and enable access rights or rights to restrict the processing, erasure or digital oblivion of data without requiring repeated direct user interaction. In this way, they would simplify the consent process. The approaches were analyzed from the point of view of how to create transparency through automated creation of an overview of the access rights of various applications; how to let the user decide individually in advance who should receive what data and for what purpose; how to enable users to take control by providing an overview of usage and how—in terms of self-protection of their data—to motivate consumers to control their data by exercising their rights to information. The examined projects showed significant differences, both in terms of their technical approach and economic implementation. They also differ in terms of how extensive the effect of the application is. For example, a specific approach may only have one focus (e.g., pure user education), or it can combine several different purposes.

How About a Consent Agent?

A particularly outstanding innovation would, of course, be a real consent agent, a tool that implements and executes the user's privacy preferences according to his specifications. In the ideal, such a "privacy-bot in the service of man" would not only be able to give consent in the sense of the data subject, but would also contain a database of declarations issued. This functionality would bring users much closer to a state of having a good overview and control in the data area.

But of course, such tools would need to fully comply with the GDPR, which is the supreme point of orientation in the area of data protection and the respective legislation for at least the next 5 to 10 years. And as the GDPR includes specific stipulations for consent, there are certain challenges for that new idea of user control. For example, to meet the conditions of GDPR Article 4, Paragraph 11, a clear confirmative action is needed from the consent assistant. Similar to any data controller under the rules of the GDPR, a privacy agent/data protection assistant would also need to explain the circumstances of any planned or programmed consent in easily accessible and understandable language to its user, including the type of data affected, purposes, recipients or categories of recipients. Developers of privacy assistance tools will also have to face the challenge that blanket consents are not valid. So it would not be possible for a user to generally instruct their privacy app to provide consent in a multitude of cases, for instance "for all cases where location data is requested by a Belgian app" or "for all apps requesting car-related location data" or other groupings like that.

For the functioning of a privacy assistant, it would be required that data protection declarations and privacy policies become machine-readable. The tools would need to assess what the vendor of any "opposite side" offers concerning the use of personal data from the user. But also, any automated translation of data protection instructions for a consent statement (for example, in the form of a list whose empty fields must be activated by the user) must be subject to verification in each individual case. This requirement might decrease the effective usability and convenience of consent agents. Difficulties might, for instance, arise if the data protection information references contract-related purposes, or if a consent statement is generated from this on an automated basis according to user preferences. However, no consent is required for contractual purposes, only transparent information. If the consent assistant is used in future to assist in the conclusion of contracts, then civil law and data protection law will have to be separated. Under civil law, consensual statements of intent are required for the formation of a contract, and as *essentialia negotii* of a purchase contract, this also includes the definition of the subject and the contracting parties. From a data protection perspective, data may be processed without consent if it is necessary for contractual purposes. Nevertheless, transparent information about the data processing (such as processing for contract-related purposes) must be provided. In designing the consent assistant, care must be taken to ensure that this separation is clear to the user.

Additionally, from a data protection perspective, consent always includes a right to revocation. With regard to the exercise of the right to revocation, any new system should offer users a self-management function, so that users can change, correct and delete their consent at any time. Thus, the requirements for a revocation from GDPR Article 7, Paragraph 3 can be met. Problems that could arise in connection with the new right to data portability (Article 20 DGPR) would then be bypassed. The necessary accuracy of the data may be ensured by the system if the consent assistant is able to prevent those types of data access in which the recipient, purpose and scope of the specific personal data do not match. The potential recipients have access to the records of users only on the condition that the right combination of legitimate recipients and processing purposes are present. In cases of deviations, the consent assistant must also be able, in dynamic form, to request and obtain the consent of the user.

As part of the design of the consent assistant, the coupling prohibition and free consent by the relevant parties must be observed to a special degree. All the circumstances must be taken into account, as well as whether the person can actually see in full the marketing and/or scoring purposes for which the personal data is used. This self-determination can be difficult to determine in certain individual cases. But the more purposes are interrelated, or the more data recipients are involved, the more likely the confusion for the person concerned. When using a consent assistant, the impression must also not be created that, as a result, the data processing is complete—especially if, for example, additional processing is planned for a legitimate reason. The legal requirements of such 'dynamic consent' must be examined separately.

The consent assistant should be able to automatically ensure that consent is not granted for an indefinite period, but that data access will automatically be prevented—either when the purpose of use no longer applies, or if after an appropriate time the user is asked if he/she wants to maintain his/her consent. In this case, the restrictions on data storage (GDPR Article 5, Paragraph 1e) and data minimization are fulfilled (GDPR Article 5, Paragraph 1c), since the person concerned decides what data is processed and that access is subject to the declaration of consent together with the category of recipients. The person responsible for the data processing must provide the consent on an informed basis. He must provide the information prior to collection of the data and must be able to show proof of consent. To support the transparent design of the choices (purpose, recipient, data) and in the interest of an informed and unequivocal expression of will, a consent assistant could—perhaps even should—apply visual elements (GDPR Recital 58). For complex data processing with different purposes or recipients, the representation could well be anything but transparent, even if a consent assistant were used. Regarding this, it could be examined to what extent the so-called one-pager mentioned above would be useful as a transparent summary of the consent given.

Overall, it should be noted that the legal requirements of informed consent and consent platforms would be difficult to establish without additional insights into economic behavior. Providing transparent information is a necessary—but not a sufficient—condition for an accurate assessment of data protection risks. In this respect, the emotions and cognitive abilities of the user are just as important, if not more important. In other words: the legal framework for informed consent can only be assessed and structured appropriately if the actual willingness of users to actively deal with the protection of their privacy is taken into account. Finally, it must be noted that it is always crucial to ensure the future achievement of the General Data Protection Regulation's aim of putting in place a uniform and high level of protection for natural persons, by applying an equivalent level of protection for the rights and freedoms of natural persons with regard to the processing of their personal data in all member states. In this context, it is particularly useful to consider uniform codes of conduct or guidelines. In practical application, a consent assistant with transparent design possibilities can contribute to the level of protection. Users have more control over their options, since data can be collected directly from them with their active participation and can be limited in time. However, technical developments must constantly be critically examined against a background of automated decisions, the possibilities of profiling and a change of purpose.

Towards a Layered Approach

In order to approach real informational self-determination, a high level of information is indispensable. Certainly, this will not be attainable using just one of the possible methods. Rather, the solution can be sought in combining them. In order to meet the different requirements for the provision of information for the intended data processing, a multi-layer model should be used. The aim is to have the best-informed consumers. The multiple layers are in the best interest of consumer policy—and they should also be in the best interest of the economy, because they are proof for companies that they have sufficiently fulfilled their information obligations under data protection law.

The bottom layer in the proposed model consists of a machine-readable data protection declaration. To this end, it is necessary for the legislator to lay down guidelines for the structure of such declarations in order to ensure uniformity. This also facilitates the use of privacy bots that can capture and evaluate the content of a privacy statement. As a second layer, the first visible-to-the-user layer builds up on the machine-readable base layer. It is the full declaration on compliance with data protection law, written by lawyers and read only by other lawyers (usually those of the competitors). This layer corresponds to the "long" version of the data protection information that we know today and which, on its own, brings no innovation what-soever. But this well-known—and well-hated—standard element is supplemented by the third and fourth layers. The third consists of the above mentioned one-pager. Reading it requires less effort and perseverance than the "full" explanation. But the effort is already considerably more manageable if a single page is to be read instead of ten or more pages. If the content of a page is already too much for the reader, the reader can go back—or better: step up—to the fourth level. It is the crowning conclusion of the information pyramid, so to speak. It consists of symbols or little pictures like those used in the license system of Creative Commons. Even current law would not need to be re-designed. The GDPR already mentions such "standardized icons" in Article 12, Paragraph 7. The icons should give "a meaningful overview of the intended processing.in an easily visible, intelligible and clearly legible manner". But since the final version of the adopted new European data protection law does not contain concrete examples of icons, work remains to be done, supported by politics and research. Initial proposals have been made in the past and new ones are being added (Specht 2018). However, we still have important tasks ahead of us, especially in the area of design and implementation.

References

BMJV. (2016). Federal Ministry of Justice and Consumer Protection—"One-Pager"—Muster für transparente Datenschutzhinweise. www.bmjv.de/DE/Themen/FokusThemen/OnePager/OnePager_node.html. Accessed April 10, 2018.
Bundesverfassungsgericht. (1983). Federal Constitutional Court, Judgment of the Second Senate of 15 Dec 1983 –1 BvR 209/83–, para 172.
ConPolicy. (2018). *Better informed? Results from behavioral science on the effectiveness of the privacy-one-pager approach and further solutions for data protection.* www.conpolicy.de/en/news-detail/better-informed-results-from-behavioral-science-on-the-effectiveness-of-the-privacy-one-pager-appro. Accessed April 12, 2018.
Horn, N., Riechert. A., & Müller, C. (2017). *New ways of providing consent in data protection—technical, legal and economic challenges.* https://stiftungdatenschutz.org/english/project-consent-pims. Accessed April 10, 2018.
Specht. (2018). Informationsvermittlung durch standardisierte Bildsymbole—Ein Weg aus dem Privacy Paradox? In *Specht/Werry/Werry, Handbuch Datenrecht in der Digitalisierung 2018.*
Steinmüller. (1971). *Grundfragen des Datenschutzes, Gutachten im Auftrag des Bundesministeriums des Innern.* German Bundestag printed paper VI/3826, pp. 5–224 www.dipbt.bundestag.de/doc/btd/06/038/0603826.pdf. Accessed April 10, 2018.

Frederick Richter, LL.M. is director of the German Foundation for Data Protection (Stiftung Datenschutz). After studying law at the University of Hamburg, Richter completed a master's degree in information law at the Universities of Vienna and Hanover. After completing his legal clerkship in Berlin, he was admitted to the bar in 2005 and worked as a research assistant to a member of the German Bundestag. From 2008 to 2010, he was a consultant and data protection officer of the Federation of German Industries (BDI). From 2010 to 2013 he was a consultant in the German Bundestag on copyright and network policy. In 2013, he was appointed founding director of the Stiftung Datenschutz. Frederick is a member of the IAPP and sits on the advisory boards of the projects AUDITOR for data protection certification and ABiDa—Assessing Big Data at the University of Münster. He is a member of the Data Protection and Ethics Panel of the AXA Group and permanent author of the journal Privacy in Germany (PinG).

Chapter 8
Erosion of Civil Rights in a Digital Society—Maintaining the Democratic Society

Jimmy Schulz

Vigilance Is Required

A Democratic, Federal and Free State Under the Rule of Law

After the Second World War, Germany was in need of a new constitution that would protect the people from the state and guarantee their fundamental rights. Therefore, the Basic Law for the Federal Republic of Germany was promulgated on 23 May 1949 in Bonn, Germany. At the same time, the Federal Republic of Germany was founded. This constitution substantiates the nature of the Federal Republic of Germany as a democratic, federal and free state under the rule of law. The essential characteristics are the Fundamental Rights, named in Article 1 to Article 19, which guarantee freedom and equality to German citizens. They are binding for all three powers: legislative, judiciary and executive power. Article 19, Paragraph 2 says: "*In no case may the essence of a basic right be affected*" (Tomuschat and Currie 2014). With this important paragraph, the German constitution protects itself from over-ambitious politicians. The authors of the Basic Law for the Federal Republic of Germany surely had no idea how digital transformation would affect our society. But that does not matter. Fundamental Rights are always valid, regardless of whether we need protection from the state in analogue life or in cyberspace.

With the progress of digital transformation, more and more decision-makers recognize challenges and possibilities of increased networking. But possibilities are not always positive, some of the new opportunities even have the power to destroy. To be able to focus on the positive ones and to avoid temptation, staunchness and conscience are the key required characteristics of decision makers.

J. Schulz (✉)
Committee on the Digital Agenda of the National Parliament of the Federal Republic of Germany, Berlin, Germany
e-mail: jimmy.schulz@bundestag.de

© Springer Nature Switzerland AG 2020
D. Feldner (ed.), *Redesigning Organizations*,
https://doi.org/10.1007/978-3-030-27957-8_8

The Big Eavesdropping Operation

Twenty years ago, in 1998, politicians in Germany decided to change one of the Fundamental Rights with the purpose of improving police work. Technical developments led to new possibilities, which brought forth a change in Fundamental Rights. This change affected Article 13: Inviolability of the home. The paragraphs 3–6 were added. These additions empowered executive authorities to use measures of acoustical surveillance in any home in which the suspect is supposedly staying, but only "pursuant to judicial order" and only under strict conditions. Furthermore, the idea of checks and balances is clearly reflected in this change. The executive power gets the right of acoustical surveillance, but only in consideration with the judiciary power. Additionally, the legislative power has to be informed regularly, which is specified in Article 13, paragraph 6.

Nevertheless, 20 years ago, this issue caused a huge controversy. A leading figure of the opponents of this Fundamental Right change was the former Federal Minister of Justice from the Liberal Democratic Party FDP, Sabine Leutheusser-Schnarrenberger. She tried to prevent the change of the law in her role as Federal Minister. When she realized she could not stop the law from passing, she consequently resigned from her post. But she did not stop fighting and brought an action before the Federal Constitutional Court (*GE: Bundesverfassungsgericht*) in Germany. With her liberal supporters—Gerhart Baum, the former Federal Minister of the Interior and Burkhard Hirsch, the former Vice-President of the German Federal Parliament—she finally succeeded in 2004, when the court delivered the judgment that large parts of the law violated human dignity and were therefore unconstitutional. The judges did not declare the changes in Article 13 themselves as unconstitutional, but numerous regulations in the Code of Criminal Procedures based on the changes of Article 13. Furthermore, surveillance should only be ordered on the suspicion of particularly serious crimes. Additionally, conversations between close relatives may only be intercepted if all involved parties are suspects and the conversation has criminally relevant content. If these conditions are not fulfilled, the corresponding records are not only worthless as evidence, but are not allowed to be made at all. In order to establish constitutionality in the conduct of surveillance, surveillance must be actively pursued by an official who, if necessary, stops monitoring as soon as the conditions specified by the court cease to exist. Any form of automatically recorded surveillance is considered non-constitutional. In summary, the right of the state to intrude into citizens' privacy is limited to situations that may pose significant risks for the community (1 BvR 2378/98 2004).

We can summarize that 20 years ago, when the process of digital transformation was still at its very beginning, politicians' ideas for using new technology for surveillance found fertile ground. It was the beginning of a steadily increasing number of comprehensive proposals to limit freedom.

But the plans from 20 years ago might seem kind of innocent compared to today's ideas. This development shows that we are still at the beginning of a revolution that might cause a huge impact on our free and democratic society. The following examples will underline this concern.

The Basic Law in Times of Digital Transformation—Under Constant Fire

Digital transformation is speeding up. New technologies like autonomous driving, hybrid humans, smart living and smart homes, new applications to simplify life, artificial intelligence in general and even digitized clothing are booming. A new generation of people, the so-called digital natives, are growing up with all these things, taking them for granted. But will this new generation also consider their civil rights as important, or will they become used to the fact that they are "transparent"? What is our duty now?

Data Retention

One of the key terms is data retention. For years, experts have emphasized that storing all communication without links to terrorism or even an involvement in crime is absolutely the wrong way to go. This does not lead to greater safety, but only to less privacy. With these data, it is possible to create a detailed profile of people. The state has access to information about the websites that citizens visit, who they called and where they were called from. Data retention can be used to identify social relationships and to provide a comprehensive background about people's private lives. Politicians who support this procedure consider every citizen a suspect.

In cooperation with supporters of civil rights, several politicians from the liberal party brought an action before the Federal Constitutional Court in Germany to stop data retention. At the moment, due to an unclear jurisdiction and several constitutional complaints, the law was put on hold until there is a judgment from the Federal Constitutional Court. Maybe it is easy to store all these data and then just look for what is needed to fight crime. It would even be easier to have a saliva sample of every person in a database and install surveillance software in every smartphone or computer (and car). Just because it is easy and possible, does not make it the right way to go. Unfortunately, the right way to go is complicated and expensive. Many people believe this is a price we should be willing to pay. About 40% of the German citizens are concerned that the state has steadily increased monitoring them as a result of technological development (Statista 2016).

Governmental Malware—Spyware Made by the Government

Another issue is a disturbing new law that was passed by the Conservatives and the Social Democrats in 2017 with regards to online searches and surveillance of telecommunication. The way this extension of state power passed the legislative process was considered controversial by many citizens and media: Two unrelated draft laws were in the middle of the normal legislative process when the Committee on Legal Affairs and Consumer Protection of the German Bundestag silently included (Deutscher Bundestag 2017) these far-reaching surveillance instruments into these draft laws. The "Gesetz zur effektiveren und praxistauglicheren Ausgestaltung des Strafverfahrens" (BT-Drucksache 18/11277) (*EN: Law on the more effective and practicable design of criminal proceedings*) and the "Gesetz zur Änderung des Strafgesetzbuchs, des Jugendgerichtsgesetzes, der Strafprozessordnung und weiterer Gesetze" (BT-Drucksache 18/11272) (*EN: Law amending the Criminal Code, the German Juvenile Court Act, the Code of Criminal Procedure and other laws*) normally would not be associated with online searches and surveillance of telecommunications. However, following the decision of the Committee on Legal Affairs and Consumer Protection (BT-Drucksache 18/12785) adopted by the plenary, the amendments had been incorporated into the laws. Therefore, the legal foundation has been extended to allow the so-called "Quellen-TKÜ" (*GE: Quellen-Telekommunikationsüberwachung*), which means lawful interception at the device-level by malware, before any end-to-end encryption can be applied. Also, online searches were added, which provides the executive powers the right to search computers and smartphones by installing malware.

By pursuing this legislative procedure, silently including far-reaching restrictions on privacy in the middle of the legislative process, a public debate nearly failed to appear. The government avoided the standard process, which consists of three readings in plenary and the involvement of the German Bundesrat. In the German Bundesrat, the federal states participate in the legislation of the federation. The Federal Data Protection Commissioner was also not involved (cf. Beuth and Biermann 2017; Grunert 2017).

With this new law, the state granted itself the right to use spy software on smartphones and computers of suspects—not only to prevent terrorism, but also to detect, for example, counterfeiting of documents or tax evasion. Nevertheless, online searches should only be applied if the alleged offense is particularly serious. But this is not the case with the use of spy software for the purpose of telecommunication surveillance. The problem is that the software used for telecommunication surveillance is able to monitor much more information than lawfully permitted.

The "Quellen-TKÜ" has the goal of lawfully intercepting encrypted communication. The other part, the online searches, goes a significant step further and implies the ability to search through all data—for example, data that is stored on a smartphone (or accessible via the cloud). The Quellen-TKÜ is supposed to be similar to the interception of a telephone call, and the online search is comparable to the search of an apartment. However, this important separation, which the legislator has envisaged, is extremely difficult to implement from a technical point of view. Unfortunately, past reports have shown that the state has been using software that was not covered by the legal framework, because the trojan used had significantly more capabilities than permitted by law.

In addition, the installation of this malicious software obviously requires the exploitation of existing vulnerabilities in IT systems. This constitutes a weak point for overall IT security, as the state participates in the trade and the dissemination of IT security vulnerabilities and prevents their effective remediation. This can lead to major collateral damages to innocent citizens, as well as companies. The use of backdoors and the participation of the state in digital black and grey markets in order to acquire knowledge about security vulnerabilities or so-called zero-day exploits from third parties are incompatible with the basic values of our liberal-democratic order.

What's Next? Public Surveillance Ideas

Data retention and governmental malware are two examples of these kind of ideas that should make "evil" cyberspace safer, better or less complicated. But in the end, they restrict the freedom of citizens. There are many other suggestions, such as face recognition in public, which has already been tested by the federal government at Berlin-Südkreuz, a heavily frequented train station in the middle of Germany's capital city. Another suggestion is the possibility of recognizing emotions via software. Some politicians think that terror attacks could be prevented—for example, at an airport— by identifying potential terrorists via monitoring of "dangerous" facial expressions. One can imagine many different reasons that a person might look stressed, angry or in any other form "dangerous" at an airport, such as if a flight is cancelled at the last minute. Such a surveillance of emotions would significantly curtail citizen liberties and constitute a huge loss of freedom.

Freedom of Speech—Online and Offline

In a democratic society, freedom of speech is fundamental. In Germany, it is written down in Article 5 of the constitution: Freedom of expression, arts and sciences. It says in paragraph 1: "*Every person shall have the right freely to express and disseminate his opinions in speech, writing and pictures, and to inform himself without hindrance from generally accessible sources. Freedom of the press and freedom of reporting by means of broadcasts and films shall be guaranteed. There shall be no censorship*" (Tomuschat and Currie 2014). Today, many people get their information through social media platforms. They use them to get updates and news from their friends or to share their thoughts and opinions. This novelty that allows everybody to express him- or herself online through a blog or a social media profile has advantages for a great majority of people. But some people use this chance to distribute lies, hate speech and even illegal content. This has led German politicians to propose the rules to require that social media companies like Facebook and Twitter quickly remove this kind of content from their sites, or face having to pay high fines. This model is now also being discussed on the European level. About 45% of the German citizens believe the so-called German Netzwerkdurchsetzungsgesetz or "NetzDG" (*EN: Network Enforcement Act*) constitutes a serious threat to the freedom of expression of the public (Civey 2018). It therefore should not become a role model. The concern about this model is that there is a risk that, in case of doubt, the companies concerned will opt for deletion of online content in order to avoid paying high fines.

Moreover, it is unacceptable to allow companies to act like judges and decide which content is covered by the right to freedom of expression and which is not. This is primarily a governmental task and will cost money, because more judges with a special training are needed.

Some European politicians even promote the idea of obligating online services to monitor and filter content, even before it's uploaded, by introducing so-called "Upload Filters". This would mean that private companies decide which content is allowed to be distributed online—taking a significant step towards censorship.

Bavarian Surveillance Fantasies Threatening a Whole Country

The past has shown that many ideas curtailing civil rights came from the conservative side of the political spectrum. Many of these ideas have their origin in Bavaria. The conservative party, called CSU (*DE: Christlich Soziale Union*), has governed Bavaria since 1957. This is a local Bavarian party, which builds a union with the CDU (*DE: Christlich Demokratische Union Deutschlands*) on the federal level. The CSU had a majority in the Bavarian Parliament in 2018, meaning they could easily make far-reaching political decisions. Furthermore, a new party appeared and gained strength against the background of the refugee crisis in 2015 in Germany. The AfD (*Alternative für Deutschland*) is another very conservative and nationalistic party that seems to be attractive to very conservative, perhaps some of them former CSU, voters. Therefore, the CSU had to regain their attention, especially with regard to the Bavarian State Elections on 14 October 2018. Additionally, the former Bavarian Prime Minister was nominated as German Federal Interior Minister half a year before the elections. In his first speech in the German Bundestag as Federal Interior Minister, he announced the desire to make Bavaria a role model for Germany as a whole.

Police with Secret Service Tasks

In 2018, the Bavarian government (CSU-led) drafted a law for the Bavarian parliament (CSU majority) that many have equated with turning the police force into an intelligence service. The plans of the so-called "Gesetz zur Neuordnung des bayerischen Polizeirechts" (BayLT-Drucksache 17/20425) (*EN: Law on the reorganization of the Bavarian police law*) lead to a weakening of judiciary power. As was the case 20 years ago, surveillance only was permitted "pursuant to judicial order" under strict conditions. The CSU wants to change this. Additionally, the law would allow the police to make videos while people participate in peaceful assemblies, butting up against Article 8 of the Federal Constitution, which guarantees the "Freedom of assembly". The draft law includes additional controversial ideas. It would allow police to spy on computers and smartphones if they think there is an imminent danger. The police could also change and delete information on these computers under special circumstances. Since the foundation of the Federal Republic of Germany and, thereby, the establishment of the German Basic Law, no police force ever had similar powers. Additionally, Members of the Bavarian Parliament submitted amendments to make this law even stricter—for example, by obligating IT companies to implement vulnerabilities in their products that could be used by the state. People all over Germany (more than 30,000 citizens in Munich, the capital of Bavaria) demonstrated against this law (Süddeutsche Zeitung 2018).

Recommended Steps to Protect Civil Rights

Transforming Possibilities into Chances

New technologies offer many possibilities to make police work more efficient but also to restrict the civil rights of a country's citizens. There are many controversial ideas popping up that are discussed, decided upon, implemented, concretized and sometimes even withdrawn. Without a doubt, the digital transformation has the potential to lead people to another age of democracy, with another view on civil rights and transparency—this is already happening.

The previous explanations focused on the dangers to civil rights while a democratic digital transformation is happening. But citizens should not be scared when it comes to the future and technical innovations. There is a lot of power in this digital transformation, along with many positive ideas that can help save and improve lives. Politicians and citizens would do well to focus more on these beneficial ideas. The question is: How can they succeed?

Investing in Education—Sensitizing People—Enlightening Politicians

To seize the full potential the digital transformation has to offer for society, there are three tasks that can be focused on: providing a better (also digital) education, raising awareness and being persistent. The future generations must be prepared to handle new technologies at school and older generations should never stop learning. In particular, multipliers like politicians or teachers should refresh their knowledge of history and explain the reasons and the importance of fundamental rights to everyone. Furthermore, today's stakeholders have the responsibility to deal with the consequences of their decisions. There need to be social debates about civil rights in the digital world, and multipliers need to be able to counter the attitude of people who say, "Surveillance is ok. I can accept it because I have nothing to hide".

A Right to Encryption

One of the basic rights in a digitized society is online privacy. On the one side, Article 10, paragraph 1 of Germany's Fundamental Rights states that "*the privacy of correspondence, posts and telecommunications shall be inviolable*" (Tomuschat and Currie 2014). If people send letters, they regularly use an envelope. But emails are often sent unencrypted, which is comparable to a postcard—anyone could read it. For that reason, a right to encryption is needed. This implies that all providers of telecommunication services should be obligated to offer the standard version of their communication service (end-to-end) encrypted.

Good Ideas from Politics

Luckily, there are also many positive political initiatives. Two of them are particularly noteworthy. As early as April 2016, the European General Data Protection Regulation (GDPR) (EU 2016/679) was adopted, and it became effective on 25 May 2018 after a two-year transition period. It establishes a harmonized data protection framework in all 28 Member States of the European Union. This regulation enables all individuals in the European Union to have control over their personal data via several instruments. The GDPR is based on several principles. One is the principle of explicit consent as the foundation for data collection and processing (opt-in), which also strengthens the data protection authorities as it provides for the possibility to impose severe sanctions in the case of data protection infringements—such as fines of up to €20 million, or 4% of the annual worldwide turnover of a company, depending on which sum is higher.

Furthermore, negotiations of the so-called ePrivacy Regulation, are taking place on the European level. The regulation's goal is to ensure "*the protection of fundamental rights and freedoms [...] in particular, the rights to respect for private life and communications and the protection of natural persons with regard to the processing of personal data*" (COM/2017/010 final—2017/03 (COD), Article 1). It will complement the GDPR with regard to electronic communications data. The ePrivacy Regulation could ensure that the same high privacy standards apply for so-called Over-the-Top communications services (e.g., widely used messenger services), as well as "traditional" telecom operators in the future.

The GDPR and the ePrivacy Regulation are important steps to ensure that all over Europe, people and businesses profit from a harmonized set of rules that strengthen the sovereignty over their own data. Data sovereignty is fundamental, if we want to ensure the autonomy of each citizen in the digital world.

Courage First—Concerns Second

"Digital first—Bedenken second" (*EN: digital first, concerns second*)—this claim was used by the Free Democratic Party (FDP) to promote innovative ideas for the digital sphere in the campaign for the Federal Elections in 2017 in Germany. This is the attitude toward new technologies that leads to great innovation. There is not a contradiction between this claim and the former explanations, as long as the fundamental rights of all people are guaranteed, and they are empowered to profit from the opportunities offered by digital transformation.

This is the guiding principle on which an international ethical fundament could be built on. The internet is a good thing after all!

References

BayLT-Drucksache 17/20425 vom 30.01.2018, Gesetzentwurf der Staatsregierung für ein Gesetz zur Neuordnung des bayerischen Polizeirechts (PAG-Neuordnungsgesetz), URL: http://www.bayern.landtag.de/www/ElanTextAblage_WP17/Drucksachen/Basisdrucksachen/0000013000/0000013038.pdf Accessed May 08, 2018.

Beuth, P., & Biermann, K. (2017). Staatstrojaner. Dein trojanischer Freund und Helfer, DIE ZEIT online, 22.06.2017, URL: https://www.zeit.de/digital/datenschutz/2017-06/staatstrojaner-gesetz-bundestag-beschluss Accessed May 08, 2018.

BT-Drucksache 18/11272 vom 22.02.2017, Gesetzentwurf der Bundesregierung: Entwurf eines Gesetzes zur Änderung des Strafgesetzbuchs, des Jugendgerichtsgesetzes, der Strafprozessordnung und weiterer Gesetze, URL: http://dip21.bundestag.de/dip21/btd/18/112/1811272.pdf Accessed May 08, 2018.

BT-Drucksache 18/11277 vom 22.02.2017, Gesetzentwurf der Bundesregierung: Entwurf eines Gesetzes zur effektiveren und praxistauglicheren Ausgestaltung des Strafverfahrens, URL: http://dip21.bundestag.de/dip21/btd/18/112/1811277.pdf Accessed May 08, 2018.

BT-Drucksache 18/12785 vom 20.06.2017, Beschlussempfehlung und Bericht des Ausschusses für Recht und Verbraucherschutz, URL: http://dip21.bundestag.de/dip21/btd/18/127/1812785.pdf. Accessed May 08, 2018.

BVerfG, Urteil des Ersten Senats vom 03. März 2004, 1 BvR 2378/98—Rn. (1-373), bverfg.de, URL: http://www.bverfg.de/entscheidungen/rs20040303_1bvr237898.html Accessed May 08, 2018.

Commission proposal for a regulation of the European Parliament and of the Council concerning the respect for private life and the protection of personal data in electronic communications and repealing Directive 2002/58/EC (Regulation on Privacy and Electronic Communications), COM/2017/010 final—2017/03 (COD), 10.01.2017, URL: http://eur-lex.europa.eu/legal-content/EN/ALL/?uri=COM:2017:10:FIN Accessed May 08, 2018.

Deutscher Bundestag, Bundestag gibt Strafermittlern neue Instrumente in die Hand, Deutscher Bundestag, 2017, URL: https://www.bundestag.de/dokumente/textarchiv/2017/kw25-de-aenderung-stgb/511182 Accessed May 08, 2018.

Grunert, M. (2017). Bundestrojaner. Durch die Hintertür zur Online-Überwachung, faz, 22.06.2017, URL: http://www.faz.net/aktuell/politik/online-durchsuchung-quellen-tkue-bundestrojaner-wird-gesetz-15071053.html Accessed May 08, 2018.

NetzDG: Einschränkung der Meinungsfreiheit? - Umfrageegebnisse 03.08.2018, Civey 2018. URL: https://civey.com/umfragen/netzdg-einschraenkung-der-meinungsfreiheit.

Regulation (EU) 2016/679 of the European Parliament and of the Council of 27 April 2016 on the protection of natural persons with regard to the processing of personal data and on the free movement of such data, and repealing Directive 95/46/EC (General Data Protection Regulation), 27.04.2016, URL: https://eur-lex.europa.eu/legal-content/EN/ALL/?uri=celex: 32016R0679 Accessed May 08, 2018.

Menschen demonstrieren gegen Polizeigesetz, Süddeutsche Zeitung, 10.05.2018, URL: https:// www.sueddeutsche.de/muenchen/massenproteste-in-muenchen-menschen-demonstrieren-gegen-polizeigesetz-1.3974427 Accessed August, 08 2018.

Tomuschat, C., & Currie, D. P. (2014). (translators), Basic Law for the Federal Republic of Germany, Juris. URL: https://www.gesetze-im-internet.de/englisch_gg/index.html Accessed May 08, 2018.

Umfrage in Deutschland zur Befürchtung, dass der Staat die Bürger überwacht, Statista/ IfD Allensbach (ACTA 2016), 2016, URL: https://de.statista.com/statistik/daten/studie/282285/umfrage/staatliche-ueberwachung-der-buerger–befuerchtung-in-deutschland/ Accessed: August 08, 2018.

Jimmy Schulz is a German internet entrepreneur and politician of the Free Democratic Party (FDP) from the district of Munich, Bavaria. After finishing his secondary education at the Ottobrunner Gymnasium, he studied political science at the University of Texas at Austin and in Munich, Germany. His passion for IT started at school, during which he worked at various IT companies and later, in 1995, founded CyberSolutions GmbH, which entered the stock market in 2000. He is currently CEO of CyberSolutions Ltd., which has its company seat in Riemerling/Hohenbrunn. Jimmy Schulz has been actively fighting for civil rights and internet freedom for more than 20 years. From 2009 to 2013, he was a member of the German Parliament. He was spokesperson for the FDP in the commission of enquiry: "Internet and digital society". From 2014 to 2016, he was a member of the ICANN At-large Advisory Committee (ALAC) and Vice Chairman of the Internet Society ISOC Germany (2015–2017). Since 2018, he has been a member of the Presidium of ISOC Germany. In 2017, he was again elected as a member of the German Parliament and is now chairing the Committee on the Digital Agenda.

Chapter 9
Transatlantic Cyber Forum—Cooperating on Borderless Cyber Security Challenges

Sven Herpig and Julia Schuetze

Introduction

This chapter aims to contribute to the discussion around the qualities of effective policy-making models and formal and informal institutions for dealing with complex issues, such as what the cyber security for democratic processes might look like and what organizational features they may have. It first describes how different some of today's challenges are, such as cyber operations against democracies, and why traditional institutions may struggle to find policy solutions for them. Subsequently, the chapter analyses the extent to which new forms of policy-making models have been applied. Ultimately, it evaluates the Transatlantic Cyber Forum as a recently established institution that includes some new features.

Legal and policy solutions always seem to be two steps behind the current reality of digitization. Cyber security challenges in particular are deemed highly complex, and they develop and change quickly. They affect a diverse group of people, and more importantly, solutions to the problem require responsibility and expertise from different sectors. Cyber security is a global issue that, nonetheless, can have direct and local effects. These features make the problem very hard to deal with in traditional policy-making and decision-making structures. Current political institutions are overwhelmed, and we have seen them struggle with the issue until today. The inability to overcome such challenges can erode the public's trust in the government. This is especially problematic for processes like elections, which derive their legitimacy from a high level of public trust.

S. Herpig
International Cyber Security Policy, Stiftung Neue Verantwortung (SNV), Berlin, Germany
e-mail: sherpig@stiftung-nv.de

J. Schuetze (✉)
Transatlantic Cyber Forum, Stiftung Neue Verantwortung (SNV), Berlin, Germany
e-mail: jschuetze@stiftung-nv.de

© Springer Nature Switzerland AG 2020
D. Feldner (ed.), *Redesigning Organizations*,
https://doi.org/10.1007/978-3-030-27957-8_9

A simple security standard recommended by the executive branch could have prevented the 2015 cyber operation against the German Parliament; however, the legislative branch did not implement the recommendation well enough, most likely because differences in security system organizations across the branches of government led to a failure in knowledge transfer. Moreover, intergovernmental responses appeared uncoordinated due to a combined lack of trust and processes in place. Similarly, in the 2016 U.S. election, "the lack of a trusting relationship between the FBI and the DNC allowed an active cyber operation to continue for months. It appears that there was neither a comprehensive communication strategy on the side of the FBI nor a trusted relationship between members of the FBI and the DNC" (Herpig 2017a, b: 15).

The election process has digitalized, and while new technologies are beneficial in this context as they improve efficiency and inclusion, they can lead to serious security vulnerabilities that threaten democracy, if the corresponding challenges are left unaddressed. Cyberattacks during elections can succeed for many reasons, and security failures may stem from a myriad of reasons, including vulnerability in a company's software, unawareness of staffers who fall for phishing emails, an active cyberattack where security infrastructure does not hold up, or failed communication channels. The list goes on. The many entry points for security failures makes solutions highly dependent on different stakeholders and their behaviors. A constitutionally created division of powers means a separation of the federal and state level—this is what actors, such as political institutions know; however, the new threat may create a need to adapt mechanisms that still ensure security. Thus, handling cyber security presents a problem for the traditional political system since it exacerbates state and federal frictions, as Shackelford et al. (2017: 631) argued, and creates problems for voting security. Moreover, private companies and individuals could become reliable for security purposes or serve as first responders when an attack occurs. Thus, institutions need to find a way to integrate their roles into the policy solution. On top of that, there is an international relations perspective as cyber operations gain momentum as a preferred tool of international conflict resolution for several nation states. This makes the challenge extremely difficult to address, not least because the solution relies on incorporating the roles, responsibilities and expertise of diverse stakeholders at all levels from local to international.

Governments need to use a variety of policy instruments to find an effective solution that reflects the interconnectedness of the problem. Kambiz (2017) found that "in the complex world of today, important policy and business decisions are still made with a 17th Century reductionist mindset and approach. Yet, complex challenges such as climate change, poverty, public health, security, energy futures, and sustainability transcend any single science, discipline, or agency. Rather, they require integration of social, economic, cultural, political, and environmental concerns to achieve acceptable and sustainable outcomes". Thus far, however, traditional institutions still work on solutions separately in their field or disciplines, which makes it harder to develop a coherent strategic plan for tackling cyber operations against democracies effectively. Gnad (2016) argued that in a world where there are difficult and highly complex challenges, such as cyber security—an unknown terrain for most states—the social stability and political order is challenged. All of this inevitably leads to the question of whether or not we need to adapt or redesign existing institutions to tackle the challenges that arise.

Features of Multi-stakeholder Models

Theoretical research has discussed what features institutions would need in order to be able to better tackle challenges like cyber operations against democracies. Fuerth and Faber (2012) talked about complex national missions and noted that relevant foundational systems, such as the electoral system, are vulnerable because of cyberspace (seen as a global network of IT-systems.). Those vulnerabilities, Fuerth and Faber (2012: 31) argued, can only be addressed by a whole-of-nation planning and synchronized execution approach. Challenges of such a complex nature call for a more diverse set of actors to engage in meeting the challenge. One example could be what Torfing et al. (2012: 14, in Fowler 2014: 3) categorized as multi-stakeholder initiatives (MSI). An MSI is a form of "interactive governance", the complex process through which a plurality of actors with diverging interests interact in order to formulate, promote and achieve common objectives by means of mobilizing, exchanging and deploying a range of ideas, rules and resources. MSIs—alongside public-private partnership, collective impact initiatives and others—aim to solve local problems caused by global governance failure, which is the case in cyber security (Abbassi et al. 2013). Kambiz (2017) argued that solutions to cyber security challenges would entail the "synthesis of diverse knowledge and perspectives in a transparent and unifying decision-making process, engaging stakeholders with competing interests, perspectives, and agendas under uncertain and often adversarial conditions" (Kambiz 2017: 14).

Furthermore, it is globally recognized that cyber security needs to be handled collectively with all stakeholders. As early as 2011, a special meeting by the United Nations recognized this as a "necessity for member states, the private sector, civil society organizations and law enforcement agencies to work in concert to manage the risks of our increasing interconnectivity" (UN DESA 2011). It is therefore even more important that the national cyber security architecture is effective and ensures a good workflow with a diverse set of actors. Kania and Kramer (2011) also agree that MSIs can be a type of collaboration, among others, to tackle the local problem. They have further identified that if actors tackle such a complex issue alone, the isolated approach is ineffective, since the problems arise from the interplay of governmental, civil society and commercial activities rather than the behavior of one specific sector alone (Kania and Kramer 2011: 38). In the related complex issue of Internet Governance, Pohle (2016) looked at the discourse and deliberations of the UN Working Group on Enhanced Cooperation (WGEC). The objective of this multi-stakeholder group was to overcome controversies on the role of governments in Internet Governance, which have persisted since the World Summit on the Information Society (WSIS). Pohle found that multi-stakeholder processes all contribute to the joint (though frequently contentious) production of discourse and a shared understanding of the issues at stake. "Overall, the production processes within multi-stakeholder groups can be considered as attempts at social ordering in Internet Governance because these processes generate discourses and create institutions which add to the shape and materiality of Internet Governance rules and procedures" (Pohle 2016). Thus, to solve those problems, discourse orientation could be useful to achieve a common understanding of a problem since no individual or group has all the answers, as there are multiple "truths" depending on one's past experiences and current reality. Hence, diverse insights and alternative points of view are imperative. Cyber security governance could be inspired by issues of Internet Governance and its corresponding proposed solutions. If an MSI transcends the national level to include international stakeholders, it certainly benefits from new perspectives and best practices. The downside is that some of those solutions might be too country-specific, and therefore difficult to adapt and adopt by others. Additionally, some challenges might be easier to solve on the international level due to the nature of cyberspace itself.

What this literature asks is that policy-making should be cross-sector, open-minded, adaptive and discourse-oriented. When it comes to the question of which countries might cooperate in solving the shared challenge of security in the era of digitization, the United States as leading technological power and Germany—with its role within the European Union and history of strong privacy and security safeguards—come to mind, among others. In theory, the U.S. and Germany would make for a capable alliance in solving those challenges. However, with the Snowden revelations, the CIA's and NSA's unfortunate role in the WannaCry and NotPetya (Herpig 2017a, b) outbreaks, as well as Trump's rather disruptive presidency, the relationship between those two governments has been far from perfect as of late.

Existing Cooperation Models

When we look at the approaches that have been implemented to tackle cyber security policy challenges, we can see thus far established cooperation models for tackling common problems. There are international, transnational, regional, EU, multilateral and bilateral cooperations. Additionally, there are national best practices that are adopted by other states. Moreover, we have seen some examples that take into account different stakeholders and aim to establish them in traditional institutional settings.

An example for international cooperation in the field of cyber security is the United Nations Group of Governmental Experts (UN GGE) on cyber norms (CCD-COCE 2017b). The goal was to establish and agree on norms for state behavior in cyberspace. After the last round of UN GGE talks ended without consensus, it is currently unclear where it will be heading. For the majority of the time, this group of experts consisted of governmental experts with diplomacy backgrounds. This has been a critical view as "experts from technical backgrounds can be 'left behind' in the sometimes intense diplomatic negotiations that accompany a GGE" (Lewis

2016: 5). Moreover, the problems and issues that led to unsuccessful negotiations in 2017 were very narrowly focused on judicial definitions. "Leaders, managers, and policy makers are often frustrated by a lack of consensus and collaboration on challenging issues—so they end up blaming outside factors or each other" (Kambiz 2017: 5). This can be challenging to overcome.

Transnational cooperation extends beyond government stakeholders and includes non-state actors such as companies and civil society organizations. Two of the main developments in this area are currently undertaken by Microsoft and the Carnegie Endowment for International Peace (CEIP). Microsoft is pushing for its "Digital Geneva Convention" (Smith 2017), which consists of three pillars: governments should agree to refrain from cyberattacks, companies should sign a Tech Accord consisting of principles to protect citizens and, lastly, to set up an independent attribution council for cyberattacks. While Microsoft's approach appears to be even broader than the UN's, CEIP is limiting its scope to a specific area and target group. Its approach aims to shield the international financial markets from cyber threats by getting the G20 states to agree to refraining from cyberattacks against financial institutions (CEIP 2018).

Two examples for regional cooperation in cyber security are the Tallinn manual drafted by the NATO Cooperative Cyber Defence Centre of Excellence (CCD COE) (CCDOC 2017a) and a set of confidence building measures developed by the Organization for Security and Co-operation in Europe (OSCE) (OSCE 2018). The Tallinn manual analyses the applicability of international law to cyberspace. OSCE's confidence building measures are designed to prevent or resolve an unintentional conflict caused by cyberattacks.

The EU level has also just recently picked up pace in promoting cyber security cooperation to its member states. One example here is the Directive on security of network and information systems (NIS Directive) (EC 2018), which consists of several aspects, such as setting up a cooperation group on cyber security, as well as a network of Computer Security Incident Response Teams (CSIRTs).

Multilateral cyber security cooperation describes partnerships between various states such as the European Government CERTs (EGC) group (EGC 2014). It consists of Computer Emergency Response Teams (CERTs) of a subset of EU states that share privileged information with each other. Similar to multilateral cyber security cooperation, there is bilateral cooperation, such as the U.S.-China agreement on cyber crime (Rollins 2015).

Last but not least, there are national best practices that become adopted by other countries. One example here would be the German *IT-Grundschutz*, an extensive set of security measures for soft- and hardware, which has been translated and adopted by Estonia (RIA 2016).

This shows that there already are steps taken and realization that the problem needs to be defined from different viewpoints, and that the solutions should include tasks for a different set of stakeholders; however, it is still very government-dominated. Moreover, it shows that traditional means are still useful and have the potential to be adapted as those new problems have local and global affects and, therefore, call for solutions that consider all those levels.

A case study that aims to combine the transnational and bilateral cooperation components with national best practices, but also uses more features of the multi-stakeholder model focusing on intersectoral work, is the Transatlantic Cyber Forum.

A Forum Was Born

In addition to the signal intelligence practices, the Snowden documents indirectly revealed that there was little to no fallback mechanism when it comes to bilateral exchanges on the political, legal, societal and technical aspects of privacy and IT security, essentially: cyber security policy between Germany and the U.S. The bilateral relations were not resilient. Thus, to enable a solution-finding process, today we cannot only rely on inter-governmental work. When the governments do not talk to each other, there is little in the way of transatlantic exchange on current developments and best practices regarding cyber security policies. Without that, it becomes increasingly difficult to tackle global security challenges arising from digitization. After all, because it is a common problem, solving it together bears a certain value and might strengthen the transatlantic relationship at the same time. This shows that a transatlantic exchange of ideas on how to improve cyber security policies is needed. The Transatlantic Cyber Forum (TCF) was established on this basis with transatlantic, independent and equal funding from the American William and Flora Hewlett and the German Robert Bosch foundations. It was clear from the beginning that a sustainable and resilient partnership cannot be achieved through a one-off conference. The underlying project was therefore designed with an initial two-year timespan to set up an expert network. It became home to lawyers, hackers, political scientists, former government employees from academia, civil society and the private sector, among others. It is still crucial to keep in touch with the respective governments—as they are the target audience of any cyber security policy proposal—but not so closely that a future disruption of the bilateral relationship would also degrade the cohesion of the network. Due to the many facets of cyber security, the network needed to draw from various backgrounds and sectors (see Fig. 9.1). As of the time of this writing (April 2018), the core network consists of 96 experts, out of which 39 are German, 50 are American and seven come from other EU countries. Academia is represented with 20 experts, civil society and think tanks with 45 and the private sector with 31.

There was no particular challenge that the network was supposed to work on—it was created as an expert network that spots and works on the most topical issues of cyber security with a comparative methodology, aiming to deliver good practices. The task is to learn from each other and create something that can be of use on both sides of the Atlantic. This process resulted in the establishment of three different working groups on the topics of "encryption policy and government hacking", "cyber defense and political IT-infrastructures" and "intelligence governance and oversight innovation" (SNV 2018b).

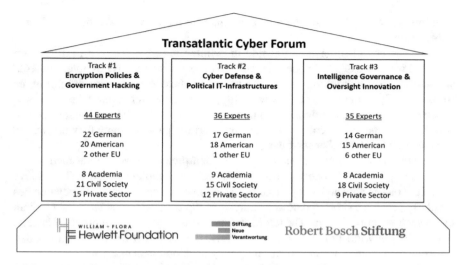

Fig. 9.1   Transatlantic cyber forum, management structure. *Own Source*, 2018

With the German foundation *Stiftung Neue Verantwortung* hosting the TCF, Berlin became the physical home of the network.

The way that cooperation works with the TCF ticks several of the international cooperation boxes while, at the same time, using features of the multi-stakeholder working model. It is composed of representatives of non-state actors from different countries, mainly Germany and the United States, and it has a transnational character. At the same time, TCF aims at creating good practices to be adopted by states. It draws from existing best practices and, therefore, also qualifies as cyber security cooperation by national best practices. TCF tries to influence policy-making mainly in Germany and the United States but—as will be mentioned later on—considers extending to other EU countries individually or collectively. TCF as a robust and active network of cyber security experts might not be unique, but it is certainly a less traditional approach toward cooperation.

Features of the TCF as a multi-stakeholder model would be that it is intersectional, interdisciplinary, discourse oriented and learning posture, adaptability mindset as in aiming to understand to what extent existing structures could be changed to be able to deal with the problems in a better way.

Establishing Trust

Trust is crucial for the cohesion of any multi-stakeholder process. Trust-building has therefore been a central component of the TCF setup. Every expert who joined the network has been invited only after an exchange with the TCF core team at the SNV.

It is important that both sides know and agree on the network's vision, goals and expectations.

Transparency also plays a vital role in maintaining trust. Thus, all information and working processes have been designed to be as transparent as possible. The TCF website features everything except the names of those experts who did not agree to be publicly named. Within the working groups, however, a list of members is regularly circulated. Additionally, all members get invited to the workshops and can participate in the online collaboration of the working papers, where they are able to see the comments of other members as well.

Constant information exchange has been identified as an additional method for maintaining trust and cohesion within this multi-stakeholder process. To achieve this, TCF has established a couple of information exchange formats. Policy debates regularly provide a short update and analysis of topical developments in the German cyber security policy arena. The target audience is the non-German group of experts within the network. For the German experts, the TCF has established a regular, informal exchange of ideas and current policy developments in Berlin.

The TCF has advocated for its role as being a point of contact for American stakeholders (experts, journalists, etc.) inside and outside the network trying to get in touch with relevant German, and partially EU, stakeholders to learn more about the current developments in the field. The same goes for German stakeholders looking for contacts in the United States and information about the current developments in cyber security policy on that side of the Atlantic. One example for its role as intermediary has been the clarification of Germany's position toward backdoors and hard- and software products. After a French-German exchange on cyber security on the EU level, their conclusion was incorrectly translated and interpreted, and the TCF reached out to U.S. (and international) experts and media to correct this misrepresentation (Herpig 2017a, b) as it might have had an undesired political impact on the other side of the Atlantic.

Defining a Common Agenda

Creating an expert network that shares information about policy developments and events is one thing. Having the experts truly engage in collaborative work is a much harder goal to achieve. It is therefore crucial to choose common challenges and develop an agenda that most of the experts agree on. Getting this buy-in is vital for the sustainability of the network. To accomplish this, the TCF started with bilateral talks in Berlin and a preliminary pre-project meeting in Washington D.C., to see what current issues the experts would like to work on. Only topics that were mentioned on both sides of the Atlantic made the cut, in order to facilitate a truly transatlantic cooperation. The second step in this methodology is to draft a problem analysis paper for each topic and get the experts together for a face-to-face workshop with possible remote participation. Having the working group members discuss the problem analysis is vital, as sometimes the underlying problems or the order of problems to be solved can only be found via open discourse between different stakeholders. Then, the goal to jointly develop a way forward as a group can be established. This requires the facilitators of this cooperation to not have a fixed solution in mind and allows the working group to set the agenda. Crowd-sourcing the work frame is useful for identifying a shared challenge that matters and can be solved by the working group. Additionally, it increases the buy-in and therefore the participation of the individual working group members during the upcoming policy development.

Working Group on Cyber Defense and Political IT Infrastructures

Political institutions have always been in the crosshair of intelligence agencies. The cyberattacks directed against the German parliament in 2015 and the Democratic National Committee (DNC) during the runup to the 2016 U.S. elections (SNV 2018a) showed that this behavior would not stop in cyberspace either. The latter, however, highlights a disturbing development. It shows that election and campaigning infrastructures are increasingly vulnerable to cyberattacks. This strikes directly at the heart of any democracy. TCF was set up only months after the DNC hack and nine months prior to the upcoming German elections. It was therefore apparent that the cyber security of political IT infrastructures would be another issue that the TCF would tackle. The election and campaigning systems in both states are very different from each other. The U.S. system is highly digitized, with several states even using voting machines without paper ballots for accountability. Furthermore, elections in the United States are clearly a state and not a federal issue. In Germany on the other hand, the campaigning and elections are much less digitized and still rely on paper ballots. Results are digitized later on, but there is always a paper ballot for accountability and recounting. The federal level takes a larger role when it comes to the security of election infrastructure. Since the working group was set up, U.S. policymakers have been debating possible measures to better protect their elections. In Germany, there was a brief discussion leading up to its 2017 elections, with some additional safeguards being implemented, but nothing beyond that. Again, the working group decided to focus on a problem that is a shared challenge for both sides and where different measures have been taken in the past to tackle it. Based on the first problem analysis paper—which was more focused on actively deterring further cyberattacks against political IT infrastructures (SNV 2018a)—the working group decided to discuss increased resilience for a data-driven electoral process in democracies and provide corresponding recommendations to the respective governments. Those recommendations are developed to help government identify and understand the different responsibilities and, therefore, a variety of possible policy instruments that would address the overall security of the democratic process.

Collaboration Is Key

For the project's transatlantic security cooperation, collaboration is key. This collaboration heavily relies, similar to the trust mechanisms mentioned earlier, on digital communication. A key component for the collaboration is certainly the face-to-face workshop. However, even those workshops were digitally infused with meticulously planned remote participation for those experts who could not join in person. Remote participation was not merely limited to a video chat with the working group, but simultaneous online collaboration in which the white board and other moderation tools and techniques were mirrored through shared documents. The goal was to make remote participation feel as non-remote as possible.

In addition to remote participation in the workshops, all working group members are encouraged and enabled to actively participate in the drafting of the input and policy recommendation papers. The initial versions of the policy products are drafted by the TCF core team based on the outcome of the first round of workshops and additional suggestions by individual working group members. The first consolidated draft is then shared online with the entire working group for collaboration. Members can add suggestions and even comment on input from other others. This dynamic process fosters collaboration, transparency, buy-in and, ultimately, the quality of the policy papers and the sustainability of the working groups.

At any stage of the process, the experts can decide to participate or not and in what capacity. They decide if their involvement is publicly reflected in the papers and on the website or not. To the working group itself, every process is presented in a transparent manner, enabling individuals to assess them and decide how and when to engage.

Joint Way Forward

The title of this collection of articles is "Redesigning Institutions—Concepts for the Digital Connected Society". Rethinking traditional institutional concepts to find joint solutions in a fast-changing field—such as cyber security—with shared challenges faced by every government around the world seems prudent. The Snowden revelations, increasing hostilities in cyberspace and the disruptive politics of the Trump presidency were a wake-up call for the transatlantic community to build more resilient networks, share information and collaborate on finding good solutions. And while the authors do not believe that the TCF is a truly novel approach, it reflects how trust building and collaboration can be achieved in the digital connected society of today. The core values TCF as an institution is built upon are independence, transparency and sustainability. The workflow uses many features that adhere to multistakeholder policy-making. This includes aspects such as regular communication, interdisciplinarity and intersectional discussions to achieve and maintain those core values. Moreover, it focuses on learning and includes a mindset that aims to engage in a deeper understanding of how the institutions today may have to adapt to better deal with the problem. Unique, too, is the openness to emergent outcomes, which allows for solutions that emerge from the discourse between the different participants.

References

Abbassi, P., et al. (2013). *Securing the net: Global governance in the digital domain global governance 2022*. Available at http://www.ggfutures.net/fileadmin/user_upload/publications/130826_GG2022_Cyber_Report_web.pdf. Accessed September 27, 2017.

CCDCOCE. (2017a). *Tallinn Manual Process*. CCDOCE (online). Available at https://ccdcoe.org/tallinn-manual.html. Accessed: April 11, 2018.

CCDCOCE. (2017b). *Back to square one? The fifth UN GGE fails to submit a conclusive report at the UN General Assembly*. CCDOCE (online). Available at https://ccdcoe.org/back-square-one-fifth-un-gge-fails-submit-conclusive-report-un-general-assembly.html. Accessed April 11, 2018.

CEIP. (2018). *Cyber policy initiative*. CEIP (online). Available at: http://carnegieendowment.org/programs/cyber/. Accessed April 11, 2018.

EC. (2018). *The Directive on security of network and information systems (NIS Directive)*. European Commission (online). Available at https://ec.europa.eu/digital-single-market/en/network-and-information-security-nis-directive. Accessed April 11, 2018.

EGC. (2014). *European Government CERTs (EGC) group*. EGC (online). Available at http://www.egc-group.org/index.html. Accessed April 11, 2018.

Fowler, A. (2014). *Innovation in institutional collaboration—The role of interlocutors*. ISS Working Paper Series/General Series, 584, pp. 1–29. Available at https://repub.eur.nl/pub/51129. Accessed September 27, 2017.

Fuerth, L. S., & Faber, E. (2012). *Anticipatory governance practical upgrades equipping the executive branch to cope with increasing speed and complexity of major challenges, forward engagement*. Available at https://dl.dropboxusercontent.com/u/44303479/Anticipatory_Governance_Practical_Upgrades.pdf. Accessed September 27, 2017.

Gnad, O. (2016). 'Wie strategiefähig ist deutsche Politik?' In Internationale Sicherheit im 21. Jahrhundert—Deutschlands internationale Verantwortung (pp. 125–140). Bonn: Bundeszentrale für Politische Bildung..

Herpig, S. (2017a). *Learning from America's mistakes*. Handelsblatt (online). Available at https://global.handelsblatt.com/opinion/learning-from-americas-mistakes-723816. Accessed April 11, 2018.

Herpig, S. (2017b). *Encryption backdoors on the EU-Level. Stiftung Neue Verantwortung* (online). Available at https://www.stiftung-nv.de/de/publikation/transatlantic-cyber-forum-policy-debates#%E2%80%9DMaerz3%E2%80%9D. Accessed April 11, 2018.

Kambiz, M. (2017). *An introduction to multi-stakeholder decision making. Multi-stakeholder decision making for complex problems* (pp. 3–14). https://www.worldscientific.com/doi/suppl/10.1142/9294/suppl_file/9294_chap01.pdf.

Kania, J., & Kramer, M. (2011). *Collective impact, Stanford social innovation review*. Available at https://ssir.org/articles/entry/collective_impact. Accessed September 27, 2017.

Lewis, J. (2016). *Report of the international security cyber issues workshop series*. UNIDIR (online). Available at http://www.unidir.org/files/publications/pdfs/report-of-the-international-security-cyber-issues-workshop-series-en-656.pdf. Accessed April 11, 2018.

OSCE. (2018). *Cyber/ICT security*. OSCE (online). Available at https://www.osce.org/secretariat/cyber-ict-security. Accessed April 11, 2018.

Pohle, J. (2016). Multi-stakeholder governance processes as production sites: enhanced cooperation "in the making". *Internet Policy Review, (online)* 5(3). Available at http://policyreview.info/articles/analysis/multistakeholder-governance-processes-production-sites-enhanced-cooperation-making. Accessed April 11, 2018.

RIA. (2016). *Three-level IT baseline security system ISKE*. Republic of Estonia Information Systems Authority (online). Available at https://www.ria.ee/en/iske-en.html. Accessed April 11, 2018.

Rollins, J. W. (2015). *U.S.–China Cyber Agreement*. Federation of American Scientists (online). Available at https://fas.org/sgp/crs/row/IN10376.pdf. Accessed April 11, 2018.

Shackelford, S., et al. (2017). Making democracy harder to hack'. *University of Michigan Journal of Law Reform, 50*(3), 629–668. Available at https://www.schneier.com/academic/paperfiles/Making_Democracy_Harder_to_Hack.pdf. Accessed September 27, 2017.

Smith, B. (2017). *The need for a digital Geneva convention*. Microsoft (online). Available at https://blogs.microsoft.com/on-the-issues/2017/02/14/need-digital-geneva-convention/. Accessed April 11, 2018.

SNV. (2018a). *International cyber security*. Stiftung Neue Verantwortung (online). Available at https://www.stiftung-nv.de/en/project/international-cyber-security-policy. Accessed April 11, 2018.

SNV. (2018b). *Defending political IT-infrastructures problem analysis*. Stiftung Neue Verantwortung (online). Available at https://www.stiftung-nv.de/sites/default/files/tcf-defending_political_lt-infrastructures-problem_analysis.pdf. Accessed April 11, 2018.

UN DESA. (2011). *Cyber security: A global issue demanding a global approach*, United Nations. Available at http://www.un.org/en/development/desa/news/ecosoc/cybersecurity-demands-global-approach.html. Accessed September 27, 2017.

Dr. Sven Herpig is head of international cyber security policy at Stiftung Neue Verantwortung. This includes the transatlantic expert network Transatlantic Cyber Forum (TCF), the EU Cyber Direct (EUCD) project funded by the European Commission as well as an ongoing analysis of German cyber security policies. Sven's current focal areas include the attack surface of machine learning with regards to national security, geopolitical responses to cyber operations, government hacking and vulnerability management. Sven served as expert on IT security for the Committee of the Interior and Homeland of the German parliament and for the European Union study on Legal Frameworks for Hacking by Law Enforcement. He presented his work inter alia at the US Congressional Cybersecurity Caucus, the European Parliament, the German parliament and the United Nations Interregional Crime and Justice Research Institute. Sven regularly appears in German and international media. Before Sven joined the Stiftung Neue Verantwortung, he was working with Germany's federal government for several years. First, he worked with the IT security staff at the Federal Foreign Office. He then became deputy of the cyber security and society unit at the Federal Office for Information Security. His PhD analyzed the strategic implications of cyber security and cyber operations for the state.

Julia Schuetze is project manager for international cyber security policy at Stiftung Neue Verantwortung. She works on cyber diplomacy of the European Union with the United States and Japan as part of the EU Cyber Direct project. Specifically her research focus is joint responses to malicious cyber activities. Her work on the project Transatlantic Cyber Forum resulted in a publication on "Securing elections in Cyberspace" together with Dr. Sven Herpig. As part of her role at SNV she has spoken at the Congressional Cybersecurity Caucus in the United States, has facilitated workshops on Germany's cybersecurity architecture with the foreign office, has organized two cybersecurity conferences with the Bundesakademie für Sicherheitspolitik and spoke at different public events including re:publica. Her work has been published or cited by news outlets, such as WirtschaftsWoche Der Tagesspiegel, F.A.Z and the BBC. Prior to SNV, she worked at Wikimedia Deutschland e.V. and has researched at the Berkman Klein Center at Harvard University. Already in her bachelors, she started to focus on cybersecurity policy at University of Stirling. In this context she wrote her bachelor thesis on 'Germany's cyber security awareness programme: Lessons from the US', which was tutored by the German Federal Office for Information Security. She is a Cybersecurity Policy Fellow at New America Foundation.

Chapter 10
Redesigning Corporate Responsibility How Digitalization Changes the Role Companies Need to Play for Positive Impacts on Society

Nicolai Andersen

Introduction

Let us imagine a girl of 11. For about one year, she has owned a smartphone with theoretically unlimited and unrestricted access to the Internet. Her parents have decided that the right time to be "old enough" for a smart-phone would be the transition from primary school to secondary school. They saw it as their responsibility to allow her to be part of a digitalized social life on WhatsApp, Snapchat, etc. Again, they saw it as their responsibility to allow her to become digitally literate. And they saw it as their responsibility to allow her to make use of the convenience of digitally assisted life.

Positive developments occurred. The 11-year-old has contributed great works of digital art to the family photo collection. She recently helped her family to understand the entire Greek and Roman mythology through instant research in the ruins of ancient Greece. And she is able to ask her parents in real-time for advice when she wants to buy a dress or tries to fix her broken bicycle.

But of course, negative developments also occurred. Like so many other kids, she also spends more time chatting with her friends than actually talking to them. She developed an addiction to mobile games, sacrificing her time on more creative leisure activities. And she learned the hard way that there is no such thing as a free lunch, that getting something for free requires giving something in return—in most cases in the digital world, this means personal data.

Today's functionalities on the 11-year-old's smartphone are not the functionalities of tomorrow. Looking into the future, technological progress on her smartphone will occur that could be considered positive or negative—depending on how you look

Published December 2018 at Deloitte. Perspectives, https://www2.deloitte.com/de/de/pages/innovation/contents/redesigning-corporate-responsibility.html

N. Andersen (✉)
EMEA Lead Innovation and Deloitte Garage at Deloitte, Hamburg, Germany
e-mail: nicandersen@deloitte.de

at it. Many researchers believe that digital technologies continue to develop at the pace of "Moore's Law" or faster, which would mean an exponential development (Shingles et al. 2016). Or in easier words: The change we saw within the last 12 years, is the change we will see in the next six years. Looking back 12 years ago takes us to the time before the market entry of the iPhone, thus before we all could even imagine what apps would do to our day-to-day lives. The same change, which we cannot imagine today, will occur to our lives in the next six years. Before the 11-year-old turns 18.

Is it just the responsibility of parents to make sure that Digitalization affects their daughters and sons only in positive and not in negative ways? Is it the consumers, employees and citizens that have the responsibility to create positive impacts on their society through Digitalization? Or is it the responsibility of governments to regulate Digitalization in a way that it only creates positive impacts on society? And what roles should companies play?

The intention of this article is to take a closer look on the responsibility of companies for the positive impacts of Digitalization on society, considering the increasing speed of change and growth of complexity of digital technologies.

Corporate (Social) Responsibility

Viewing companies as responsible for the positive impacts of their business is not a new concept. It has long been discussed and is, in parts, being executed by companies under the umbrella term "Corporate Social Responsibility" (CSR). In one of its wider definitions, CSR is seen as "the continuing commitment by business to behave ethically and contribute to economic development while improving the quality of life of the workforce and their families as well as of the local community and society at large" (Moir 2001).

There have been numerous discussions on why companies actually should invest in CSR programs, including the view of CSR as a part of creating stakeholder value and thus shareholder value (Hübscher 2015). Benefits from acting socially responsible include—among others—talent attraction, public image, process efficiency and employee loyalty (Shingles et al. 2016), as well as creating new market opportunities, enabling proactive regulatory relationships and building resilient, sustainable supply chains (Mennel and Wong 2015, see Fig. 10.1).

Despite these positive impacts on a company's own benefits, investing strategically in social responsibility is not a given. A Deloitte study examining the social impact practices of the 2014 Fortune 500 global public companies revealed four business archetypes:

There may be differences in the percentages in each of the archetypes, if you take different samples, depending on the size and country of origin of the companies. Some companies just have a different corporate social performance depending on specific environments, stakeholders and local issues (Moir 2001). But it still has to

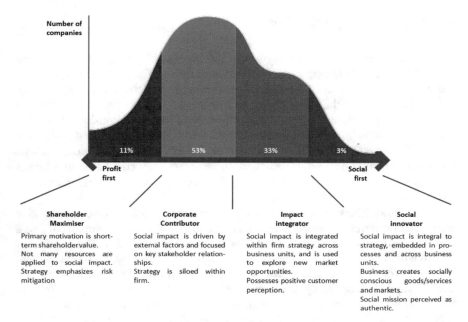

Fig. 10.1 Business archetypes of fortune 500 companies concerning their 2014 social impact strategies. *Source* Mennel and Wong (2015)

be concluded that CSR—for the majority of companies—is not perceived as being of the highest importance for a company's strategy.

The reason for that may reside in the two dichotomies that, according to Hübscher (2015), lead to CSR being treated in a rather reactive and residual way, versus treating CSR strategically:

- The dichotomy between responsibility for the economy versus responsibility for society
- The dichotomy between goals for the economy versus goals for society

Only if these dichotomies are cleared, Hübscher argues, would companies consider CSR as way to create positive shareholder value and would, thus, invest in it in a noticeable way.

As Digitalization is completely redefining not only the products and services landscape, but is leading to radical changes in the economy, society, politics and even our values and beliefs (Gärtner and Heinrich 2018), does this create a chance to also clear these dichotomies and radically change the way companies approach their own responsibility?

The Impact of Digitalization

Uncountable publications have been written and uncountable public discussions have been held about the good impacts and the bad impacts of Digitalization. Digitalization is driven mainly by the combination of the increase in available data and the ability to access and process this data, leading to new ways to produce, to consume and to work (Gärtner and Heinrich 2018). These "new ways" are often considered as innovations. And these innovations, in many cases, lead to opportunities and challenges at the same time (Mühlner et al. 2017). Take, for example, intelligent algorithms processing granular data on communication patterns: They could lead to a higher crime prevention rate, but could restrict the freedom and privacy of individuals and could affect the culture of a society.

Digital Innovations have already changed our lives in both positive and negative ways. Positive developments of Digitalization undoubtedly are the comfort of being able to access information and services 24 h a day, 7 days a week. The comfort of individualizing what, how and when we consume products, services and information. The comfort of being able to afford more because of decreasing product costs and increasing price transparency. The comfort of taking part in each other's lives through easily sharing visual and acoustical experiences (e.g., my daughter's video message to her Grandma from one end of the world to the other).

At the same time, there are current developments that are undoubtedly negative: The case of the last U.S. presidential election has shown how the processing of granular personal related data in connection with automated content generators and in connection with so-called "Fake News" have led to content bubbles of perceived truth. These may have actually influenced voters' behaviors and, ultimately, may have affected the outcome of the election (Voigt 2018).

A root cause of this—but also a separate negative development—can be seen in the so-called "digital divide": the gap between those who take part in digitalization and those who do not (see Fig. 10.2). There is a yearly survey in Germany analyzing the degree of Digitalization of the German population, taking into account four categories (D21 2018):

- Digital access (Internet use at home/at work, available equipment)

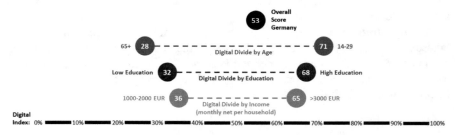

Fig. 10.2 Degree of digitalization of the German population in 2017 and selected gaps. *Source* D21 (2018)

- Digital use (Variety of applications used, average usage time of the Internet)
- Digital competence (Knowledge about digital topics (eHealth, cloud, …), technical competence)
- Digital openness (Mindset toward the use of the Internet and digital devices)
  These categories lead to a score between 1 and 100, where people with a score <20 are considered as "Offliners" and people with a score >80 as "Technology Enthusiasts". The 2017 score for Germany is 53, which does not say much without taking a look at the gaps between certain parts of the population:

If there are positive and negative changes to our lives today through Digitalization, how about the development in the future? How will our lives be affected when the exponential advancement of technology, as discussed before, leads to new applications of Digitalization that we can hardly imagine? Will the future be as positive, as described by the visionaries of the Silicon Valley, who see technology as the way to solve humanity's greatest challenges (Diamandis and Kotler 2012)? "The financial services industry, for example, might explore new ways for Blockchain to democratize banking, enable micro-transactions, and simplify philanthropic donations. The consumer food industry could potentially leverage biotechnology to change the health benefits profile and affordability of their products. The entertainment industry might partner with educational leaders to leverage advances in augmented and virtual reality to revolutionize learning and education. By supporting the maker movement and exploring new ways to leverage 3D printing, manufacturers could help provide affordable housing and basic necessities to the world's underserved populations. Hospitals and the health care industry have opportunities to use digital medicine to reinvent and democratize prevention, diagnosis, and treatment" (Shingles et al. 2016).

These future visions sound desirable, and in the stated examples, maybe do not even have negative downsides. But how about a future vision of every human having a chip implanted in the brain, connected to our mobile phones. No, not for mind-reading and -writing purposes. This functionality may still be far out in the future. But in the nearer future, we could already be technically able to offer a very useful functionality that could save lives: The 11-year-old walks down the street, only focusing on the screen of her mobile phone, not watching the world around her. She does not notice the red light crossing the street and the truck approaching at high speed. Her mobile phone could detect the approaching vehicle, predict the likelihood of an accident and then send a signal to the chip in the girl´s brain that triggers an impulse for her to jump backwards. Surely a useful functionality, but is this a positive future or a negative future?

From CSR to CDR

Digitalization undoubtedly creates opportunities for positive futures for societies. These opportunities cannot and should not be realized by governments alone, but should also be realized by companies—with the side effect of capitalizing on the business opportunities of these positive futures (Shingles et al. 2016). But at the same time, Digitalization creates an increasing degree of responsibility for politics and the economy to prevent negative side effects on our societies (Capurro 2017).

These considerations have led to a discussion in the recent past, about whether there is a general responsibility for companies resulting from Digitalization: Corporate Digital Responsibility (CDR). There is no common definition for CDR yet. CDR is seen as differently as:

- Extension of classical CSR into Digitalization: The responsibility of companies to act with discernment within and outside their boundaries when applying digital business processes, creating digital services and products and interacting with employees, business partners and society (Mühlner et al. 2017)
- Application of ethics in Digitalization: The responsibility of companies to embed ethical considerations at company, individual and societal levels (Raivio 2018)
- Creating trust of societies toward Digitalization: The responsibility of companies to create transparency on the use of data, algorithms and bots to increase the level of societal trust in Digitalization (Osburg 2017)
- Creating trust of consumers toward Digitalization: The responsibility of companies to keep and increase the level of trust consumers have in the use of digital applications (Thorun 2018)
- Solving problems through Digitalization: The responsibility of companies to help leverage digital technologies not only for their own benefit but for driving greater good in society (Shingles et al. 2016).

In summary, the existing definitions of Corporate Digital Responsibility agree in the aspect that CDR is not just using digital technology to be more efficient and effective in managing CSR. But the definitions seem to differ in two dimensions:

- Stakeholder Dimension: Just the consumer of a company versus stakeholders of a company (consumers, employees, business partners) versus wider group of stakeholders (society in general)
- Impact Dimension: Primarily preventing negative developments of a company's actions versus primarily achieving positive developments through a company's action

  For this article—and I suggest also for any further discussion and implementation of Corporate Digital Responsibility in companies—CDR should be considered in the widest possible definition: Corporate Digital Responsibility is the strategy and execution of a company to prevent negative impacts and achieve positive developments from Digitalization on the entire society.

While this definition may be academically easy to take on, it is in reality an umbrella term for at least four completely different mindsets regarding CDR (see Fig. 10.3).

Looking at the four different types of "CDR Mindsets", however, the dimensions lead to two different possible trade-offs:

- Stakeholder Dimension: The trade-offs between the responsibility of a company for its own consumers versus the responsibility of a company for society in general.

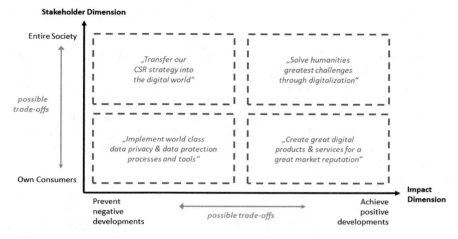

Fig. 10.3 Different mindsets on corporate digital responsibility. *Source* Own illustration, 2018

The example of e-commerce illustrates this quite well: Delivery of goods with the lowest possible costs in the supply chain creates value for consumers, because all-in prices for consumption will go down, ultimately leading to a higher standard of living for the consumers. But due to the negative effects of last-mile delivery (e.g., increase in city traffic, decrease of achievable income for delivery drivers), negative effects are created for the standard of living of the overall society.

- Impact Dimension: The trade-offs between the responsibility of a company to create positive impacts versus the responsibility of a company to prevent negative impacts from happening. Examples of this—which leads us to the ethical discussion in the next section of this article—are any kind of digital applications that help to optimize daily life situations by processing individual personal data. Our car may warn us about an icy part of the road just ahead of us, taken from the data of another car on the same street. This would motivate us to slow down, ultimately leading to a decreased probability of a car crash. But this requires cars to send individualized granular data to a central processing mechanism, decreasing data privacy and increasing the risk of data-protection violations.

We need to be aware of these trade-offs in general, but especially in specific daily life situations, to be able to take the first steps at implementing Corporate Digital Responsibility programs on the company and—more importantly—the country level.

Corporate Digital Responsibility in Our Daily Lives

To be able to debate the CDR trade-offs in daily life situations, we need to take one step back to define a framework for our digital lives. According to Mühlner et al.

(2017), we need to distinguish four different aspects of Digitalization that are causing challenges:

- Datafication: The ability to generate and process an increasing amount of granular data. This leads to the possibility of generating specific insights for a higher degree of individualization, with the downside of a centralization of these insights outside of the control of the individuals.
- Automation: The ability to make automated decisions based on algorithms. This leads to the possibility of quicker and more fact-based decisions, with the downside of possibly losing control over the question of what is right or wrong (e.g., discrimination based on facts).
- Connection: The ability to exchange and combine data from "things" (e.g., sensors). This leads to the opportunity to virtualize and remotely control actions, with the downside of losing a sense of responsibility for the effects of the action.
- Interaction: The ability to have machines work together with humans. This leads to the possibility of fulfilling tasks more comfortably and easily—and even less dangerously—with the downside of possibly eliminating jobs and/or personal relationships.
  These trade-offs in the four aspects of Digitalization according to Müller are actually existing ethical trade-offs applied to the new realities and/or opportunities caused by Digitalization:
- Trade-offs in values:

  - Accessibility versus Privacy (e.g., mobility data to optimize public transport)
  - Individualization versus Privacy (e.g., user behavior data to optimize products and services)
  - Customer Experience versus Objectivity (e.g., nudging to motivate behavior)

- Trade-offs in interests:

  - Insights versus Privacy (e.g., pharmaceutical/medical research on personal health data)
  - Security versus Privacy (e.g., crime prevention through tracking and storing of personal data)

- Trade-offs in consequences:

  - Short-Term Benefits versus Long-Term Risks (e.g., automation of tasks in work profiles)
  - Option 1 versus Option 2 (e.g., prioritization of digital infrastructure investments).

To understand and discuss these trade-offs, Müller suggests viewing them in the context of various areas of living: Learning and Education, Health and Personal Care, Communication, Mobility and Logistics, Work Life and Private Life (see Fig. 10.4).

There are uncountable trade-off decisions in daily lives. In many cases, companies have to make decisions for their consumers, for their employees and for their

	Datification	Automation	Connection	Interaction
Learning & Education	Smart Learning optimizes individual methods vs. Collection of personal data	Algorithms assign learners into optimized clusters vs. Sorting by machines		
Health & Personal Care			Insurer collect data from wearables for individualized offers vs. Data protection	Support of health care personnel through robots vs. In personal care
Communication		Filtering of information based on indivudal patterns vs. Self-determination	Communication with anyone anytime vs. Loss of „real life" social contacts	
Mobility & Logistics	Collection of connected-car data to optimize traffic flow vs. Movement profiles	Prevention of human errors in traffic vs. Machines taking ethical decision		
Work Life				Elimination of errors caused by humans vs. Elimination of jobs
Private Life			Watching of children through wearables / connected toys vs. Loss of privacy	

Fig. 10.4 Exemplary trade-off decisions caused by digitalization in various areas of daily lives. *Source* Müller and Andersen (2017)

business partners, with impacts on wider stakeholder groups or society in general. Or they have to help consumers, employees and business partners to make decisions in the full knowledge of possible consequences. Referring back to the definition of CDR presented in the last chapter, I would like to extend this view even further: Companies should see themselves as responsible for helping all members of a society—including governments—to make the right decisions regarding trade-offs caused by Digitalization that affect our daily lives.

For humans personally this would mean…

…in their role as a consumer: the responsibility of companies to develop digital products that increase the quality of my life. And the responsibility to explain to me, in easily understandable words, what data and algorithms they use for what purpose, what advantages this brings to me and what risks.

…in their role as an employee: the responsibility of companies to make my job—with the help of digital assistants—as easy as possible and to pay me a fair salary. And the responsibility to keep on educating me so that I can switch into a different job profile even at a higher age, when my original job has been replaced by a machine.

…in their role as a citizen: the responsibility of companies to make as much information available to me as possible to enable me to make self-determined decisions. And the responsibility to value the functioning of a free and open society based on a democratic system higher than the value of the company's own stock price.

…and last but not least—ending the consideration where I started off at the beginning of this article—in the role of a person as a parent: the responsibility of companies to close the gaps of the digital divide and improve overall digital literacy, especially of the younger generations. The education system cannot be blamed for not being able to teach our children every aspect of Digitalization, given the increasing speed and complexity of changes through Digitalization, as discussed earlier. I would like to see the education system teach my daughter the principles of humanity to be able to consider ethical questions for herself. Companies should take responsibility for protecting my daughter from the negative impacts of Digitalization through open and honest explanations of context and consequences. And at the same time, they should take responsibility for getting my daughter excited about the opportunities Digitalization is creating.

Corporate Digital Responsibility can improve the overall well-being of societies through Digitalization. This requires a complete redesign of institutions and—in order to achieve this—an entirely different mindset in politics and society regarding the role of companies, and inside the companies regarding their responsibility.

It is not the sole responsibility of me as father to help my daughter benefit from Digitalization. It is the responsibility of all of us.

References

Capurro, R. (2017). Digitale Ethik. In: *Homo Digitalis – Beiträge zur Ontologie, Anthropologie und Ethik der digitalen Technik.* Berlin: Springer.

D21. (2018). Initiative D21 (Hrsg.): *D21 Digital Index 2017/2018 – jährliches Lagebild zur digitalen Gesellschaft.* www.initiatived21.de/app/uploads/2018/01/d21-digital-index_2017_2018.pdf.

Diamandis, P., & Kotler, S. (2012). *Abundance—The future is better than you think.* Free Press.

Gärtner, C., & Heinrich, C. (2018). Vorwort. In *Fallstudien zur Digitalen Transformation.* Wiesbaden: Springer Gabler.

Hübscher, M. (2015). Understanding CSV: Ein neues Narrativ des Kapitalismus? In *zfwu, 16/2,* (pp. 203–218). Reiner Hampp Verlag.

Mennel, J., & Wong, N. (2015). *Driving corporate growth through social impact.* www2.deloitte.com/content/dam/Deloitte/us/Documents/strategy/us-strategy-operations-social-impact-corporate-archetypes.pdf.

Moir, L. (2001). What do we mean by corporate social responsibility? In *Corporate governance,* (Vol. 1(2), pp. 16–22). Bingley: Emerald Publishing.

Mühlner, J., et al. (2017). *Corporate digital responsibility—Unternehmensverantwortung in einer digitalen Welt.* www.charta-digitale-vernetzung.de/app/uploads/2018/01/20170504_Forum_Europrofession_CDR_Unternehmensverantwortung_in_einer_digitalen_Welt_M%C3%BCChlner_final_16-9.pdf.

Müller, L.-S., & Andersen, N. (2017). *Denkimpuls Digitale Ethik: Warum wir uns mit Digitaler Ethik beschäftigen sollten—Ein Denkmuster.* www.initiatived21.de/app/uploads/2017/08/01-2_denkimpulse_ag-ethik_digitale-ethik-ein-denkmuster_final.pdf.

Osburg, T. (2017). Sustainability in a digital world needs trust. In *Sustainability in a digital world* (pp. 3–19). Basel: Springer International Publishing.

Raivio, T. (2018). *The future of work & corporate digital responsibility.* www.csreurope.org/sites/default/files/uploads/Corporate%20Digital%20Responsibility%20Factsheet.pdf.

Shingles, M., Briggs, B., & O'Dwyer, J. (2016). Social impact of exponential technologies. In *Tech trends 2016—Innovation in the digital era* (pp. 112–125). Deloitte University Press.

Thorun, C. (2018). Corporate digital responsibility—Unternehmerische Verantwortung in der digitalen Welt. In *Fallstudien zur Digitalen Transformation* (pp. 174–191). Wiesbaden: Springer Gabler.

Voigt, M. (2018). Digital trump-card? Digitale transformation in der Wähleransprache. In *Fallstudien zur Digitalen Transformation* (pp. 161–170). Wiesbaden: Springer Gabler.

Nicolai Andersen was first exposed to Artificial Intelligence at the age of 15, when he tried to teach his Atari 1040 ST to think like a human through a BASIC program. That did not work out. In the 1990s, he studied Business & Engineering at the Karlsruhe Institute of Technology and took courses in Genetic Algorithms for a second attempt at creating real AI. Again, it did not work out. In the meantime, he is heading The Deloitte Garage and serving as Deloitte's Chief Innovation Officer in the EMEA region. Through these roles, he is exposed to the question of how AI and other Tech Trends might replace our jobs—or how they could actually create new business opportunities. Even more importantly, he is exposed to the following question: How can you transform a traditional corporate environment into a platform for innovation? This is much more about the human side of change than its technological side. Nicolai thinks it is very important to build the future on the fundament of a well-educated society. He is a member of the board of "Initiative D21" and leads their group working on "Ethics in the digital world". In his free time, he also likes to build fantastic desirable futures together with his four daughters and tons of Legos. And he loves to let his thoughts take journeys while he plays music or runs through the mud.

Chapter 11
The Algorithmic Society

Agnieszka M. Walorska

Introduction

Although Artificial Intelligence (AI) research has its roots back in the mid-20th century, the year 2011 seems to have been a breakthrough year marking the end of the long AI winter. In that year, IBM's Watson had beaten the best human players in Jeopardy, Google's self-driving car prototype had driven more than 100,000 miles, and Apple had introduced the personal agent Siri (O'Reilly 2018, p. 232). Public interest in Artificial Intelligence has been constantly growing since then—as can best be shown by the increase in the number of Google search queries (see Fig. 11.1).

Looking deeper into the AI-related searches, another interesting observation can be made: while the number of searches for "AI potential" stayed roughly constant, the interest in the phrase "AI risk" increased significantly, especially since 2016 (with a noticeable spike at the end of 2011, just around the time Siri was introduced; see Fig. 11.2).

The current discourse on *superintelligence*—triggered by Nick Bostrom's book with the same title, published in 2014—bears a fair share of responsibility for this development. Since then, prominent figures have made their warning voices heard among the general public: Stephen Hawking, "The development of full artificial intelligence could spell the end of the human race" (Cellan-Jones 2014); Elon Musk, AI is "a fundamental existential risk for human civilisation" (Sulleyman 2017); and, recently, Henry Kissinger, "Philosophically, intellectually—in every way—human society is unprepared for the rise of artificial intelligence", (Kissinger 2018).

While it's not the intention to downplay the importance of the debate on superintelligence, it is here suggested to don't consider it to be the most pressing aspect of the AI discourse. As Andrew Ng verbalized it in a *WIRED* interview, "I think that hundreds of years from now if people invent a technology that we haven't heard of

A. M. Walorska (✉)
Creative Construction Heroes, Berlin, Germany
e-mail: agnieszka@creativeconstruction.de

© Springer Nature Switzerland AG 2020   149
D. Feldner (ed.), *Redesigning Organizations*,
https://doi.org/10.1007/978-3-030-27957-8_11

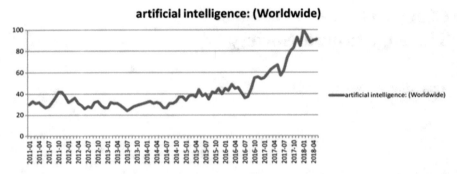

Fig. 11.1 Trends visualization of increasing interest in artificial intelligence. Authors own graphic based on data from Google Trends (2018)

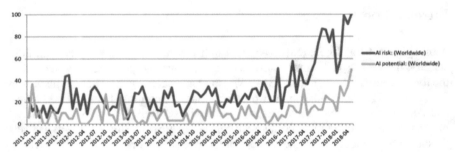

Fig. 11.2 Trends comparison of interest in AI risk versus AI potential. Authors own graphic based on data from Google Trends (2018)

yet, maybe a computer could turn evil. But the future is so uncertain. I don't know what's going to happen five years from now. The reason I say that I don't worry about AI turning evil is the same reason I don't worry about overpopulation on Mars. Hundreds of years from now I hope we've colonized Mars. But we've never set foot on the planet so how can we productively worry about this problem now?" (Garling 2015).

While *superintelligence*, or even *AGI* (artificial general intelligence), is still far away, the *narrow AI* with its exceedingly powerful algorithms is already here and plays an increasing role in shaping our societies, economies, and politics. The algorithms themselves are not a new phenomenon—they have been used for decades. The current interest in them, though, is due to the vast amounts of data—each day, 2.5 quintillion bytes of data are being generated, 90% of the world's data has been created in the last two years (IBM 2017)—and the need to process and understand it.

The Risks of an Algorithmic Society

In 2008, the mathematician Cathy O'Neil was working for a hedge fund and saw firsthand how algorithms contribute to a financial crisis. Since then, she has become one of their loudest critics, calling the harmful mathematical models *weapons of math destruction*. "The math-powered applications powering the data economy were based on choices made by fallible human beings. Some of these choices were no doubt made with the best of intentions. Nevertheless, many of these models encoded human prejudice, misunderstanding, and bias into the software systems that increasingly managed our lives", (O'Neil 2016, p. 3).

Algorithmic Bias

The problem of *algorithmic bias* is becoming more and more significant as the technology is already spreading to critical areas such as hiring, insurance, media, justice, and policing, and as more people who lack a deep technical understanding are utilizing it in their work. A particularly problematic area where bias can occur is in risk assessment models, which may, for example, determine a person's chances of obtaining a loan, being granted parole, or being hired. We want to think of algorithmic decisions as neutral, efficient, and free of bias, but for each learning algorithm, the output is determined by the data input. The first type of algorithmic bias is the *sample selection bias,* occurring when the training data over-represents a certain population. One of the most striking examples of this kind of bias is the malfunction of most of the existing facial recognition software. In 2015, Google came under heavy criticism after its algorithm labeled a black couple as "gorillas", (BBC 2015). Interestingly, Google still didn't manage to refine its algorithm to prevent this from happening again, so it erased gorillas, and some other primates, from the service's vocabulary (Simonite Simonite 2018a). While most existing facial recognition software can recognize the gender of white women with at least 95% accuracy, it errs 10 times more often when analysing pictures of dark-skinned women. Also, when it comes to correctly identifying a specific person in a photograph, the algorithms do very well when the person in question is Caucasian, while making far more errors with black or Asian people (Simonite 2018b). If such biased facial recognition software is used by the police, it might not do any better than the biased eyewitnesses do (Johnson 1984). That's just one of the most obvious examples of the algorithmic bias emerging from incomplete training data. Other examples might not be that easy to spot, such as most of the *implicit biases*, in which the algorithm correlates available information with race, gender, disability, or sexuality. Many algorithms used in recruiting and HR, for example, contain both the *selection bias* and the *implicit bias* towards women and minorities. Being trained on data going back to the time when women were barely existent in the professional space, the algorithms naturally favor men for hire or promotion. The *implicit bias* may also lead to correlating a candidate's name or extracurricular activities with information not disclosed in the application, like age, race, or class, which can lead to a discriminatory action related to such factors.

The most common algorithmic bias though is the *emergent bias,* which learns from every interaction to reconfirm the user's mindset. We encounter it in almost every digital interaction—Google Search, Facebook News Feed, YouTube Autoplay—and we have a more popular name for it: "the filter bubble". The 2016 US presidential elections were the most prominent example of the damage that can be done by the *emergent bias.* With 61% of millennials (and 39% of baby boomers) relying primarily on news gathered from social media (Gottfried and Barthel 2015)—where the algorithms show each person more of what they respond to positively—and though confirming their biases, the truth became relative and the "fake news" flourished.

Opacity

All these biases can be corrected given the algorithms and the data they run on are transparent enough. But this is not always the case. As algorithms get smarter, they're also becoming incomprehensible "black boxes", systems in which the inputs and outputs are known, but the process of transforming one to the other is unknown. Currently we encounter three levels of algorithmic opacity. In most of the everyday cases, like receiving recommendations by Foursquare or getting directions from Google Maps, we trust the algorithms even though we don't have the capacity to understand them, while being convinced that others understand them. This kind of opacity usually doesn't bother us, as it provides immediate benefits without much of a downside—most of the time. But at the next level, the design of the algorithm is kept secret even from the experts capable of understanding it—like in the case of the algorithms driving Google Search or the Facebook News Feed. The third level could be called *opacity by design.* In 2016, the chip maker Nvidia brought an autonomous vehicle onto the road. What was remarkable about this car was that its algorithm wasn't programmed to follow specific instructions defined by its creators, but to teach itself to drive by observing a human doing it. There is only a little bit of concern as long as everything goes right. But if such a car is involved in an accident (like in the recent case of an Uber vehicle—Nvidia's customer), it's extremely difficult to isolate the specific malfunction. On this level, the algorithm's behavior is opaque even to its own creators.

With medical diagnostics, investment decisions, and potentially even military decisions being driven by algorithms *opaque by design,* the increasing demands for disclosure and regulation are not surprising. But are we going to be able to disclose and regulate "black boxes" without sacrificing their benefits? Will the great question of the twenty-first century be "Whose black box do you trust?" as stated by John Mattison, the Chief Medical Information Officer of the health provider Kaiser Permanente (O'Reilly 2018, p. 224).

Surveillance and Control

The dystopian *Black Mirror* episode called "Nosedive" premiered in October 2016, presenting a terrifying vision of the world in which everyone is being rated by others for every interaction and the public ranking based on these ratings has a major impact for their socioeconomic status. This dystopia is already becoming reality in China, where the government started working on a social credit system back in 2014 and plans to make it mandatory by 2020. The rating criteria range from credit history to behavior and preferences. Sesame Credit, an affiliate of Alibaba, is one of the companies providing a social credit solution and does not reveal the exact logic behind the algorithm. The company's Technology Director, Li Yingyun, gives a glimpse of it though: "Someone who plays video games for ten hours a day, for example, would be considered an idle person. Someone who frequently buys diapers would be considered as probably a parent, who on balance is more likely to have a sense of responsibility", (Botsman 2017). Many of the now still "voluntary" participants with low scores are already experiencing major disadvantages of the system: nearly 11 million are not allowed to take a flight and 4 million are suspended from trains (Locker 2018). China's blending of a communist surveillance state with rapid technological progress makes it a pioneer in the new era of *algocratic governance*—a post-bureaucratic system of management by algorithms (Aneesh 2002).

While there seems not to be an immediate threat of implementing such a type of state-driven *algocracy* in western democracies, more and more companies are introducing *algorithmic management* to optimize internal processes. "The algorithm is the new shift boss", states Tim O'Reilly in his 2017 book, *WTF? What's the Future and Why It's Up to Us*—and not just for the "gig economy" (O'Reilly 2018, p. 198) workers providing services to companies like Uber, Lyft, or Deliveroo. Companies with more traditional employment models—like Starbucks, McDonalds, and Walmart—utilize scheduling software to ensure employees are booked to work only when they are needed. While letting companies cut costs and operate more efficiently, it results in substantial disadvantages for the employees: irregular schedules that make it impossible to handle childcare, education, or any private matter, or "clopening" (the same employee closing the store late in the evening and returning early in the morning to open it) that leads to lasting sleep-deprivation (O'Neil 2016, pp. 123–124).

Automation of Jobs

While *algorithmic management* is already replacing the job of the shift manager, it will surely not be the only job to be replaced by an algorithm. Companies like Uber make no secret of the fact that the gig workers are just a logical step towards the automation of most of the unskilled work. This threat isn't new. Back in 1964, 35 scientists and activists addressed a letter to President Lyndon B. Johnson warning of the "cybernation revolution" leading to a "separate nation of the poor, the unskilled, the jobless", (Pauling et al. 1964). Only this time around, most experts agree that this "cybernation revolution" is already happening. Estimates of the degree of automation range from about 30% (PWC 2018) to almost 50% (Frey and Osborne 2013) within the next 20 years. Unlike the industrial revolution of the 19th century, the *technological unemployment* of the 21st century will not only affect low-skilled workers—the machines are coming for the highly-skilled white-collar jobs, as well. A programme called StatsMonkey is generating automated sports reporting in a compelling narrative indistinguishable from human writing (Ford 2016, p. 94), IBM Watson and Google Brain are making diagnoses with an accuracy similar to or better than human, "the world's first robot lawyer" DoNotPay successfully contests 160,000 parking tickets in London and New York (Gibbs 2016), and Google's and Microsoft's software learns how to programme itself (Simonite 2017). "Now comes the second machine age. Computers and other digital advances are doing for mental power—the ability to use our brains to understand and shape our environments—what the steam engine and its descendants did for muscle power", (Brynjolfsson and McAfee 2014, pp. 7–8).

The Potentials of an Algorithmic Society

The risks mentioned above are outside of the question—we need to be aware of them and work on eliminating, or at least reducing, their negative consequences. But that's only one perspective on our algorithmic society—the one we tend to overemphasise due to our very human shortcomings. Our views are skewed not only by the *availability heuristic* (Tversky and Kahnemann 1973)—which makes us estimate the probability or frequency of an event based on the ease with which it comes to our mind—but also by the *negativity bias*, which makes us dread losses more than we look forward to gains. It's no wonder, then, that spectacular and publicly debated algorithmic failures like the lethal crash of the autonomous Uber car and the outrage about Cambridge Analytica make us develop a sheer "progressophobia" (Pinker 2018, pp. 41–48). No doubt, AI is an epochal development in many dimensions and, as with almost every epochal development, it gives rise to extremely negative reactions. But if we handle it wisely, it can bring humanity immense benefits, from the global to the personal level (McCorduck 2015, p. 51).

Human Augmentation

As already mentioned above, more and more tasks are being replaced by intelligent algorithms and the obsolescence of human work becomes a serious threat. But a closer look at many of these tasks reveals that human work is not necessarily being replaced, but rather augmented, by having algorithms perform tasks humans would not be able

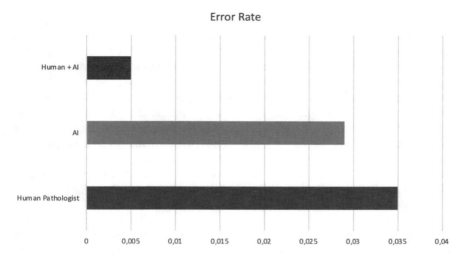

Fig. 11.3 Error rate in identifying breast cancer. Authors own graphic based on the study by PathAI, 2016

or willing to perform. Machines are good at long, monotonous tasks like analyzing vast amounts of data for patterns and trends, making themselves particularly useful in areas like medical diagnostics. Just one of the examples is an algorithm using image recognition to identify breast cancer. While the algorithm's success rate of 92% was inferior to the 96% success rate of human pathologists, the best result, a stunning 99.5%. was reached by augmenting the human evaluation with algorithmic analysis (Kontzer 2016) (see Fig. 11.3).

Medical diagnostics is by far not the only example for successful augmentation—similar examples can be found in areas like education, farming, law, and many others.

Meaningful Work

According to a Gallup study from 2017, only 15% of full-time employees across 155 countries are "engaged" in their jobs (highly involved and enthusiastic about the content and environment of their work). Although the percentage of engaged employees varies considerably depending on the country and region (ranging between 6% in China, 10% in Western Europe, and 31% in North America), the number of disengaged employees significantly exceeds the number of engaged ones in all countries evaluated in the study (Gallup 2017). One of the reasons for this might be that "the job" is "an artificial construct, in which work is managed and parceled out by corporations and other institutions, to which individuals must apply to participate in doing the work", (O'Reilly 2018, p. 301). Wouldn't it then be a desirable state in which the machines free us from such kind of dehumanizing or simply boring work? With the routine tasks carried out by algorithms and robots, wouldn't the human touch become more valuable? Certainly, even if most of the workers are not engaged in their jobs, they are highly dependent on them. That's why the future of algorithmic societies depends on which opportunities they provide to those who lose their jobs—not only with regard to how they will fund themselves and their children, but also how they will spend the time they once spent at work. While the first issue might be resolved by an unconditional basic income, the answer to the second one might be even more difficult (Levi 2015, p. 235). On the other hand, the reduction of working hours for the same salary might be an even more natural way of dealing with task automation. In 1870, the average American man used to work 62 h per week; since about the 1960s, this number is down to about 40 h (O'Reilly 2018, p. 304) without a significant proportion of the population complaining about the abundance of spare time (see Fig. 11.4).

Why should it be any different if by 2050 we only worked 18 h per week? John Maynard Keynes was already convinced in 1930, that by the time his grandchildren had grown up the average work week would occupy only 15 h. This estimation turned out to be a little too optimistic. The generation of Keynes's grandchildren is already retired and used to work rather 40 and more hours a week, but his prediction might be true for the grandchildren of his grandchildren. If this amount of paid work would

Fig. 11.4 Working hours 1840–2015. Author's own graphic, but the data source is McKinsey & Company (2017)

be sufficient to satisfy our economic needs, we could use the newly gained free time to engage in activities we consider valuable and fulfilling, but not necessarily financially beneficial.

Equality and Diversity

Among these valuable and fulfilling—but not financially rewarding—activities are many of the unpaid tasks currently carried out by women: childcare, eldercare, community work. Activities with high value to the society but very low or non-financial prestige. With algorithms taking over more and more typically male jobs, these social activities might become both more esteemed and distributed more equally between men and women. The majority of men over the generations have been taught and appreciated for performing predictable, repetitive, and emotionless tasks—tasks that machines are, by their nature, far better at. The typical female work, by contrast, is the opposite of it, focusing on intuition, empathy, and emotional intelligence—attributes that are uniquely humane and very difficult to automate (see Fig. 11.5).

According to these numbers, today's typical women's work might become more prevalent in the future, potentially resulting in a shift in the division of labor. While this kind of shift may currently sound difficult to imagine, it might actually have a

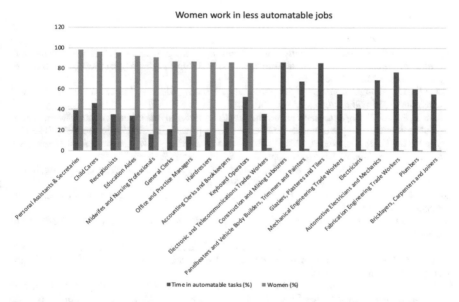

Fig. 11.5 Women work in less automatable jobs. Author's own graphic, but the data source is Hanrahan and Evlin (2017)

positive impact on society: fair division in unpaid work and family duties and more esteem and higher wages for "pink-collar-jobs".

Conclusion

Humans often encounter technological progress with scepticism and anxiety. During the Industrial Revolution, the Luddites and other groups organized to destroy machines that were automating their jobs; when the first railway opened, people feared using it, as they believed the human body wouldn't survive the speed of 30 miles an hour; and driving a car in 19th century England required a second person to precede the vehicle on foot waving a red flag (The Motor Miscellany 1865). Almost every technology though, despite the initial difficulties, contributed to a significant rise in the quality of life over the long term. Almost everyone now knows how to drive a car, and we've created regulations that make us use them in a reasonably safely way. Why shouldn't we manage the same with algorithms? *"Technology is [...] giving us ways to do harm and to do well; it's amplifying both.[...] but the fact that we also have a new choice each time is a new good"*, (Kevin Kelly quoted by Edge 2014).

References

Aneesh, A. (2002). *Technologically coded authority: The post-industrial decline in bureaucratic hierarchies.* http://web.stanford.edu/class/sts175/NewFiles/Algocratic%20Governance. pdf. Accessed May 30, 2018.

BBC. (2015). *Google apologises for photos App's Racist Blunder.* http://www.bbc.com/news/technology-33347866/. Accessed May 30, 2018.

Botsman, R. (2017). *Big data meets big brother as China moves to rate its citizens.* http://www.wired.co.uk/article/chinese-government-social-credit-score-privacy-invasion/. Accessed May 30, 2018.

Brynjolfsson, E., & McAfee, A. (2014). *The second machine age: Work, progress, and prosperity in a time of brilliant technologies.* New York: W. W. Norton & Company.

Cellan-Jones, R. (2014). *Stephen Hawking Warns artificial intelligence could end mankind.* http://www.bbc.com/news/technology-30290540/. Accessed May 30, 2018.

Edge. (2014). *The Technium.* https://www.edge.org/conversation/kevin_kelly-the-technium/. Accessed May 30, 2018.

Ford, M. (2016). *Rise of the robots: Technology and the threat of a jobless future.* New York: Basic Books.

Frey, C., & Osborne, M. (2013). *The future of employment: How susceptible are jobs to computerisation?* https://www.oxfordmartin.ox.ac.uk/downloads/academic/The_Future_of_Employment. pdf/. Accessed May 30, 2018.

Gallup. (2017). *State of the global workplace.* http://www.managerlenchanteur.org/wp-content/uploads/Gallup-State-of-the-Global-Workplace-Report-2017_Executive-Summary. pdf/. Accessed May 30, 2018.

Garling, C. (2015). *Andrew Ng: Why 'Deep Learning' is a mandate for humans, not just machines.* https://www.wired.com/brandlab/2015/05/andrew-ng-deep-learning-mandate-humans-not-just-machines/. Accessed May 30, 2018.

Gibbs, S. (2016). *Chatbot Lawyer overturns 160,000 parking tickets in London and New York*. https://www.theguardian.com/technology/2016/jun/28/chatbot-ai-lawyer-donotpay-parking-tickets-london-new-york/. Accessed May 30, 2018.

Gottfried, J., & Barthel, M. (2015). *How Millennials' political news habits differ from those of Gen Xers and Baby Boomers*. http://www.pewresearch.org/fact-tank/2015/06/01/political-news-habits-by-generation/. Accessed May 30, 2018.

Hanrahan, C., & Evlin, L. (2017). *Artificial intelligence: Men's jobs face higher risk of automation than women, low-paid workers also at risk*. http://www.abc.net.au/news/2017-08-09/ai-automation-men-and-lower-paid-workers/8741518/. Accessed May 30, 2018.

IBM. (2017). *10 key marketing trends for 2017 and ideas for exceeding customer expectations*. https://www-01.ibm.com/common/ssi/cgi-bin/ssialias?htmlfid=WRL12345USEN/. Accessed May 30, 2018.

Johnson, S. (1984). Cross-racial identification errors in criminal cases. *Cornell Law Review, 69*, 933–987.

Kissinger, H. (2018). *How the enlightenment ends*. https://www.theatlantic.com/magazine/archive/2018/06/henry-kissinger-ai-could-mean-the-end-of-human-history/559124/. Accessed May 30, 2018.

Levi, M. (2015). Human responsibility. In J. Brockman (Ed.), *What to think about machines that think* (p. 235). New York: Harper Perennial.

Locker, M. (2018). *China's terrifying "Social Credit" surveillance system is expanding*. https://www.fastcompany.com/40563225/chinas-terrifying-social-credit-surveillance-system-is-expanding/. Accessed May 30, 2018.

McCorduck, P. (2015). An epochal human event. In J. Brockman (Ed.), *What to think about machines that think* (pp. 51–53). New York: Harper Perennial.

McKinsey & Company. (2017). Jobs lost, jobs gained: Workforce transitions in a time of automation. https://www.mckinsey.com/~/media/McKinsey/Global%20Themes/Future%20of%20Organizations/What%20the%20future%20of%20work%20will%20mean%20for%20jobs%20skills%20and%20wages/MGI-Jobs-Lost-Jobs-Gained-Report-December-6-2017.ashx/. Accessed May 30, 2018.

Kontzer, T. (2016). *Deep learning drops error rate for breast cancer diagnoses by 85%*. https://blogs.nvidia.com/blog/2016/09/19/deep-learning-breast-cancer-diagnosis/. Accessed May 30, 2018.

O'Neil, C. (2016). *Weapons of math destruction*. New York: Crown Publishing Group.

O'Reilly, T. (2018). *WTF? What's the future and why it's up to us*. London: Random House Business Books.

Pauling, L., et al. (1964). *The triple revolution*. Santa Barbara.

Pinker, S. (2018). *Enlightenment now: The case for reason, science, humanism, and progress*. New York: Penguin Books Limited/Viking.

PWC. (2018). *How will automation impact jobs?* https://www.pwc.co.uk/services/economics-policy/insights/the-impact-of-automation-on-jobs.html/. Accessed May 30, 2018.

Simonite, T. (2017). *AI software learns to make AI software*. https://www.technologyreview.com/s/603381/ai-software-learns-to-make-ai-software/. Accessed May 30, 2018.

Simonite, T. (2018a). *When it comes to gorillas, Google photos remains blind*. https://www.wired.com/story/when-it-comes-to-gorillas-google-photos-remains-blind/. Accessed May 30, 2018.

Simonite, T. (2018b). *How coders are fighting bias in facial recognition software*. https://www.wired.com/story/how-coders-are-fighting-bias-in-facial-recognition-software/. Accessed May 30, 2018.

Sulleyman, A. (2017). *Elon Musk: AI is a "Fundamental Existential Risk for Human Civilisation" and creators must slow down*. https://www.independent.co.uk/life-style/gadgets-and-tech/news/elon-musk-ai-human-civilisation-existential-risk-artificial-intelligence-creator-slow-down-tesla-a7845491.html/. Accessed May 30, 2018.

The Motor Miscellany. (1865). *The locomotive act*. https://sites.google.com/site/motormiscellany/motoring/law-and-the-motorist/locomotive-act-1865/. Accessed May 30, 2018.

Tversky, A., & Kahneman, D. (1973). Availability: A heuristic for judging frequency and probability. *Cognitive Psychology, 4,* 2017–2032.

Agnieszka M. Walorska is a founder of CREATIVE CONSTRUCTION, kickstartAI and hidence. As one of the very first employees at one of the biggest exit-success stories on the German market (studiVZ), Agnieszka has played a crucial role in the explosive growth of this social network. She is the author of publications and sought-after expert and keynote speaker on the influence of Artificial Intelligence on User Experience and Design, Agile Management and Digital Innovation. She led successful Innovation and Customer Experience projects across several industries, including utilities, banking and media. Agnieszka studied social sciences and politics at Warsaw University and Humboldt-University of Berlin and received a scholarship from Studienstiftung and Hertie Foundation. When she is not busy solving innovation problems, she is training for her next IronMan.

How Should One Respond to Political Risks?

Chapter 12
Smart Cities and Smart Regions—The Future of Public Services—Solidarity and Economic Strength Through Smart Regions and Smart Cities

Katherina Reiche

Introduction

Digitalization is permanently transforming the economy and our society. Although it poses a major challenge, it also presents an opportunity for cities and regions to become safer, more efficient, and more attractive for residents and businesses. Local public utilities are seizing this opportunity—without losing sight of the challenges posed by digitalization or other megatrends, such as climate change. The strategic course for the digital age is clear: cities and regions are becoming smart in order to improve quality of life. And many actors are already on the move. The large corporations of the digital world, which have already upended many markets in the analog world, are hoping for billions in business. The smart city market in Germany was valued at €20.4 billion in 2017, but it is expected to increase to some €43.8 billion in the near future. However, do large corporations really have to manage smart cities and regions? And is this even desirable—particularly in light of data privacy scandals and the concerns of the public? Shouldn't it be local and regional actors that actively strengthen public services—which, after all, are the basis for equal living conditions, economic success and social cohesion—with digital means, to make people's lives better?

Local public utilities are taking action. They want to contribute to the development of municipalities to become connected cities and regions. They aim to capture the economic and social opportunities offered by digitalization for the benefit of citizens. The first section of this paper examines the action areas and paths in which local public utilities can be active in the future. They face the challenges of four megatrends, primarily regarding infrastructure upgrades. The second chapter examines the opportunities digital technologies can offer them in practical terms; this also

K. Reiche (✉)
German Association of Local Public Utilities (Verband Kommunaler Unternehmen e.V.),
Berlin, Germany
e-mail: hgf@vku.de

© Springer Nature Switzerland AG 2020
D. Feldner (ed.), *Redesigning Organizations*,
https://doi.org/10.1007/978-3-030-27957-8_12

explains what we mean by smart cities and regions. But for this vision to become reality, local public utilities require an appropriate, reliable framework. This is discussed in the third chapter. The actions outlined here will enable them to capture the economic and social opportunities offered by digitalization for the benefit of citizens, thus creating a foundation upon which smart cities and regions can build.

Four Megatrends Will Shape the Future of Cities and Regions

Four megatrends will define the cities and regions of the future: demographic and social change, urbanization, climate change and, of course, digitalization. These long-term trends affect every area of society and the economy and will change them fundamentally. As such, they define the action areas and paths in which local public actors will move in the future (Quadriga 2017). These megatrends pose challenges for municipalities and the local public utilities, because they place new demands on the public services that they provide and, hence, on their infrastructure. At the same time, they also offer opportunities to supply energy, water/wastewater, telecommunications and waste management/municipal cleaning, which local actors can take advantage of, thus advancing toward becoming smart cities and regions. New services will also emerge, because the tasks of the local public utilities are not static—they focus on the needs and expectations of the residents as well as technological change.

Urbanization and Demographic Change Require Adaptation of Infrastructures

The trend toward urbanization is clear: population centers and their surrounding regions will continue to grow, while rural areas will shrink. The latter is already happening today in the eastern and central regions of Germany (BBSR 2017). These rural-to-urban migration flows require changes to the infrastructure for public services: in growing regions, infrastructures will have to be established or expanded—for example, water pipes and sewage conduits. In shrinking regions, in contrast, capacities will have to be cut back and new ways of utilizing them explored (Quadriga 2017). These rural-to-urban migration flows are being augmented by demographic change: life expectancy in Germany has been growing steadily for decades, while the birth rate has been declining. Germany is aging, and the excess mortality rate is another reason why the number of inhabitants of the country is shrinking. In 2060, Germany will have around 68–73 million inhabitants instead of the current 82 million (Statistisches Bundesamt 2015). These trends have consequences: Municipalities and their local public utilities have to adapt, plan and invest in the long term. The specific local circumstances are decisive when assessing the exact needs for infrastructure-related construction projects and the associated costs.

The Energy Transition and Resilient Infrastructures Help to Protect Against Climate Change

Safe, economic and renewable: local public utilities aim to create a new energy world. Municipal utilities are following a two-pronged strategy: First of all, they are minimizing the emission of greenhouse gases, to contribute towards achieving climate targets. The energy transition is an important tool for achieving this. The transition to energy from renewable sources requires a complete transformation of the energy system: from a centralized to a decentralized energy infrastructure. The German government aims to increase the share of renewable energy to 65% by 2030 (decided by the fourth Merkel cabinet). To achieve the necessary infrastructure transformation, local public utilities are backing digital technologies such as smart grids and smart meters (see Chap. 3).

Secondly, local public utilities are already facing the consequences of climate change, including extreme weather events. Heat waves increase energy consumption for air conditioning and refrigeration. The water and wastewater industry is adapting to extreme rainfall scenarios with new concepts for wastewater disposal and flood protection. The aim is to create flexible, adaptable structures that enable municipalities to respond to new climatic conditions. In case of an emergency, it must be possible to intervene quickly. Furthermore, the structures must remain operable even if individual elements fail. The overall goal is to strengthen the resilience of critical infrastructures, to prepare them for climate change (Rottmann and Grüttner 2016).

Social Change: The Desire for Security and a Change in Consumer Behavior

Those who wish to shape life in the cities and regions of Germany must include the people and actively address changes of the society. This is the only way changes will be accepted by the public. In addition to individualization, the need for security is growing. For example, according to a poll, three quarters of Germans said that security is increasingly important to them. Places two and three can be interpreted as the wish for stability and support in a familiar environment: values expressed in the poll, such as home (63%) and trust (58%), are gaining importance (GfK 2017). These desires likely have many sources. One possible explanation is the negative consequences of social change: a trend towards income polarization, a lack of educational opportunities and spatial segregation (Deutscher Städtetag 2015; Rottmann and Grüttner 2016).

Moreover, people are also changing in their role as consumers. Almost nine out of ten people in Germany (88%) can imagine lending things and utilizing sharing schemes (Verbraucherzentrale Bundesverband 2015). Their value system is clearly changing away from possession and toward usage. In addition, market research has shown that a familiar environment, regional origin, price-performance ratio and eco-friendliness are important factors in purchase decisions (Nielsen 2016).

Furthermore, experts predict that digitalization will change people's perception of their roles as customers or citizens; the first changes are already apparent. On the one hand, their market position is improving tangibly: Access to the internet, with search engines and comparison websites, as well as social networks and messaging services, are helping them become better informed and connected. Their quality standards are also rising: they expect individualized products, delivery within 24 h and free updates (Bundesministerium für Wirtschaft und Energie 2017). These demands also affect local public utilities—essentially, all service providers.

On the other hand, people know that they are no longer anonymous in the digital world. They pay with their data. But while individuals are becoming increasingly transparent (consciously or unconsciously), companies are keeping a low profile. Hardly anyone knows which data they are making available to whom and what they are doing with those data. Critics warn that, at the moment, customers have lost ownership of their data (Osburg 2017). In turn, the loss of transparency and privacy is causing a loss of trust, which is an essential precondition for the digital economy to function (Osburg 2017). In fact, Facebook and other internet giants are struggling with real confidence issues (see GPRA 2017), and a clear legal framework for handling data[1] would help to rectify the situation (see Chap. 3).

Furthermore, digitalization also poses other questions, including ethical ones. What rules should algorithms follow in their decisions? How are programming parameters decided? We will need transparent criteria for these questions in the future. This will also help to create acceptance for new products and services. One important prerequisite is that the practical benefits of digitalization be perceptible to the public. Local public utilities enjoy an advantage here, because their digital products and services are linked directly and closely to the places where people live and work. One example: sensors in water channels measure the rise in water levels after heavy rainfall. This means local transport companies can be alerted to the threat of flooding in underground stations. Their customers could also be notified of precautionary station closings, for example, via text message. This would reduce the risk of traffic chaos, and people would be able to look for other ways home in advance.

Societal change raises many questions in general. Local public utilities have answers, because they are deeply rooted in the region and understand the local situation. They are trusted by the public as a competent and reliable contact, with the technical expertise to resolve problems on site. They provide public services in a

---

[1] With the General Data Protection Regulation, which came into force in May 2018, the European Union has shown that it intends to counter this loss of control. It is too early at the moment (May 2018) to assess the actual effects of the GDPR and people's perception of its effectiveness.

safe, reliable manner: 24 h a day, 365 days a year. As a recent survey showed, 72% of people trust municipal utilities—an excellent result, putting the utilities in fourth place on the trust barometer (Forsa 2018).

These characteristics undoubtedly set the local public utilities apart from internet companies in distant Silicon Valley. The former reflect the values of the people, who seek greater security. This is a decisive advantage that the utilities should and want to exploit in the competition for markets in smart cities and regions. When developing digital products and services, they have to consider the new market position of customers, their focus on usage and the importance of regional aspects in influencing people's decisions.

Digitalization: Creating Infrastructures to Capture the Economic and Social Benefits of Digital Transformation

Germany needs high-performance digital infrastructures for the gigabit society. In Germany, small and mid-sized enterprises (SMEs) are responsible for more than half of all value created (Federal Ministry for Economic Affairs and Energy 2018). To remain competitive in the future, SMEs in particular need access to high-performance digital infrastructures—independently of whether the company is located in a metropolis like Hamburg, or in a small village like Dingolshausen. Residents also need access to digital infrastructure, in order to ensure that they can participate in and benefit from the digital transformation.

Therefore, a prerequisite is to provide broadband coverage to all regions throughout the country. A fiber-optics broadband network is essential to giving everyone access to the internet and ensuring equal living conditions in the digital world. It will also help to strengthen social cohesion to a considerable extent. Therefore, the digital infrastructure itself must be considered a public service.

Rural areas must remain a focal point—for example, more than half of Germany's residents live there (definition and calculation by BMEL 2016). In rural areas, only fast internet access can provide the technical means of working from home. This would render stressful, time-consuming commutes obsolete and enable employees to spend more time with their families.

A lot remains to be done in rural areas. One hundred and fifty committed local public utilities in Germany are dedicated to building out the broadband network. Other providers still advocate for the outdated vectoring technology and are installing copper cables—even though the demand for broadband will continue to grow. The advance of the Internet of Things (IoT) alone—essentially, connected devices and machines—will increase data traffic by approximately 50% by 2025 (Boston Consulting 2018), in addition to the growing demand for mobile internet by other sectors of society. As such, the 5G mobile communications standard will play a key role. But for the high-speed network to develop its full potential, 5G transmitters must be connected to the broadband network. This means the importance of a broadband connection cannot be ignored. At the same time, 5G offers thinly populated, difficult-to-reach rural areas the chance to bridge the last mile from broadband cable to households and businesses (Internet Economy Foundation/Roland Berger 2018). The infrastructure is supplemented by cloud computing, data centers and wireless solutions on different frequencies e.g. LoRaWan. As such, the highest priority at the moment involves creating a broadband infrastructure as the foundation for smart cities and regions. It is the key precondition for capitalizing on the economic and social advantages of digitalization (see Chap. 3).

Summary So Far: Smart Cities and Regions Must Build on the Right Infrastructure

To capture the opportunities offered to people by digitalization, the infrastructures for public services must be adapted. They are the foundation of smart cities and regions. The megatrends of urbanization and demographic change require infrastructures to be expanded or dismantled, while climate change demands increased resilience. The energy transition will require a complete retrofit. And only digital infrastructures can provide access to the digital world. They will decide on the competitiveness of companies and the social engagement of every resident. A ubiquitous broadband network is the key. Equal living conditions are decisive for social cohesion. These infrastructures will make it possible for cities and rural regions to seize the opportunities offered by digitalization: digital technologies will help to tackle the challenges of these megatrends and will improve the lives of people in smart cities and regions (see Chap. 3). What's more, the necessary adjustments to the infrastructures vary from place to place. The local public utilities have the required expertise in building and maintaining infrastructures and also know the local situation and needs. Social acceptance is an important prerequisite for the transformation—whether for smart cities and regions or the new products and services. It is a question of gaining people's trust in the digital world. As competent, reliable, local actors deeply rooted in their region, local public utilities have a clear advantage over internet corporations. Their vision of the digital transformation counters the digital disruption from Silicon Valley: actively shaping the digital transformation of cities and municipalities to the benefit of the locals.

How Smart Cities and Smart Regions Can Seize Digital Opportunities

Smart cities and regions are created by connecting people, administrations, businesses and public services in an intelligent manner. Together they make up the smart nation of Germany—a cosmopolitan, socially lively, economically prosperous nation. It can successfully reduce the risks posed by megatrends and seize the opportunities offered by digitalization, as the following examples show.

Smart Cities and Regions Build on Efficient Processes

Digital technologies capture efficiency potential in all corporate processes, starting with internal HR and resource planning and the opportunity for intercommunity collaboration via platforms. Digital technologies will enable public utilities to perform better. This benefits residents, as well. After all, local public utilities provide a broad spectrum of public services.

In general, every digital tool serves a variety of use cases, as demonstrated by the Internet of Things and AI, for example. Waste management companies and water companies can both benefit from the Internet of Things. Sensors can measure the fill levels of waste bins for waste management companies and send a signal when the bins have to be emptied, thus saving unnecessary trips. As a result, route planning can be optimized to save costs. Consequently, the general traffic levels decrease, thus protecting the environment and the climate. The water and wastewater industries, in turn, can install sensors in pipes and digital meters, to enable smart management of the water supply and wastewater disposal, using the data collected. This will help to deal with periods of heavy rainfall.

Artificial intelligence (AI) is also promising. Testing AI is likely to reap benefits in the long term. It could be used to optimize gas turbines autonomously, to ensure better monitoring of smart electricity grids and for predictive maintenance of plants and equipment. It is likely to play a key role in the supply and disposal industries, in particular—for example, using smart waste-sorting robots. They will not only save companies time, money and effort, but also improve the recycling economy in general, helping to save the environment.

These examples might cause some readers to fear job losses. But in fact, many new digital technologies require two things of employees in local public utilities in particular: new skills and qualifications. New jobs that demand new skills will also be created. That is why it is important to prepare today's trainees, students and employees to handle the tasks of tomorrow. Digital education will undeniably become a key task in the future for businesses, regional governments and, of course, local public actors.

Smart Cities and Regions Are Developing New Services and Products

Like all other sectors of society, local authorities and the municipal utilities have to cope with the stronger market position of well-informed, connected customers: digital services, the orientation of local administration to online services, as well as energy and logistics solutions are in demand (Eco 2017). In smart cities and regions, people and businesses communicate with local authorities digitally and will obtain services via their user accounts. In this respect, local government administrations in Germany have a lot catching up to do. The broad opinion was that city administrations in Germany were a "digital services wasteland" (EFI 2016). Local public utilities must also adapt their portfolios of products and services, along with their business models, to their customers' wishes if they hope to remain competitive. Their new or improved products and services should offer people genuine added value that improves their quality of life. That is what characterizes smart cities and regions.

Two examples from the transport sector illustrate the benefits of digital technologies and how they can help to counter the megatrends of climate change and urbanization.

IoT technologies are useful instruments for intelligent traffic control to suit various needs. The German public utility Stadtwerke München is already testing sensors installed in road surfaces, which can direct people to free parking spaces via an app. The time-consuming search for a parking space has thus become a thing of the past, while also reducing traffic congestion and environmental pollution. After all, 30–40% of urban traffic is caused by people searching for parking spaces. But that is only the beginning. Sensors in the road or in street lamps can also record the volume of traffic and warn traffic control centers and drivers of possible traffic jams in advance. The former can then take action—either automatically or manually—while alternative routes can be proposed to drivers. This is how the Internet of Things can improve the control of all traffic flows. Smart sensors offer other potential use cases, as well: they can make street lamps shine brighter when pedestrians, cyclists or cars approach. This lowers energy costs for municipalities. Or they can record environmental data and report excess particulate pollution or emissions—two problems that are likely to worsen in population centers in light of the trend toward urbanization. In general, this technology can help protect the climate, the environment and individual health, thus improving the quality of life. People in population centers, in particular, will benefit from intelligent traffic control.

Self-driving vehicles will also open up new possibilities. German cities, such as Berlin and Duisburg, are already using autonomous shuttle buses to boost local public transport. The public utility companies in Arnsberg and Menden in the Sauerland region of the country develop various mobility concepts, including self-driving vehicles to improve local public transport. This will open up new prospects for the automotive supply industry with its 43,000 employees in the rural region of southern Westphalia. The technology harbors potential for rural areas most of all, since it can individualize local public transport. People in rural areas could take a self-driving bus even late at night. Autonomous driving on-demand transport schemes are also conceivable. The public utility company Stadtwerke Augsburg is testing self-driving electric buses. In general, autonomous driving schemes can offer an additional service that is currently not available in many rural areas, due to the lack of demand. Local people would gain mobility, and rural areas would become more attractive. And in the best case, it would also be cost-efficient.[2]

Smart Cities and Regions Are Fighting Climate Change

Smart cities and regions can utilize digital technologies to accelerate the energy transition, in order to contribute toward fighting climate change. The switch to renewable energy technologies is in full swing. But the volatility of energy from sun, wind and biomass remains a challenge. Structural change—from centralized to decentralized energy supply—must also be reflected at the grid level. This change can be achieved by digitalizing electricity grids: intelligent energy networks, or smart grids, ensure that local production facilities, networks and storage facilities, together with consumers, interact in optimum fashion through intelligent control systems, thus compensating for the volatility of renewable energy technologies. Hence, smart grids are the necessary prerequisite to harmonize the aims of security of supply, economic efficiency and environmental sustainability.

Public debate often focuses on the major transmission lines that transport the electricity generated from renewable sources over long distances at low losses. Distribution grids are all too often forgotten in such cases, even though they transport electricity directly to businesses and consumers. There are currently 25,000 km of transmission lines, compared with 1.7 million km of distribution grids. They will play a major role in the structural transformation to smart grids. Expansion, modernization and digitalization lie in the hands of the operators of local public distribution grids. If distribution grids were used more efficiently, the need for expansion could be more than halved, representing a financial savings of €400 million (Federal Ministry for Economic Affairs and Energy 2014).

Smart grids help smart cities and regions to increase flexibility in their electricity grids. In combination with smart metering technology, they enable commerce and industry, as well as general consumers, to tailor their consumption to the electricity supply. If a lot of electricity is available, SME manufacturers can ramp up production, while at home, the washing machine starts its daily routine. Both consumer groups would be rewarded financially for their flexibility, and the grid load would be reduced, in turn improving the stability of the electricity supply. These types of solutions are already being tested in SINTEG (Smart Energy Showcases—Digital Agenda for the Energy Transition) model regions.

---

[2]This is the concept for self-driving schemes for smart cities and regions. Of course, this vision currently presupposes further technical and pre-commercial development, which would result in a reduction in procurement and maintenance costs.

Furthermore, smart grids are essential for a breakthrough in e-mobility and, therefore, will be decisive for the success of the transport transformation, which in turn will be decisive for the success of the energy transformation. Public debate should concentrate more on expanding the charging infrastructure and the required number of charging stations. However, it is just as important to ensure that charging stations are connected to the distribution grid. Therefore the expansion of both should go hand-in-hand. Grid overload, which could occur by charging a large number of cars at the same time, would be balanced out in a smart grid. In addition, inductive charging would be a good bet. Electric cars could not only be charged with electricity from charging points in the roads, but could also feed electricity back into the grid when needed. They would become mobile electricity storage devices that could contribute to the security of supply.

These approaches show that local public utilities are creating the foundation for the energy transition with smart grids. Smart cities and regions will supply people with electricity, heating and mobility by linking e-mobility, power plants and storage facilities in a smart energy grid. Initial approaches have already demonstrated the integrated solutions developed by public utility companies in the construction or renovation of residential areas. Here, district heating, power generation from renewable sources, power storage facilities, landlord-to-tenant electricity, smart homes, e-mobility, car-sharing and virtual power stations are all interconnected intelligently. People are able to enjoy the benefits of this new energy world and take advantage of safe, economical and clean energy generation, which also contributes to climate protection.

Implementation: Smart Cities and Regions Need Legal Certainty and a Consistent Strategy

To turn the vision of smart cities and regions into reality, we need more than just a digital infrastructure. The diversity of our cities and communities demands diverse solutions. Regional digital strategies identify specific needs and strengths on the ground, thus helping to develop appropriate solutions—for the specific site and to deal with the consequences of megatrends. Cooperation is essential to develop and implement solutions. It allows cities and communities to combine their relative strengths. Through cooperation, they capture synergies that benefit all parties. This is the reason why smart cities and regions need the possibility to build networks, exchange and cooperate with local public actors, startups, the businesses, researchers and, of course, the people themselves. That is also why discussion and experimentation spaces and platforms are important: they connect the knowledge of stakeholders with data from cities and regions. Therefore innovation centers are important meeting places.

Local data are merged in a shared local platform—the urban data room. Data include geographical, cadastral, mobility, energy and environmental data, as well as social and economic data. The aim is to provide, exchange and utilize local data securely, transparently and independently for the development of new products and services. Data should resolve problems on the ground and, hence, offer genuine real value. In addition to interfaces, clear governance rules that follow the principles of data security and data sovereignty are important. Data are only shared when necessary; there are sensible, comprehensible and—above all—transparent rules as to who is allowed to access which data (see Quadriga 2017).

This requires a consistent legal framework that clearly regulates how data are used and handled, such as the data local public utilities use to create new products and services. A "Data Act" should classify both open data and fee-based data. Not all data should be available free of charge. The ultimate aim of a Data Act would be to create legal certainty for people and businesses, thus ensuring planning reliability for decision-makers in local public utilities. For example, a Data Act would have to regulate how local public utilities can use data to develop new products and services (see Quadriga 2017).

Local public utilities see digitalization as a transformation, not a disruption. Digital progress needs to serve people. Innovations are accepted when users perceive their benefits. Digital products and services must add tangible value for the residents. Local public utilities in smart cities and regions must always develop products, services and business models with a focus on how they benefit the people and whether they resolve local issues or improve people's lives.

In general, local public utilities are using digital tools to offer their services more efficiently and more competitively. With a view to climate change, urbanization and demographic change, their solutions should also be as resilient as possible, universally applicable and scalable. They must work independently of population density and also take sustainability into account. The needs of future generations must not be forgotten. Citizens will be able to experience the benefits of products and services created from local data first hand, for their local region and their daily lives. Local public utilities can help them gain trust in the digital transformation.

Summary

Four megatrends pose a challenge for cities and regions: demographic and social change, urbanization, climate change and, of course, digitalization. To mitigate the negative effects of these megatrends and capitalize on the opportunities offered by digitalization, local public utilities need to adapt their infrastructures for public services and, above all, upgrade the digital infrastructure.

Firstly, a sound digital infrastructure will ensure the competitiveness of businesses and the social participation of every resident in the digital age. Its quality is decisive: comprehensive access to broadband, the aim of local public utilities, is essential to ensuring that everyone has equal access to the digital world. In turn, this equal access is decisive for social cohesion. Therefore, the digital infrastructure itself must be seen as part of the public service.

Digital infrastructures will create the prerequisites for smart cities and regions, enabling them to capture the economic and social advantages offered by digitalization for local residents. They link citizens, government administration, local businesses and public services. Innovation centers for interchange, experimentation and cooperation are important, as is the urban data room where all local data are collected. This can be distinguished from private providers through clear rules of governance and the principles of data ownership and sovereignty. With this approach, smart cities and regions can create a solid foundation for encouraging people's trust in the digital world—a trust that is being shaken by the loss of privacy and lack of transparency. Generally speaking, people place a high value on security. This poses an opportunity for local public utilities, since they are able to cater to the residents' desire for reliability. In fact, the majority of people place great trust in them. As competent, reliable local partners rooted in the local region, they have a clear advantage over often-suspect internet corporations.

Data will also help local public actors to develop regional digital strategies, for example by identifying local challenges and strengths and finding solutions that fit the local scenarios. This means local public utilities can improve their products and services or even create new ones. They can utilize the opportunities offered by digital technologies to solve problems on the local level and give local people genuine added value: savings of time, energy and costs, as well as greater efficiency, and this will make the energy transition a success. The new world of energy in smart cities and regions is based on smart power grids that link e-mobility, power plants and power storage facilities intelligently, to supply local people with electricity, heating and mobility generated from renewable energy sources. Residents will enjoy safe, economical and clean energy generation, which will also contribute to climate protection. With an eye to the megatrend of urbanization, this will help cities the most.

Smart regions can use digital technologies to create new products and services. One example: autonomous vehicles to improve local public transport. Mobility would improve, and rural areas would be more attractive, thanks to their improved accessibility. Smart cities and regions can employ new, digitally-based products and services to mitigate the negative consequences of megatrends—provided that digital infrastructures and clear data laws are in place. At the same time, smart cities and regions can use them to improve Germany's competitiveness, as well as increase social cohesion. Local public utilities are countering the digital disruption from Silicon Valley with a digital transformation on their own terms. They are shaping the digital transformation by seizing the social and economic opportunities presented by digitalization for the benefit of their citizens.

Bibliography

Association of German Cities. (2015). Integrated urban development planning and urban development management. Position paper of the Association of German Cities.

BBSR (Federal Institute for Research on Building, Urban Affairs and Spatial Development). (2017). Wachsen und Schrumpfen von Städten und Gemeinden 2010 bis 2015 im bundesweiten Vergleich.

BMEL (Federal Ministry of Food and Agriculture). (2016). Ländliche Regionen verstehen. Fakten und Hintergründe zum Leben und Arbeiten in ländlichen Regionen.

Eco—Association of the Internet Industry. (2017). The German smart city market 2017–2022. Facts and figures.

EFI—Commission of Experts for Research and Innovation. (2016). Digitale service-Wüste in deutschen Amtsstuben.

Federal Ministry for Economic Affairs and Energy. (2014). Moderne Verteilernetze für Deutschland.

Federal Ministry for Economic Affairs and Energy. (2017). Mittelstand-Digital Themenheft Digitale Geschäftsmodelle.

Federal Ministry for Economic Affairs and Energy. (2018). Wirtschaftsmotor Mittelstand. Zahlen und Fakten zu den deutschen KMU.

Federal Statistical Office. (2015). Germany's population by 2060. Results of the 13th coordinated population projection.

Federation of German Consumer Organizations. (2015). Sharing economy: The consumers' view in Germany.

Forsa (2018): Vertrauensranking im Auftrag von RTL/n-tv-Trendbarometer.

German Public Relations Consultancies Association: Likes, Lügen, Lethargie—GPRA-Vertrauensindex 2017. Last Accessed May 28, 2018.

GfK Verein (2017): Werte-Studie 2016/2017.

Internet Economy Foundation/Roland Berger. (2018). Success factor 5G. Innovation und Vielfalt für die nächste Stufe der Digitalisierung.

Nielsen. (2016). Die Herkunft zählt: Das halten die Deutschen von Marken Made in Germany.

Osburg, T. (2017). *Sustainability in a digital world*. CSR, sustainability, ethics & governance.

Quadriga—Studie im Auftrag des Verbands kommunaler Unternehmen. (2017). Digital. Kommunal. Deutschland. Smart Nation durch Smart Regions.

Rottmann, O., & Grüttner, A. (2016). Smart cities—Handlungsfelder und Konzepte: Studie zum 9. Mitteldeutschen Energiegespräch.

Society for Consumer Research. (2017). *Focus topics—Values: Desire for security increases further*. Last Accessed April 27, 2018 at 13.45.

Katherina Reiche is the CEO of the German Association of Local Public Utilities (Verband Kommunaler Unternehmen e.V.), an association representing more than 1500 companies that provide electricity, gas, water, sewage, and transportation services to the public in local communities. She was born in Luckenwalde, in the former East Germany, and, after obtaining a degree in chemistry from Potsdam University, she worked at universities in the US (Clarkson University) and Finland (Turku). At the age of 25, she was elected to the German Bundestag, where she specialized in education and science policy, and later became deputy chair of the CDU parliamentary group. From 2009 until 2013, she served as Parliamentary State Secretary at the Ministry for the Environment, Nature Conservation, and Nuclear Safety; from 2013 until 2015, she held the same function at the Ministry of Transport and Digital Infrastructure. In September 2015, she left the German Bundestag to assume the role of CEO of the German Association of Local Public Utilities. Since 2016, she has been President of the European Centre of Employers and Enterprises providing Public Services, CEEP. During her entire career, she has dealt with questions concerning research, sustainable development, digitalization, infrastructure, and energy. She sits on a number of advisory boards of German energy companies, scientific institutes, and the Atlantik-Brücke, an association to promote transatlantic cooperation. She is married with three children. She was appointed by Chancellor Merkel to serve on the German "coal committee" on growth, structural change, and employment to develop a concept for Germany's energy transition from fossil fuels and nuclear power to renewable energies.

Chapter 13
China's Authoritarian Internet and Digital Orientalism

Maximilian Mayer

Introduction

Globally, more than 4.1 billion people use social media, smartphone applications and other Internet-based services. That represents a 54% penetration by end of 2017. We are witnessing a transformational epoch, as information societies have become a reality (Castells 1997). But concerns about the ramifications of mushrooming digital technologies are growing. The relentless changes in education, industrial production, communication attitudes, workspaces and entertainment, to mention just a few, cause many observers to worry (Morozov 2011; Lynch 2016). Against this background, news reports and popular writings typically link China's Internet revolution to concerns with digital surveillance and manipulation. The specter of an Orwellian society has become a dominant theme, rendering almost any digital trend in China as inherently dangerous, uniquely dictatorial and ethically questionable.

This chapter explores the meaning of widespread anxieties about the authoritarian potentials of digital technologies. It argues that the responses to Chinese digital innovations tend to fall in a self-referential trap. Drawing on Edward Said's problematization of how Western discourses portray other cultures and societies as different and problematic (Said 1985; Palat 2000), today's intellectual reflex is akin to "digital orientalism". My focus is on German perceptions about China's digitalization to illustrate the problem of stereotypes. The negative image of digital China is interesting, because it mirrors techno-skepticism and buttresses the overemphasis on risks and regulation in a European context.

The following aims to clarify the challenges that spring from China's digitalization for European societies and governments. By studying China's Internet revolution without resorting to dystopian registers, European populaces and policy makers may refine their understanding of *global* trends of multiplying "authoritarian practice[s]

M. Mayer (✉)
International Studies, University of Nottingham Ningbo China, Ningbo, China
e-mail: maximilian.mayer@nottingham.edu.cn

© Springer Nature Switzerland AG 2020
D. Feldner (ed.), *Redesigning Organizations*,
https://doi.org/10.1007/978-3-030-27957-8_13

in the field of digital communication technologies" (Michaelsen and Glasius 2018). The experiences related to China's rapid construction of digital ecosystems carry dire warnings but also crucial insights: for instance, how to embrace beneficial outcomes and perhaps to regain a competitive edge in digital innovations. The act of balancing that defines a differentiated view then dampens techno-skepticism in European democracies, without normalizing modes of authoritarian digitalization. So, before taking a closer look at European views on digital developments in China, it is worth reconsidering for a moment Europe's own digital conundrum.

Europe's Digital Recession

Digitalization reshapes our life worlds in unforeseen ways. Digital technologies are also implicated in the reorganization of global markets. And Europe has lost momentum in this competitive process. From AI development and emerging data platform companies to glamorous fintech "unicorns", few European countries are "both highly digitally advanced and exhibit high momentum" (Chakravorti 2015; Chakravorti and Chaturvedi 2017). The boom of Chinese AI startups, in contrast, surpassed the U.S. in 2017 with a value of $15.2 billion (Snow 2018). China is leading in fintech developments both financially and technologically (Creehan and Borst 2017). In the first half of 2018, Ant Financial alone, a payment system that is part of the Alibaba business empire, raised $14 billion (Yang 2018a). While China's Internet giants increase their global presence, Europe economies have few indigenous giant Internet platforms. Both social media startups and established service sectors have difficulties propagating attractive visions of digital services, products and lifestyles.

Europe is even further behind in the field of AI. For example, between 2012 and 2017, European AI startups were only a minuscule share of major AI company acquisitions (CBInsights 2018; Yang 2018b). Europe's share of platform companies is 15.3%; their share of market capitalization is less than 2%, while none of the largest 20 Internet companies resides in Europe (see Table 13.1). As a result, novel

Table 13.1 Global distribution of platform and Internet companies

	Share of platform companies (%) in 2015	Platform enterprises' market capitalization (in Billion USD) in 2015	Market capitalization of the largest 20 internet companies (in Billion USD) in May 2018
North America	36.3	3.12	4319 (all in the U.S.)
Asia	46.5	0.93	986 (all in China)
Europe	15.3	0.18	--

Source Peter C. Evans and Annabelle Gawer (2016), p. 10; and https://www.statista.com/statistics/277483/market-value-of-the-largest-internet-companies-worldwide/

digital worlds, applications and services emerging abroad put a mounting pressure on domestic companies.

Europe's politicians began to demand "disruptive innovations", even in the fields of security and military technology (Vincenti 2018). However, bridging the widening gap with respect to digital and AI development is extremely difficult. Europe responds to the rapid progress in the U.S. and China with very limited investments. Many young talented individuals and leading researchers have already left. How can we explain this failure that could have potentially devastating consequences for Europe's wealth and productivity (McKinsey 2017)? Obvious reasons include inadequate amounts of venture capital, a missing digital single market, the overregulation of digital businesses and the relatively slow reaction of traditional industrial companies (Zilgalvis 2014; Beise and Schäfer 2016; Scott 2017; Champion 2018). But a deeper cultural layer is perhaps more consequential for shaping this response.

Woven through this web of behavioral and cognitive hurdles is a long-standing pattern of political culture: rather than embracing both opportunities and risks, European societies display an overly critical sentiment that slows down and sometimes outright prevents experimentation with emerging technological possibilities (see Jasanoff 2011; Mager 2017). The European, and particularly the German public sphere, is characterized by highly skeptical attitudes toward Internet technologies and the disruptive social transformations they may engender (Han 2013; Borchardt 2015). The recent drafting of an Artificial Intelligence (AI) strategy by the European Commission, for instance, privileges the "legal and ethical problems raised by AI and discusses the 'legitimate concerns' the technology generates". The main EU narrative frames AI as creating ethical problems and being in need of regulation (Macaes 2018), while the Chinese government wants to make the country an "innovation center for AI" by 2030 and already uses nascent AI systems to support foreign policy-making (Chen 2018a, b). Other emerging applications face skepticism too, including smartphone payment systems, new fintech services and facial recognition techniques. Europe's fintech market, concentrated in London, was only $6 billion in 2015, while China's was $102 billion in 2015 (Guarascio 2017). Europe's digital "recession", in short, closely corresponds with a mindset that de-emphasizes the opportunities and benefits of digitalization.

Framing Internet Developments in China

Domestic discourses about digital technologies are, under conditions of globalization, intimately connected to perceptions about models of digitalization in other nations. While Silicon Valley serves as the global gold standard for innovational aspirations and policy recommendations (Keese 2016), many view China as the worst nightmare. The negative framing of China's Internet is the corollary of the pessimistic outlook mentioned above. For a German sociologist, China represents all that could go wrong with digital technologies (Welzer 2015). The most important

China think tank in Europe, the Merics, sees "IT-backed authoritarianism" buttressing the power of China's party-state (Meissner and Wübekke 2016). Joschka Fischer, former German Foreign minister, warns against a "digitally supported Leninism" (Mayntz 2018; Heilmann 2017). German media portray China as an authoritarian political system with Internet surveillance, censorship and control as a central theme. The number of news reports between 2010 and 2017 that feature the terms "China", "Internet" and "dictatorship" and that frequently used notions including "IT dictatorship", (Deutschlandfunk 2017) "Internet dictatorship", "data dictatorship" (Mayer-Kuckuk 2017) and "Big Data dictatorship" (Assheuer 2017) has increased sharply (see Table 13.2).

The fast-growing use of this peculiar triadic connection— China/Internet/dictatorship—is interesting for various reasons. A comprehensive 2008 study on German media reports about China notes that only 0.6% of all articles referred to China's Internet censorship (Richter and Gebauer 2010). A decade ago, to begin with, any combination of these terms had seemed contradictory. The Internet was linked to democracy as a generation of Californian entrepreneurs and utopian visionaries aspired to use cyberspace as a catalyst for freedom and democratization worldwide (Meng 2010; Deibert and Rohozinski 2010). The Internet, according to the hope of many observers, would help to usher in China's progressive democratization or, at least, enable activists to increase the pressure on the regime (Yang 2009).

Yet, today's China provides ample evidence for the opposite, namely, the potential realization of a "totalitarian Internet society" (Lüdke 2017). The leadership of the Communist Party indeed uses social media, Internet platforms and big data infrastructures creatively and without democratic restrictions. When access to foreign information sources is blocked, censorship automated, mutual surveillance and universal scoring is normalized via all-encompassing platforms, there appears to be no escape from "total control" (Stockmann and Gallagher 2011; Siemons 2018). Even abroad, if Chinese tourists enjoy cheap Internet access via mobile roaming, they remain virtually inside the great firewall while walking through Rome, New York or Tokyo. Hence, China came to symbolize all that went wrong. Naturally, the pessimistic and alarmed voices in Western media are growing because the consequences of China's digitalization do not only bury progressive aspirations but also contradict the values of liberal democracies, such as data privacy and freedom of speech—not to mention the "right to be forgotten" or the idea of digitally opting out.

To make sense of these developments, journalists frequently invoke Orwell's vision of a totalitarian society—articulated in his famous book *1984*—to signify the radical difference of a morally appalling and dangerous "other" (Hoffmann 2018). Yet, the question that should be raised is whether this sort of black-and-white narrative does justice to the entire cyber reality in China. Digitalization evolved through both tensions and collaborations among state agencies and private Internet companies. The result thus far is a marriage of convenience: a surveillance-obsessed state and data-hungry platform enterprises working hand-in-hand to create a "networked authoritarianism" (Deibert et al. 2010). By 2018, China had more than a

Table 13.2 The theme of China's digital dictatorship in German online news

2010	2011	2012	2013	2014	2015	2016	2017	Overall hits	Terms
180	377	415	652	1060	1310	2280	4020	36,400	"China, Dictatorship"
50	99	168	298	520	553	1100	2190	19,200	"China, Internet, Dictatorship"
13	15	32	60	79	130	408	480	3000	"China, Orwell"

Own Source 2018
Note The results of these searches might still include data noise

billion Internet users, of which the majority access online services through mobile phones. WeChat—China's leading instant online messenger and a subsidiary of Tencent—which combines multiple chat, payment, work and entertainment features within a single ecosystem, has more than one billion daily users. In short, through the widespread use of smartphone applications, unencrypted communication, digital surveillance and censorship became socially acceptable and pervasive, as well as inexpensive for the party state.

Online finance services generate another great pool of data. Hundreds of millions buy on Alibaba's platform and use Alipay, a part of Alibaba group, to electronically pay for anything from ice cream to donations for the homeless. Cash payment has all but vanished and face recognition techniques have rapidly become normal in commercial and public settings. The use of cameras, censors and cookies goes beyond data collection in policing, traffic, communication and finance (Mozur 2018). In some Chinese factories, workers' brain activities are measured to improve production processes (Chen 2018a, b). There are schools with cameras in classrooms that stream their live-feed in real time online. The Chinese state apparatus 2.0 has—at least in principle—access to all the data generated by these technologies. As a result, an ideologically vibrant regime can rely on "algorithmic governance" that is capable of what was only a dream to the East German Stasi—it has automatic, instant and unrestricted access to even the tiniest details of the financial, communicative and emotional life of its citizens. New propaganda and persuasion tactics diminished earlier hopes that internet activism could nudge China's political system to become more transparent, participatory and democratic (Creemers 2017; Repnikova and Fang 2018).

So, has Orwell's dark vision even been surpassed? The ultimate "smoking gun" for observers lies in the official plan of the National and Development and Reform Commission to establish a national "social credit system". The latter aims at rating the behavior of all persons, companies and other actors in four areas: "administrative affairs, commercial activities, social behavior, and the law enforcement system" (ASAN 2017). Rooted in traditional Confucian ideas of moral control and collective management, the social credit system draws on hitherto unavailable data sources and big data methods to govern China's economy (Creemers 2018; Meissner 2017). Reducing citizens to a single score that represents their overall trustworthiness dramatically redefines the modus of digital personhood. Individuals with a low ranking could become second class citizens overnight, barred from traveling, online access and other public services (Botsman 2017).

The wildest frontier of authoritarian digitalization, however, lies in Western China. In the remote province of Xinjiang, layered measures of a repressive police state and a system of 'educational camps' blend with the ubiquity of the latest surveillance high-tech. It is here, in China's remote West, where tech startups—far away from their glittering offices in Shanghai or Shenzhen—experiment with new face recognition, surveillance software and AI applications that are employed against minorities without public scrutiny or legal restrictions (Economist 2018; Millward 2018). China's recent AI hype is organically linked to a new type of highly intrusive and repressive techno-policies in Xinjiang.

The Orwellian Narrative Revisited

Despite these often literally unspeakable realities of digitally enhanced abuse and repression, the Chinese experiences with digital technologies cannot be purely described through a register of control. Digital China as a whole is not a dystopian case. To begin with, opinion polls show that a majority, notwithstanding online protests and offline resistance, has a positive view of government. Although some of the polls might be misleading, Chinese society displays a high score of trust in its political institutions and the achievements of the Communist party (Chen 2017; Zhao and Hu 2017; Tang 2018). Modern infrastructure and digital technologies are widely seen as a sign of the success and efficiency of one-party rule.

Moreover, against the seemingly persuasive framing of Orwellian narratives, one should not forget that digitalized services made life extremely convenient for the average (urban) person. Chinese users were extremely quick to adopt digital technologies. They did not care much for privacy and data security. The digitalization of entertainment and work environments, service sectors and education unfolded perhaps nowhere as quickly and comprehensively as in Chinese society. This observation holds across different generations, as well as urban and rural settings (Delisle et al. 2016). The majority of Chinese can no longer imagine missing the mobile phone-based options for mobility, communication, shopping, dating or gaming, while the pervasive practice of Internet censorship has been largely normalized and is—even among critical voices—a less contentious issue (Wang and Mark 2015). The benefits dominate public perception. The competitive effects of commercial online services in banking, for instance, improved the effectiveness of traditional banks (Ye 2017). In the field of charity, to name another example, the new charity law—in combination with digital platforms for NGOs—rendered a corruption-infested sector into a more transparent and professional operation.[1] Finally, platforms such as Ant-Financial may also help to change behavior towards a more environmentally friendly pattern (Yang et al. 2018; Wu 2018).

Positive outcomes of digitalization are also obvious in the vast rural areas where, despite rapid urbanization, still more than 40% of the Chinese population lives. The exploding fintech branch—which now has more clients than traditional banks—made small loans and other financial products accessible to poor families and small business for the first time in a fast, affordable and rational manner (Shen 2017). Automated financial credit scoring replaces the bureaucratic and complicated credit procedures. In 2017, one third of Alipay's users were from rural areas. JD Finance, which belongs to JD Group, China's largest online seller, operated in more than 300,000 villages to provide online financial services. Large online marketplaces empower small farmers in remote parts of China to sell directly into major cities (Zhou and Hua 2013). CreditEase, a fintech company, pioneered the leasing of machines, tractors and even livestock (Shen 2017). Online education platforms, meanwhile, offer reliable and affordable language courses for children living outside the major metropolitan centers. Digital ecosystems offer various education opportunities in

---

[1]Interviews and observations in Shanghai 2017 and 2018.

M. Mayer

third-tier cities and significantly contribute to the inclusion of migrants and rural denizens in China (Arnold and Willis 2016; Aveni and Roest 2018).

A more nuanced look at the social credit system, which appears to be an overpowering totality, is also necessary. The first observation is perhaps counter-intuitive for outsiders. A key rationale of this system is to re-establish basic trust and guaranty responsible behaviour among actors in commercial, legal, health and administrative settings. China's breakneck economic growth was always connected with major scandals ranging from food safety and corruption in the real estate sector to financial Ponzi schemes. In the early phase of e-commerce and virtual communication, too, online fraud and other cybercrimes have skyrocketed (Liang and Lu 2010). In 2010, trust among the Chinese reached its lowest level according to researchers from the Chinese Academy of Social Sciences (Dan 2013). And many experts argue that transparent rating and scoring systems, based on gathering big data from different sources, will help to restore interpersonal and public trust in cyberspace, as well as offline (Wang 2017; Shen 2016). The system's Chinese name (*shehui xinyong tixi*) indirectly conveys a Confucian connotation, alluding to "sincerity", "integrity" and "honesty" (*xin*) behind the modern word of "credit" (*xinyong*) (Meissner and Wuebekke 2016). Given the similarities to other credit ratings in the West, the legitimate motivation to restore societal trust cannot simply be ruled out. Recent polls show that Chinese users indeed value the trust- and transparency-enhancing function of different rating systems (Kostka 2018).

Second, it is doubtful whether the central government will ever be able to realize its comprehensive vision from 2014 (Chorzempa et al. 2018). Major obstacles include bureaucratic infighting about data and responsibilities and the difficulty of working closely together with private companies in order to share massive amounts of data instantly. For instance, in cities such as Rongcheng (Shandong province) that test a local version of the system, the collected data stem only from state agencies. No direct link to private companies or neighborhood committees (*shequ*) exists.[2] The practical problems for data pooling run in parallel with fragmented censorship practices. A comparison of Chinese video platforms, for example, shows that there is no monolithic system of censorship. Different private platforms have diverging approaches (Knockel 2015). A biotope of different, partly overlapping or conflicting financial, social and moral scoring systems that might never be fully integrated is more likely to emerge. Among the operators of these systems are private, public and commercial actors, including courts, railways and airlines, banks, NGOs and platform companies, as well as the police and state security apparatus (Creemers 2018; Daum 2017). While this scenario differs substantially from a single, unified social credit system, the effects of such a fragmented landscape of digital scoring are perhaps even less desirable.

What is clear, however, is that the three-way interplay between party-state, platform companies and citizen-consumers—the mediated ubiquitous datafication infrastructure—will remain a complex and partly unpredictable coevolution under the conditions of authoritarian digitalization (Qin et al. 2017; Negro 2017). Hence, the

---

[2]Results from author's field work in Rongcheng.

certainty and moral high ground that media reports routinely express when looking at China's digital advancements neither allow for an objective account nor generate productive insights. The moralizing stance is instead indicative of a tendency, widespread across Europe, to undervalue digital technologies as actual improvements. Consequently, positive sides of China's digitalization are systematically overlooked, or downplayed, and negative sides occupy the lion's share of attention. Furthermore, attempts to compare China with other countries are largely missing, although they would be necessary to replace digital Orientalism with a more comprehensive perspective.

Comparative Perspectives on Digital Surveillance

China's authoritarian digitalization needs to be contextualized by international comparison. I cannot satisfyingly undertake this task here due to space constraints. Yet, a few observations and examples shall suffice for our purpose. One option is to compare China alongside other authoritarian system such as Russia, Iran or Saudi Arabia. Another one, as suggested by Marcus Michaelsen and Marlies Glasius, is to compare the spread of a set of "authoritarian practices" across *different* political systems (Michaelsen and Glasius 2018). The latter approach reveals that there are various commonalities between China, the EU and the U.S. In democratic countries, new forms of digitally enabled governance emerged that lack accountability and legitimacy (Leggett 2014; Danaher et al. 2017). Comparing 14 different practices of surveillance, manipulation and control reveals a great deal of overlap, reaching from technical support of Internet surveillance abroad, to the massive breach and misuse of private data, to automated algorithmic censorship (albeit with different intensity) and large-scale ubiquitous data collection (Table 13.3).

Observation that surveillance, datafication and various types of "soft" manipulation—e.g., nudging, scoring (see Yeung 2017; Dencik et al. 2019; Bradshaw and Howard 2018) or even big data supported remote killing (Taibbi 2018)—are normalized across the board presents us with a tricky puzzle. Science and technology studies offer at least two theoretical lenses to explain the phenomenon of global communalities. We could speculate that similar authoritarian practices arise from the specific "affordances" of the involved technologies. Affordances are, everywhere, more or less the same regardless of the nature of political or commercial digital technology systems (Conole and Dyke 2004; Autio 2018; Dahlberg 2011). The related digital authoritarian practices can therefore easily travel between different political cultures. Others, in contrast, argue that "digital totalitarianism" (Taz 2017) is created by the mutual constitution of digital technologies, algorithmic governance *and* security discourses that evolve in democracies and authoritarian regimes alike (Lehr 2019; Amoore 2013; Crawford 2019). But, whatever the answer to this question is— whether one prefers a techno-determinist or a co-productive logic—the Orwellian narrative becomes less convincing against the backdrop of comparative reasoning.

Table 13.3 Comparison of "authoritarian digital practices" used for surveillance, manipulation and control over Netizens in the U.S., China and the EU

Examples of authoritarian digital practices	EU	USA	China
Systematic use of techniques of predictive policing			
Party-led censorship (media companies)			
Brain hacking, dopamine labs (companies)			
Automated algorithmic censorship (social media/Internet)			
Automated, data-based killing			
Public lists (with felons' wrongdoings)			
Goal to build a comprehensive national social credit system			
Algorithmically generated content bubbles in social networks			
Unrestricted or legally non-transparent access to personal private online data			
Illicit selling/misuse of private data			
Targeted political online campaigns			
Automated upload filters / blocked content			•
Support of internet surveillance abroad	•	•	•
Ubiquitous data collection by devices, intelligence services and enterprises			
Systematic filtering and blocking of major foreign information sources			

Source Authors own compilation from open-access sources

The intellectual prudence of a comparative approach challenges simplifying world views that naively see "digital China" as the "evil other". To contextualize China's realities makes parallels and similarities, as well as differences in digitalization visible. As a result, a more accurate, yet complex picture arises that pinpoints disconcerting global commonalities and transmutations. Ultimately, understanding the risks, dangers and opportunities of authoritarian digitalization is key to rethinking Sino-German cooperation in a global data economy and digitalized industrial production.

Conclusion

This chapter argues that—in portraying China—many observers fall into the trap of digital orientalism. Digital China is framed as the negative, the failure and a threat. Yet, this sort of prism is biased. It overlooks positive outcomes resulting from digitalization—and can't explain the reason for why the absolute majority of Chinese don't resist authoritarian digitalization. Indeed, regarding digital rating systems, research by Genia Kostka shows that "urban Chinese have an overwhelmingly positive view of commercial and government-run systems" (Kostka 2018). My point here, analytically speaking, is to stress the need for a thorough contextualization of the Chinese experience. One such option is to interrogate digital authoritarian practices across the world. Using a comparative approach, of course, does not imply excusing or normalizing the misuse of digital technologies in China. However, a comparative analysis reveals that digital technologies are increasingly employed in authoritarian ways in the U.S. and EU. Various similarities with China exist. It follows that democratic countries, as they critique surveillance in China, have to review their own practices more carefully.

China's digital authoritarianism has serious implications for Sino-European cooperation, both politically and economically. At the political level, competition will intensify, because China's ideology is now supported by, and intertwined with, a strong technological component and expertise attractive to other countries. Concerning trade and investment, the commercial success of digital authoritarianism doesn't just pose a huge challenge for Europe's own underdeveloped digital capitalism (Lobo 2018). Data localization requirements, deep-seated mistrust and conflicting laws will complicate the establishment of Industry 4.0 ecosystems and cross-border IT infrastructures. The relevant rules, standards and administrative procedures for data storage and security are already under negotiation. But there are further complicated moral and political issues. If, for instance, German autonomous cars are to drive in China, German companies will become directly complicit in a massive surveillance operation (Fasse and Scheuer 2016). The unfolding US-Sino conflict about Huawei products such as 5G network technology, that has greatly escalated in the recent month, puts additional pressure on European countries to reconsider their reliance on China built digital infrastructures. However, the European perspective also needs to focus on positive advances and examples of creativity. China's administrative state has digitally reinvented itself, and there are further lessons for European policymakers from the Chinese use of digital technologies in public security and transport, health, education, competition, inclusive finance and entrepreneurship. Assessing the fruits of China's massive and multi-faceted drive for digital innovations requires an open-minded and balanced analysis.

If it is true that the process of digitalization is still in an early stage, then attempts to determine whether its results will be overwhelmingly beneficial or harmful are premature. Meanwhile, the discursive shift away from the Internet democratization thesis to growing anxieties about expanding digital authoritarianism indicates that the pendulum swings back. The Orwellian narratives about China's Internet revolution express a legitimate concern: the utopian story of the Internet's beginning might turn completely and irreversibly into a nightmare. And this nightmare might very well become a reality—perhaps with a social credit system extending its effects beyond China's borders (Hoffmann 2018). But, it is worthwhile to repeat for all techno-skeptics among us that digital technologies are as much made from materials and powered by algorithms as they are social and cultural constructs. Digital technologies and Big Data have no inherent tendency that inevitably renders them either authoritarian tools or instruments of democratization. Hence, to think of positive imaginaries of the future is, at this juncture, more important than ever before (Broy and Precht 2017; Dencik 2018). In other words, we need many more systematic efforts to develop new ideas and practices of shaping and redesigning current sociotechnical formations of digital technologies in Europe, China and the U.S. in the light of our visions for a thriving and better society.

References

Articles

Autio, E., et al. (2018). Digital affordances, spatial affordances, and the genesis of entrepreneurial ecosystems. *Strategic Entrepreneurship Journal, 12*(1), 72–95.

Chakravorti, B., et al. (2015). Europe's other crisis: A digital recession. *Harvard Business Review* https://hbr.org/2015/10/europes-other-crisis-a-digital-recession.

Champion, M. (2018). Europe wants a robot army to challenge the U.S. and China on AI, Bloomberg, 25 https://www.bloomberg.com/news/articles/2018-04-25/europe-wants-a-robot-army-to-challenge-the-u-s-and-china-on-ai.

Chen, D. (2017). Local distrust and regime support: Sources and effects of political trust in China. *Political Research Quarterly, 70*(2), 314–326.

Creemers, R. (2017). Cyber China: Upgrading propaganda, public opinion work and social management for the twenty-first century. *Journal of Contemporary China, 26*(103), 85–100.

Dahlberg, L. (2011). Re-constructing digital democracy: An outline of four 'positions'. *New Media & Society, 13*(6), 855–872.

Danaher, J., et al. (2017). Algorithmic governance: Developing a research agenda through the power of collective intelligence. *Big Data and Society, 4*(2), 2053951717726554.

Deibert, R., Palfrey, J., Rohozinski, R., & Zittrain, J. (2010). Access controlled.

Deibert, R., & Rohozinski, R. (2010). Liberation versus control: The future of cyberspace. *Journal of Democracy, 21,* 43–57.

Dencik, L. (2018). Surveillance realism and the politics of imagination: Is there no alternative? *Krisis: Journal for Contemporary Philosophy, 2018*(1), 31–43.

Dencik, L., Redden, J., Hintz, A., & Warne, H. (2019). The 'golden view': Data-driven governance in the scoring society. *Internet Policy Review*, [online] *8*(2). Available at: https://policyreview.info/articles/analysis/golden-view-data-driven-governance-scoring-society [Accessed: 7 Sep 2019].

Leggett, W. (2014). The politics of behaviour change: Nudge, neoliberalism and the state. *Policy & Politics, 42*(1), 3–19.

Liang, B., & Lu, H. (2010). Internet development, censorship, and cyber crimes in China. *Journal of Contemporary Criminal Justice, 26*(1), 103–120.

Mager, A. (2017). Search engine imaginary: Visions and values in the co-production of search technology and Europe. *Social Studies of Science, 47*(2), 240–262.

Meng, B. (2010). Moving beyond democratization: A thought piece on the China internet research agenda. *International Journal of Communication, 4,* 501–508.

Michaelsen, M., & Glasius, M. (2018). Authoritarian practices in the digital age—Introduction. *International Journal of Communication, 12,* 3792.

Palat, R. A. (2000). Beyond orientalism: Decolonizing Asian studies. *Development and Society, 29*(2), 105–135.

Qin, B., Strömberg, D., & Wu, Y. (2017). Why does China allow freer social media? Protests versus surveillance and propaganda. *Journal of Economic Perspectives, 31*(1), 117–140.

Repnikova, M., & Fang. K. (2018). Authoritarian participatory persuasion 2.0: Netizens as thought work collaborators in China. *Journal of Contemporary China,* 1–17.

Said, E. W. (1985). Orientalism reconsidered. *Race & Class, 27*(2), 1–15.

Scott, M. (2017, October 17). In Europe's digital race, the winner is the United States. i, https://www.politico.eu/article/digital-innovation-europe-digital-startups-venture-capital-united-states-tech-london-berlin-paris/.

Shen, Y. (2016). Shehui zhili Chuangxin yu shehui xinyong tixi jianshe di. *Hehai daxue xuebao, 18*(3), 72–77.

Snow, J. (2018, February 14). China's AI startups scored more funding than Americas last year. *MIT Technology Review*, https://www.technologyreview.com/the-download/610271/chinas-ai-startups-scored-more-funding-than-americas-last-year/.

Stockmann, D., & Gallagher, M. E. (2011). Remote control: How the media sustain authoritarian rule in China. *Comparative Political Studies, 44*(4), 436–467.

Tang, W. (2018). The 'Surprise' of authoritarian resilience in China. *American Affairs, 2*(1) https://americanaffairsjournal.org/2018/02/surprise-authoritarian-resilience-china/.

Wu, P. (2018, January 25). Ant financial and the greening of Fintech. *The Diplomat*, https://thediplomat.com/2018/01/ant-financial-and-the-greening-of-fintech/.

Yang, Y. (2018a, July 05). China surpasses North America in attracting venture capital funding for first time as investors chase 1.4 billion consumers. *South China Morning Post*, https://www.scmp.com/tech/article/2153798/china-surpasses-north-america-attracting-venture-capital-funding-first-time.

Yang, Z., et al. (2018). Switching to green lifestyles: Behavior change of ant forest users. *International Journal of Environmental Research and Public Health, 15*(9), 1819.

Yeung, K. (2017). 'Hypernudge': Big data as a mode of regulation by design. *Information, Communication & Society 20*(1), 118–136.

Zhao, D., & Hu, W. (2017). Determinants of public trust in government: Empirical evidence from urban China. *International Review of Administrative Sciences, 83*(2), 358–377.

Zilgalvis, P. (2014). The need for an innovation principle in regulatory impact assessment: The case of finance and innovation in Europe. *Policy & Internet, 6*(4), 377–392.

Books

Amoore, L. (2013). *The politics of possibility: Risk and security beyond probability*. Durham, London: Duke University Press.

Beise, M., & Schäfer, U. (2016). *Deutschland digital: Unsere Antwort auf das Silicon Valley*. Frankfurt a.M.: Campus.

Borchardt, A. (2015). *Das Internet zwischen Diktatur und Anarchie: zehn Thesen zur Demokratisierung der digitalen Welt*. Süddeutsche Zeitung: Edition Streitschrift; Frank Schirrmacher, Technologischer Totalitarismus: Eine Debatte (edition suhrkamp).

Castells, M. (1997). *Power of identity: The information age: Economy, society, and culture*. Cambridge, MA: Blackwell Publishers, Inc.

Conole, G., & Dyke, M. (2004). What are the affordances of information and communication technologies, 113–124.

Delisle, J., Avery Goldstein, A., & Yang, G. (Eds.). (2016). *The internet, social media, and a changing China*. Philadelphia: University of Pennsylvania Press.

Han, B.-C. (2013). *Im Schwarm. Ansichten des Digitalen*. Berlin: Matthes & Seitz Verlag.

Jasanoff, S. (2011). *Designs on nature: Science and democracy in Europe and the United States*. Princeton: Princeton University Press.

Keese, C. (2016). *Silicon Germany: Wie wir die digitale Transformation schaffen*. München: Albrecht Knaus Verlag.

Lehr, P. (2019). *Counter-terrorism technologies a critical assessment*. Switzerland: Cham.

Lynch, M. P. (2016). *The Internet of us: Knowing more and understanding less in the age of big data*. New York: Liveright.

Morozov, E. (2011). *The net delusion: How not to liberate the world*. Penguin UK.

Negro, G. (2017). *The Internet in China: From infrastructure to a nascent civil society*. Cham: Springer.

Wang, D., & Mark, G. (2015). Internet censorship in China: Examining user awareness and attitudes. *ACM Transactions on Computer-Human Interaction (TOCHI), 22*(6), 31.

Welzer, H. (2015). *Die smarte Diktatur. Der Angriff auf unsere Freiheit*. S. Fischer, Frankfurt a. M.

Yang, G. (2009). *The power of the Internet in China: Citizen activism online*. New York: Columbia University Press.

Media

Assheuer, T. (2017, November 29) Die Big-Data Diktatur. *ZEIT Online*, https://www.zeit.de/2017/49/china-datenspeicherung-gesichtserkennung-big-data-ueberwachung.

Aveni, T., & Roest, J. (2018, January 11). What can mobile money make possible? China has many answers. *CGAP*, https://www.cgap.org/blog/what-can-mobile-money-make-possible-china-has-many-answers.

Botsman, R. (2017, October 21). Big data meets big brother as China moves to rate its citizens. *WIRED*, http://teomeuk.s3.amazonaws.com/wp-content/uploads/2017/11/05/Wired_Big_data_meets_Big_Brother_as_China_moves_to_rate_its_citizens.pdf.

Broy, M., & Precht, R. D. (2017, February 9). Daten essen Seele auf. *Die Zeit*, https://www.zeit.de/2017/05/digitalisierung-revolution-technik-seele-menschen-grundrechte.

CBInsights. (2018). Research briefs, Feb. 27, 2018, The Race for AI: Google, Apple In a rush to grab artificial intelligence startups.

Chen, S. (2018a, July 30). Artificial intelligence, immune to fear or favour, is helping to make China's foreign policy. *South China Morning Post*, https://www.scmp.com/news/china/society/article/2157223/artificial-intelligence-immune-fear-or-favour-helping-make-chinas.

Chen, S. (2018b, April 29). 'Forget the Facebook leak': China is mining data directly from workers' brain on an industrial scale. *South China Morning Post*, https://www.scmp.com/news/china/society/article/2143899/forget-facebook-leak-china-mining-data-directly-workers-brains.

Dan, H. (2013, February 18). Trust among Chinese 'drops to record low'. *China Daily*, http://www.chinadaily.com.cn/china/2013-02/18/content_16230755.htm.

Deutschlandfunkt Weltzeit. (2017 September 05). Auf dem Weg in die IT-Diktatur. *Podcast*, https://www.deutschlandfunkkultur.de/chinas-sozialkredit-system-auf-dem-weg-in-die-it-diktatur.979.de.html?dram:article_id=395126.

Daum, J. (2017, December 24). China through a glass, darkly, *ChinaLawTranslate*, https://www.chinalawtranslate.com/seeing-chinese-social-credit-through-a-glass-darkly/?fbclid=IwAR1Ui9GH-WMRCMn9_aXvpOa-wPFo2zX5ep0shobEy3gV6QF3vz4hJ_enkZQ&lang=en.

Fasse, M., & Scheuer, S. (2016, October 13). China Plans E-Car Surveillance, https://www.handelsblatt.com/today/companies/delicate-data-china-plans-e-car-surveillance/23541570.html?ticket=ST-2988627-QUKWfChxkL5QJcMvGKfe-ap4.

Guarascio, F. (2017, September 11). EU weighs strategy to compete in Fintech with global rivals. *Reuters*, https://www.reuters.com/article/us-eu-ecofin-fintech/eu-weighs-strategy-to-compete-in-fintech-with-global-rivals-idUSKCN1BM1XL.

Lobo, S. (2018, July 11). Das autoritäre Erfolgsmodell. *Spiegel Online*, http://www.spiegel.de/netzwelt/netzpolitik/china-wird-bei-der-digitalisierung-den-ton-angeben-kolumne-a-1217577.html.

Lüdke, S. (2017). China plant die Komplett-Überwachung—So soll sie funktionieren. Bento, http://www.bento.de/politik/china-plant-mit-dem-social-credit-system-die-komplett-ueberwachung-so-soll-sie-funktionieren-1891016/.

Macaes, B. (2018, March 19). Europe's AI delusion. *Politico*, https://www.politico.eu/article/opinion-europes-ai-delusion/.

Mayer-Kuckuk, F. (2017, December 05). Chinas neue Daten-Diktatur, *Frankfurter Rundschau*, http://www.fr.de/kultur/netz-tv-kritik-medien/netz/ueberwachung-chinas-neue-daten-diktatur-a-1400913.

Mayntz, G. (2018, March 6). Aufschwung in China Joschka Fischer blickt besorgt in Europas Zukunft. *Rundschau Online*, https://www.rundschau-online.de/politik/aufschwung-in-china-joschka-fischer-blickt-besorgt-in-europas-zukunft-29824812.

Millward, J. A. (2018, February 3). What it's like to live in a surveillance state. *The New York Times*, https://www.nytimes.com/2018/02/03/opinion/sunday/china-surveillance-state-uighurs.html.

Mozur, P. (2018, July 8). Inside China's dystopian dreams: AI, shame and lots of cameras. *The New York Times*, https://www.nytimes.com/2018/07/08/business/china-surveillance-technology.html.

Shen, L. (2017, December 5). How this Chinese Fintech Company Is innovating by leasing cows. *Fortune*, http://fortune.com/2017/12/05/china-fintech/.

Siemons, M. (2018, May 11). Die totale Kontrolle. *Frankfurter Allgemeine Zeitung*, https://www.faz.net/aktuell/feuilleton/debatten/chinas-sozialkreditsystem-die-totale-kontrolle-15575861.html?printPagedArticle=true#pageIndex_0.

Taibbi, M. (2018, July 19). How to survive America's Kill list. *Rolling Stone*, https://www.rollingstone.com/politics/politics-features/how-to-survive-americas-kill-list-699334/.

Taz. (2017). Der digitale Totalitarismus, 04.01.2017 http://www.taz.de/!5367252/.

The Economist. (2018, May 31). China has turned Xinjiang into a police state as no other. *The Economist*, China has turned Xinjiang into a police state as no other, https://www.economist.com/briefing/2018/05/31/china-has-turned-xinjiang-into-a-police-state-like-no-other.

Vincenti, D. (2018). Return of the JEDI, European disruptive technology initiative ready to launch, 14. März 2018, EURACTIV.

Wang, S. (2017). Zhima kaimen nide Xinyong you jifen? *Shanghai Xinzi hua, 7*(3), 76–78. (in Chinese).

Yang, X. (2018b, September 19). Europa ist abgemeldet. *ZEIT Online*, https://www.zeit.de/2018/39/weltkonferenz-kuenstliche-intelligenz-shanghai-technologie-china-usa/komplettansicht.

Ye, J. (2017, August 11). Big banks on notice that they're losing ground to China's Fintech giants. *South China Morning Post*, https://www.scmp.com/business/companies/article/2106105/big-banks-fear-theyre-losing-ground-chinas-fintech-giants.

Zhou, J., & Hua, X. (2013, January 18). Digital financial inclusion growing in China. *China Daily*, http://www.chinadaily.com.cn/a/201801/13/WS5a59b862a3102c394518f050.html.

Studies

Arnold, J., & Willis, S. (2016, April 5). Improving access usage of financial products and services in China. *Center for Financial Inclusion*, https://www.centerforfinancialinclusion.org/improving-access-and-usage-of-financial-products-and-services-in-china.

ASAN Institute for Policy Studies. (2017, February 28). Orwell's Nightmare: China's Social Credit System, http://en.asaninst.org/contents/orwells-nightmare-chinas-social-credit-system/.

Bradshaw, S., & Howard, P. (2018). Challenging truth and trust: A global inventory of organized social media manipulation. Oxford University Internet Institute, http://comprop.oii.ox.ac.uk/wp-content/uploads/sites/93/2018/07/ct2018.pdf.

Chakravorti, B., & Chaturvedi, R. S. (2017, July). Digital planet 2017 how competitiveness and trust in digital economies vary across the world, The Fletcher School, Tufts University July 2017, https://sites.tufts.edu/digitalplanet/files/2017/05/Digital_Planet_2017_FINAL.pdf.

Chorzempa, M., Triolo, P., & Sacks, S. (2018). 18–14 China's social credit system: A mark of progress a threat to privacy? Policy brief. Peterson Institute for International Economics, https://piie.com/system/files/documents/pb18-14.pdf.

Creehan, S., & Borst, N. (2017). Asia's Fintech Revolution, 2017, Asia Program, Federal Reserve Bank of San Francisco.

Creemers, R. (2018, May 9). China's social credit system: An evolving practice of control. Available at SSRN https://papers.ssrn.com/sol3/papers.cfm?abstract_id=3175792.

Crawford, M. B. (2019). Algorithmic governance and political legitimacy, in American affairs, Summer 2019/Vol. III, Number 2. https://americanaffairsjournal.org/2019/05/algorithmic-governance-and-political-legitimacy/.

Evans, P. C., & Gawer, A. (2016). The rise of the platform enterprise: a global survey. The emerging platform economy series No. 1 (The center for global enterprise).

Hoffmann, S. (2018). What's the problem, Australian Strategic Policy Institute, https://www.aspi.org.au/report/social-credit.

Heilmann, S. (2017, September 29). Big data reshapes China's approach to governance. *Financial Times*, https://www.merics.org/de/node/3641.

Knockel, J., et al. (2015). Every rose has its thorn: Censorship and surveillance on social video platforms in china. *5th USENIX workshop on free and open communications on the internet (FOCI 15)*.

Kostka, G. (2018, September 17). Chinas soziale Bonitätssysteme sind—noch—beliebt. *MERICS*, https://www.merics.org/de/blog/chinas-soziale-bonitaetssysteme-sind-noch-beliebt-englisch.

McKinsey. (2017). Digitally-enabled automation and artificial intelligence: Shaping the future of work in Europe's digital front-runner, https://www.mckinsey.com/~/media/mckinsey/featured% 20insights/europe/shaping%20the%20future%20of%20work%20in%20europes%20nine% 20digital%20front%20runner%20countries/shaping-the-future-of-work-in-europes-digital-front-runners.ashx.

Meissner, M., & Wübekke, J. (2016). IT—Backed authoritarianism: Information technology enhances central authority and control capacity under Xi Jinping". In H. Sebastian & S. Matthias (Eds.), *China's core executive, leadership styles, structures and processes under Xi Jinping*. Mercator Institute for China Studies, 52.

Meissner, M. (2017). China's social credit system a big-data enabled approach to market regulation with broad implications for doing business in China," Merics China Monitor, May 24, 2017.

Richter, C., & Gebauer, S. (2010). *Die China-Berichterstattung in deutschen Medien. Anhang-Band zur Studie* (p. 48). Berlin: Heinrich Böll Stiftung.

Dr. Maximilian Mayer is currently assistant professor at the University of Nottingham Ningbo China. He is research fellow at Renmin University, Beijing (2018–2020), worked as Research Professor at Tongji University, Shanghai (2015–2018) and was senior researcher at the Munich Center for Technology in Society, Technical University Munich (2018–2019). Maximilian holds a master degree from Ruhr University Bochum and obtained his Ph.D. at Bonn University. He joined Bonn University's Center for Global Studies (CGS) in October 2009 where he worked as managing assistant, senior fellow and lecturer between 2009 and 2015. His research interests include the global politics of science, innovation, and technology; China's foreign and energy policy; global energy and climate politics; theories of International Relations. Maximilian presents regularly at international conferences, publishes his research in peer-reviewed journals, and is co-editor of the two-volume work *The Global Politics of Science and Technology* (http://www.springer.com/social+sciences/political+science/book/978-3-642-55006-5) (Springer 2014), coeditor of *Art and Sovereignty in Global Politics* (https://link.springer.com/book/10.1057%2F978-1-349-95016-4) (Palgrave, 2016), and edited *Rethinking the Silk-Road: Chinas Belt and Road Initiative and Emerging Eurasian Relations* (https://www.palgrave.com/de/book/9789811059148# aboutBook) (Palgrave, 2018). Maximilian was a visiting scholar at Harvard Kennedy School, Program on Science, Technology and Society, and section co-chair of STAIR (http://www.isanet.org/ISA/Sections/STAIR.aspx) (Science, Technology, Arts and international relations) of the International Studies Association (2015–2017) and STAIR program chair (2014–2015).

Chapter 14
Digitalization and Public Policy—Conceptualizing a New Space

Aaron Maniam

World is crazier and more of it than we think
Incorrigibly plural …
—Louis MacNeice, "Snow"

Introduction

We need better conceptual explanations for the widely differing public sector experiences with digital technology—at national, regional, municipal and other levels. While perhaps not "incorrigibly plural", the range of governments' lived realities with digitalization has at least some roots in how core ideas like digital resources and digital governance are defined. This chapter examines how more nuanced and textured definitions are gradually emerging from gaps in the existing public administration literature, in the process raising important questions for policy formulation and delivery.

The Current State of Play

More and more governments are adopting digital technologies in policy development, internal administration/management and service delivery. Some writers even note that "it is difficult to think of a public problem or government service that does not

Policymaker by profession and joined the Singapore government in 2004, Author.

A. Maniam (✉)
Blavatnik School of Government, University of Oxford, Oxford, Great Britain
e-mail: Aaron.maniam@bsg.ox.ac.uk

The Birthday Collective, Singapore, Singapore

© Springer Nature Switzerland AG 2020
D. Feldner (ed.), *Redesigning Organizations*,
https://doi.org/10.1007/978-3-030-27957-8_14

involve (such technologies) in some substantial way" (Gil-Garcia et al. 2017). As early as 2006, Dunleavy et al. noted that in limiting cases, some agencies "become their websites … the electronic form of the organization increasingly defines the fundamentals of what it is and does". Today, government services are increasingly delivered via digital platforms (e.g., smart telephones, personal computers), and some websites are becoming increasingly passé.

The manifestations of public sector digitalization have been highly varied. In some countries, governments use digital technology to acquire data that enables greater control over citizens—for example, China's emerging system of social credit and "Police Cloud" (Hvistendahl 2017; Mistreanu 2018) Elsewhere, digitalization has been used to provide more citizen-centric policies and delivery—Estonia's public rhetoric suggests this, although actual policies display additional goals to balance a perceived threat from Russia (Heller 2017). Yet other countries seek to reduce government reliance on manpower and Weberian paper-based procedures, and to increase resource efficiency (the United Kingdom (UK), Australia and Singapore, inter alia). These differences in motivation are compounded by variations in efficiency: some governments digitalize at great cost, with little by way of concrete results; other more modest programmes have much higher success rates.

This breadth of empirical experience makes it unsurprising that there is still no clear, common definition of government digitalization—the pervasive adoption of digital technology in various facets of public sector life, including policy making, service delivery, internal administration and management of public organizations, and engagement of businesses, citizens and other stakeholders.

There is certainly no lack of effort to articulate such definitions. In fact, the gray literature has provided multiple descriptive definitions. Consulting firms like McKinsey and Deloitte have been prolific on digital government, particularly in the past five years. International organizations, particularly the United Nations (UN), its subsidiaries and the World Bank have weighed in. The definition from the UN Educational, Scientific and Cultural Organization (UNESCO) definition is typical: "the public sector's use of information and communication technologies (ICTs) with the aim of improving information and service delivery, encouraging citizen participation in the decision-making process and making government more accountable, transparent, and effective" (2005).

The academic literature has moved beyond description to analytical typologies. These generally take the form of frameworks or models, each emphasizing the interactions of different dimensions. Several adopt a sequential or generational approach. Chun et al.'s four-stage model (2010) begins with "digital presence", or simple, passive, information-providing websites, through to "shared governance to transform how government operates, in terms of seamless information flow and collaborative decision making". Janowski's four-stage model (2015) involves Digitization (the presence of technology in government, in a static, non-interactive way), Transformation (electronic government), Engagement (electronic governance involving non-public sector stakeholders) and Contextualization (policy-driven electronic governance). These sequential models mistakenly assume that digitalization processes follow a deterministic teleology—early stages being necessary before subsequent ones, and later stages being superior to earlier ones. Most lived reality suggests that neither of these is true. Countries or agencies can learn from others and leapfrog more rudimentary stages, while some deliberately remain at a particular stage more appropriate to their needs (e.g., foreign ministries deal primarily with information and naturally develop deep Facebook, Twitter and other social media capabilities for public diplomacy, but only basic web platforms for administrative and consular functions like passport and visa applications). All this adds to the range of empirical experiences in need of analysis and categorization.

Several scholars have attempted to address this flaw with richer frameworks. Gil-Garcia and Luna-Reyes (2006) offer a meta-examination by categorizing digital government under four headings: electronic services (e-services); e-management; e-democracy; and e-policy. Dawes (2009) defines digitalization as a dynamic framework embedded in socio-economic phenomena—an approach enhanced by Dawes and Helbig's conceptual model (2015) of digital government as a dynamic phenomenon in which policies, management and organization, technology and data interact within social, political and economic contexts. However, they are silent on exactly how such interaction occurs. This gap is addressed by Pollitt (2011), who extends Bellamy and Taylor's (1998) insight that government digitalization constantly engages with "the complexities of the political and social world in which technologies are being adopted". Pollitt describes these engagements in terms of digitalization and technology altering time, place, tasks and activities, rules, resource flows and individuals. Moon et al. (2014) extend the classification by Rosenbloom (1998), exploring how digitalization shapes three dimensions of public administration: political, managerial and legal.

Dunleavy and Margetts (2013 and forthcoming) offer arguably the most comprehensive definition in current literature. Building on an earlier concept, "Digital Era Governance" (Dunleavy et al. 2006), they suggest the idea of "Essentially Digital Governance" (EDGe), comprising 48 aspects grouped into three headings:

a. (re)integration of government functions (in several cases, reintegration followed the agency specializations advocated by some interpretations of New Public Management (NPM) reforms);
b. holistic concern with, and response to, citizen needs (what the authors call "needs-based holism");
c. widespread adoption of digital technology.

This definition encompasses similar technical elements to those in the aforementioned models. It also examines the normative issue of how EDGe, at its most expansive, qualitatively differs from preceding governance paradigms: Progressive Public Administration, which emphasized the importance of a well-trained professional bureaucracy, insulated from political, commercial and popular pressure; and NPM, which focused on transposing neoliberal market principles to running public agencies (Dunleavy 2009).

These multiple, overlapping definitions have typically been applied to assessing how digital technology is transforming one or more aspects of governance. The vast majority of public administration authors focus on transformations in different geographical areas, a sampling of which includes member states of the Organization for Economic Cooperation and Development (OECD) (Foley and Alfonso 2009); the United States (Norris and Moon 2005; Dawes 2008; Tolbert et al. 2008; Fountain 2009; Norris and Reddick 2012); Sub-Saharan Africa (Schuppan 2009); groups of countries in the Middle East (Chatfield and Alhujran 2009; Jasimuddin et al. 2017); Australia (Henman 2010); the UK (Brown et al. 2014); Brazil (Musafir and Freitas 2015); the Nordic states (Joseph and Avdic 2016); Canada (Clarke et al. 2017); Estonia (Lember et al. 2018; Margetts and Naumann 2017).

Beyond geographies, other authors examine digital technologies' transformative effects on governance in general (Hood and Margetts 2007; Margetts 2008, 2009; Dominguez et al. 2011). Yet others adopt a meta-approach, studying how digital governance as a field is evolving (Dawes 2009; Gil-Garcia et al. 2017). A small clutch of articles has moved beyond backward-looking analysis to forward-looking advocacy on how governments should digitalize. O'Reilly (2011) suggests seven design principles[1] for Gov 2.0. Dunleavy and Margetts (forthcoming) suggest nine design principles for EDGe, sub-divided into five 'Do' principles[2] providing the framework for administrative and service design, and four 'Choice' principles[3] articulating a normative framework for policy-making and service delivery.

While this literature extensively describes the range of government programs, projects, practices and success rates, what is missing is a rigorous explanation of why government practices and principles occupy such a broad range. The next section takes up this challenge.

Digitalization as a Two-Dimensional Space

A core reason for the breadth of government digital practices is how differently scholars and practitioners' approach two key issues. What follows is a broad overview, followed by a more in-depth treatment of each.

First, there is a range of definitions of the type of resources managed in a digital environment. Traditional ideas of resource scarcity are gradually being supplemented by the view that some new resources are abundant, or even generative or self-replicating from frequent use. Call this the scarcity-generativity axis; definitions of digital resources can be found across the whole spectrum.

---

[1] The seven principles are: (1) Open standards to spark innovative growth; (2) Build a simple system and let it evolve; (3) Design for participation; (4) Learn from your hackers; (5) Use data mining to harness implicit participation; (6) Lower barriers to experimentation; (7) Lead by example.

[2] The five Do principles are: (1) Deliver public services for free; (2) Use already existing digital information; (3) Do it once; (4) Grow scalable services in competition; (5) Isocratic (Do-It-Yourself) administration.

[3] The four Choice principles are: (1) Value equality of outcome over process; (2) Provide formal rights and real redress; (3) Keep the state nodal; (4) Experiential learning.

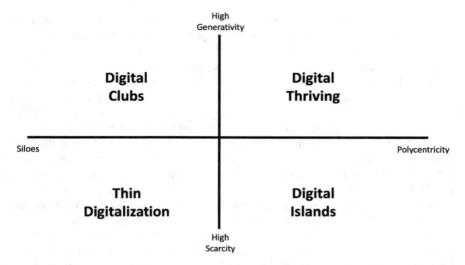

Fig. 14.1 Digitalization as a 2 × 2 space, Maniam 2018

Second, there are varied definitions of how governance needs to be structured: on the one hand, there are arguments for specialization and division into professionalized siloes; on the other hand, policymakers advocate for "Whole-of-Government", "Joined-Up Government" and "Whole-of-Society" approaches that involve more polycentric governance, transcending the orthodox boundaries of public, private and community sectors (Pollitt 2003). Call this the silo-polycentricity axis. As with resources, definitions of digital governance can be found across the entire spectrum.

Together, these two axes generate a 2 × 2 space (Fig. 14.1) with four core archetypes of digitalization, rather than a single canonical experience.

The next two sections present more detailed derivations of each axis, followed by a description of what digitalization might entail in each quadrant.

Reconceptualizing Resources: The Scarcity-Generativity Axis

It is noteworthy that the first chapters of nearly all economics textbooks are titled "Scarcity: The Central Problem of Economics", or something similar. This makes sense in the non-digital space, where much of the discipline is concerned with reconciling unlimited wants with limited resources, through optimisation processes that "maximize" some variable (utility, profits, wages, lifetime income, et al.) within constraints and parameters. Such maximizing approaches have extended to public policy, where efficient resource utilization is often cited as a key priority of governments.

The fundamental assumption of scarcity works well for physical (and therefore finite) resources like oil or land, or even non-physical but nonetheless finite resources such as time. But it is less clear if scarcity characterizes the resources increasingly important in digital economies—data, knowledge and connections underpinned by relationships. These grow rather than deplete from being used, particularly given improvements in analytic capabilities and Internet-enabled network effects.

Consider some simple examples. Data begets more data once it has been interpreted and analyzed: raw data on road usage, for instance, can generate new insights leading to new studies and models, perhaps even the collection of new data. One person's knowledge—say, in a book or chapter like this one—can catalyze new ideas, interpretations and innovation. All successful thinkers and scholars stand on the shoulders of their predecessors and antecedents; even the most ground-breaking work draws on prior research. In addition, what Putnam (1995) called the "social capital" underpinning relationships benefits from being tended and nurtured, so that trust, reciprocity and predictability are created and sustained, much like gardens generate new life from regular maintenance. Of course, data and knowledge need physical storage (although technological advancements increasingly expand these limits), and relationships take attention and time to maintain. But one could call these second-order scarcities: in and of themselves, the resources involved have no physical form and no physical quantities to be "depleted" from use, so they are not "consumed" in the literal sense.

Traditional economics is only partially helpful in understanding the nature of such resources. Drawing on Elinor Ostrom's Nobel Prize-winning studies of how common-pool resources are governed, as well as the prior work of Samuelson (1954), resources like data, knowledge and relationships would be seen as:

a. Non-rivalrous or non-subtractable—Person A's consumption of data and knowledge, or experience of a relationship, does not prevent the simultaneous consumption/experience by Person B;
b. Non-excludable—Person A could not stop Person B from consuming data or knowledge, or from being in a particular relationship.

Where the concepts of rivalry, subtractability and excludability falter is their assumption that resources do not change in the process of being utilized or experienced. Samuelson's public goods, like street-lighting, lighthouses and defense, or the rivers, fisheries and forests that formed Ostrom's key case studies, remain relatively unchanged by consumption, even if they transform due to external or systemic factors. In contrast, data, knowledge and relationships are fundamentally altered when utilized, because this very utilization results in more of each resource being created.

The term I suggest to describe such self-replicating resources is "generative" (Maniam 2016). Given the inadequacy of existing descriptions, we need a fundamentally different way of thinking about what generative resources are and how they work. Some useful ideas come from French sociologist Mauss (1966) on what he terms a "gift economy"—where things have intangible value even if their tangible worth is unclear, and are exchanged reciprocally rather than as instrumental, capitalist commodities. However, while Mauss' *process of exchanging gifts* is indeed generative, he does not offer much of an account for the generativity *of the gifts themselves* (the additional usage created for each resource).

On this latter issue, the conceptual heavy lifting is only just beginning. For now, five aspects seem particularly critical in defining this new vocabulary:

1. Generative resources are not static but ***constantly evolving***. This might seem obvious, until we remember that for the better part of history, our notion of resources has focused on substances like wood, iron and oil, which stayed (by and large) the same during a productive process. Physical states sometimes changed, but at its core, each resource remained relatively constant. In contrast, generative resources are dynamic and iterative: data feeds on itself, networks generate cycles that can be virtuous or vicious, and social capital can undergo both quantitative and qualitative transformations as a result of relationship building within a community.

2. Generative resources exhibit *"**input-output polymorphism**"*. It is difficult to tell whether data, knowledge and social capital constitute input into, or output from, productive processes. The input-output dichotomy—while useful for a world where goods and services were produced in linear, discrete processes—is much less relevant when resources and raw "materials" are less tangible. There are inevitable feedback loops between input and output. Old boundaries between production and consumption, or producers and consumers, will grow less salient over time. We will all increasingly become "prosumers" in our interactions with generative resources.

3. Generative resources suffer, not from the overuse that prompted Hardin (1968) to coin the famous term "tragedy of the commons", but from ***underuse***. Websites bereft of traffic; physical and online networks atrophying with sluggish usage; prediction algorithms with insufficient training data; neglected communities on the sidelines of cities—the lost potential and capacity in each of these are certainly tragic. But the tragedy lies in insufficient rather than excessive exploitation— sometimes by choice, in the case of relationships, and sometimes because we might lack the computational and/or cognitive power to analyze new data or knowledge. Again, underuse is related to second-order scarcities: time, attention and capacity are limited, even if the resources on which they act are generative. The sobering reality is that such second-order scarcities will not disappear easily, although they can be mitigated with technological improvement and innovation.

4. As a result of their evolving nature, generative resources have ***fuzzy and dynamic edges***, not clear boundaries. Governments usually rely on well-defined boundaries when governing a resource: knowing the boundaries of oil deposits, for instance, is key in deciding the validity of ownership claims. But where do things like data, networks and relationships start or end? If clear boundaries allow for clear principles of governance, then fuzzy and dynamic boundaries may also require governance by fuzzy and adaptive logic—broad norms underpinning the use of a resource, rather than mere physical concepts like quantity. Ostrom's work on fisheries and forests offers clues on what such norms might look like— fishermen collectively choose to adhere to norms of throwing young fish (below

a certain length) back into the water, while trees with trunks below a certain diameter are deliberately untouched by loggers. These norms can only be imperfectly defined and enforced—hence their fuzziness—and may need to evolve with time. But Ostrom's work suggests that self-organizing, self-monitoring and self-enforcing communities can actually achieve reasonably high rates of adherence within these conditions—in some cases, such communities achieve outcomes that are closer to Pareto Optimality than those achieved by either top-down state interventions or market-based solutions. This could apply equally to data and knowledge (collective norms in universities on plagiarism, for instance) or norms governing the behavior of communities, both on- and offline.

5. The intangible and evolving nature of such resources mean that there are *few clear "equilibrium" points* in the way they can be used or managed. Much as equilibria help to simplify analysis in what the positivist tradition terms "comparative statics", the real world does not usually exhibit stable or immutable equilibria. Instead, the ways in which data, knowledge and relationships are used, as well as sustained, are likely to be much more emergent and unpredictable *ex ante*. Optimization approaches may need to start giving way to design-based, behavioral approaches that emphasize iterative and experimental learning by discovery.

To be clear, my argument is that generativity will and should supplement, not supplant, scarcity in our understanding of resources. Scarcity of physical resources, or units like time, will persist. But we do need a more nuanced view of emerging resources in the digital space, and different circumstances will require us to situate ourselves at different points on the scarcity-generativity axis.

Reconceptualizing Governance: The Silo-Polycentricity Axis

In addition to resources, our conceptualization of governance must also evolve. Even before the advent of digital technology, the traditional Weberian model of government—politicians and civil servants acting with relative autonomy on businesses and citizens who are passive recipients of policy—has been giving way to more multi-sector approaches where governments steer, not row (Osborne and Gaebler 1992; Peters 2011) or are "platforms" for a broad range of actors, rather than "control towers" or singular actors functioning with unimpeded autonomy (Kettl 2008; O'Reilly 2011).

Digital technology has accelerated and deepened these trends. While some sectors continue to require governments to row, command and control (decisions on tax rates, for instance, or implementation of the highest levels of foreign policy), many do not. Education requires the involvement of parents and school stakeholders, particularly when enabled by the use of home computers and mobile applications; healthcare is no longer the sole purview of medical professionals, and patients have access to far more online information than before (for better and worse). Even the process of making policy increasingly involves citizens—as seen in the advent of deliberative democratic platforms like Citizen Juries, participatory budgeting and other participatory platforms (Fung 2006)—many of them at least complemented, at most transformed by digital platforms (Fung et al. 2013; Farrell 2012).

Even where citizens do not play active roles in policy formation, digital technology allows for the collection, connection and synthesis of unprecedentedly large amounts of data, all of which shed light on the needs, preferences and perspectives of individual citizens—what Dunleavy and Margetts (2013 and forthcoming) call "needs-based holism". This data richness allows for a degree of focus on "citizen needs" hitherto unseen, adopting "user-oriented" design principles and insights from behavioral psychology. In line with this trend, a growing number of governments are establishing government or government-linked units focused on design (e.g., Denmark), behavioralism (Australia) or both (Singapore and the UK).

Collectively, these trends substantiate arguments in the institutionalist literature that governance of complex systems must be "polycentric"—most recently by Elinor Ostrom (2009), building on earlier work by Ostrom et al. (1961):

> 'Polycentric' connotes many centers of decision-making that are formally independent of each other. Whether they actually function independently, or instead constitute an interdependent system of relations, is an empirical question in particular cases. To the extent that they take each other into account in competitive relationships, enter into various contractual and cooperative undertakings or have recourse to central mechanisms to resolve conflicts, the various political jurisdictions in a metropolitan area may function in a coherent manner with consistent and predictable patterns of interacting behavior. To the extent that this is so, they may be said to function as a 'system.'

Polycentricity means, among other things, ensuring adequate porosity between the state (government), markets (businesses) and communities of citizens, as well as decision-making that takes place as close as possible to, if not by, the stakeholders most affected. Where digital technology enables the transcending or attenuation of physical distance, or the synthesis and analysis of hitherto inaccessible (volumes of) data, the momentum toward polycentricity is likely to increase. Different societies' positions on the silo-polycentricity axis will depend on how they balance between two poles: on the one hand, governments acting as specialists and experts, separate from and independent of other stakeholders; and on the other, governments acting more in concert with counterparts in business and the community.

Bringing the Axes Together

This brings us back to the quadrants in Fig. 14.1. The boundaries of the quadrants are porous, rather than rigid, but each quadrant is a useful archetype of the kinds of digitalization that may exist in different contexts.

The top right quadrant involves high generativity and high polycentricity. Such settings involve moving beyond traditional assumptions of scarcity, harnessing the generative potential of data and online trust. Estonia's vision of an e-nation, or the most fully-developed aspects of Singapore's Smart Nation policy, seem to fall into this category. Key examples include the use of integrated systems like Estonia's central data infrastructure X-Road (enabled by a universal personal identification system), a publicly accessible "technology stack" in Singapore, and the presence of strong leadership (both countries have designated ministers in charge of digital and technology issues). I describe this as a situation of **Digital Thriving**, not just digital efficiency, since the mix of polycentricity and generative resources catalyzes an overall ecosystem vibrance, not just mechanistic optimization. Such systems are sometimes necessarily untidy—like the most dynamic ecosystems—and characterized by complexity, unpredictability and turbulence.

The lower right quadrant involves high polycentricity and low generativity (high scarcity). I describe this as a situation of **Digital Islands**: disparate pockets of deep and high-quality digitalization are not sufficiently integrated across agencies or sectors and, hence, fail to achieve the integration and needs-based holism outlined by Dunleavy et al. (2013) in their concept of EDGe. A key underlying reason for the low levels of generativity is that agencies focus on optimizing internally—at a local level, rather than at the overall system level. Such islands are characteristic of the United Kingdom's digital system, where a very modern Government Digital Service (GDS) and swiftly modernizing tax system sit alongside antiquated legacy systems that prevent fulsome digitalization in areas like pensions and social welfare payments. Overall citizen-centered approaches are aspirations that remain unrealized due to low levels of cross-agency integration.

The upper left quadrant involves high generativity and low polycentricity (a siloed approach to governance). In this situation, generativity of resources faces the countervailing effect of barriers and agency-oriented divisions. Big companies could end up protecting the data they gather, knowledge is restricted behind paywalls and relationships are confined to narrow networks (e.g., traditional Old Boys' networks)—hence my use of the term **Digital Clubs**. In these cases, artificial "fees" or membership constraints are imposed on otherwise generative goods. One might describe this as a situation of "generated scarcity", or scarcity that is contrived in order to maintain particular individual or organizational interests. Sociologist of science Merton (1968) refers to the Gospel of Matthew (25:29) when he describes such generated scarcities as examples of the "Matthew Effect", where "… whoever has will be given more, and they will have an abundance. Whoever does not have, even what they have will be taken from them".

The final quadrant, on the lower left, involves low generativity (scarcity) and low polycentricity (siloes). I term this **Thin Digitalization**: where digitalization occurs, it is episodic and sectorally narrow, rather than system-wide. This may well be a necessary starting point for many under-developed digital systems—which are still beholden to traditional scarcity—and specialist-based models of governance.

It is also plausible that even within a country, different sectors may be located at different quadrants, since the quality, pace and depth of digitalization may differ across functional areas. In Estonia, for instance, while X-Road and the e-identity system are examples of Digital Thriving, other sectors—like welfare and medical services—are far less sophisticated and could arguably be described as examples of Thin Digitalization (Lember et al. 2018).

Implications for Policy

The core implication of the four quadrants is that there is no single, canonical manifestation of digitalization in different policy spaces; instead, there are many digitalizations, depending on where countries lie on each axis. This generates several implications for both the principles that guide policymaking, and the delivery or implementation of policies.

Policy Principles

The evolving nature of generative resources, as well as their input-output polymorphism, will require changes in how we measure economic value—not just through static traditional measures like Gross Domestic Product (GDP), but through new measures that capture the catalytic effects caused by resource use. Such new measures are very much works-in-progress, but it is telling that new and more variegated measures of human welfare, like the Legatum Institute's Prosperity Index, include a sub-index on generative resources like Social Capital.

The fuzzy and dynamic edges of generative resources will have important ramifications for how we define intellectual property, which will need to be seen as much more dynamic and kaleidoscopic. Assigning precise ownership over extended periods will grow increasingly complicated; these will need to be calibrated in order to avoid the most extreme and pernicious manifestations of the Matthew Effect, and maintain the generativity of resources.

Lack of clear equilibrium points calls for a more systems-oriented approach that considers not just individual agents and nodes, but their interactions within larger ecologies. This entails analyzing the interactions among pieces of knowledge, exploring the creative potential of networks and tapping into the collaborative capacity in generative relationships. Some of this has already been proposed by scholars of complexity science and complex adaptive systems, who explore how non-linear, interdependent systems require iterative, experimental approaches that are fundamentally different from the stable, predictable systems popularized by the European Enlightenment. These links can be deepened and broadened through further research.

Policy Delivery and Implication

Tragedies of the generative commons, arising from under- rather than over-use, will need new and creative approaches to tax and regulatory policies, where some forms of exploitation will need to be encouraged rather than limited through instances of generated scarcity.

Greater public participation in public policy—in particular, more deliberation by citizens on the decisions that affect their lives and more policy that is truly "of, by and for" the people—is one key way to guard against the risk of such underuse. In many ways, generative resources are like muscles, and public participation ensures their constant use, stretching and suppleness. Indeed, such engagement can *itself* be a generative platform, leading to the creation of new knowledge and learning, civic awareness and social capital among mutually engaged citizens.

Such participation is a subset of a larger imperative to conceive of political actors—governments, businesses and citizens—as capable of constant adaptation and learning, rather than stuck in rigid, immutable maximization games. This is a core requirement for the polycentricity of systems in Ostrom's work, with substantive echoes in the fourth of Dunleavy and Margetts' (pending) Choice Principles (experiential learning), as well as the broader literature on "democratic experimentalism" (Dorf and Sabel 1998). Policymakers accustomed to stable, neat systems with clear boundaries will have to accustom themselves to more non-linear, unordered and emergent policy milieu. Government recruitment, remuneration and training schemes will have to evolve to attract, retain and provide public officials with the right skills and capabilities.

Conclusion

In "Little Gidding", one of his "Four Quartets", poet T. S. Eliot observes that

> … last year's words belong to last year's language
>
> And next year's words await another voice.

This chapter is an initial attempt to articulate a new language and voice for the emerging space of government digitalization, particularly in terms of how resources and governance are conceptualized. Future research could fruitfully focus on applying the ideas of generativity and polycentricity to specific policy areas, like education, healthcare, tax administration and multiple others. This is a conversation in its most nascent stages, and both scholars and practitioners will have to be open to further dynamism, flux and evolution in the years to come.

References

Articles and Books

Bellamy, C., & Taylor, J. (2018). *Governing in the information age*. Buckingham: Open University Press.

Chun, S., Shulman, S. W., Almazan, R. S., & E. Hovy. (2010). Government 2.0: Making connections between citizens, data and government. *Information Polity, 15*(1,2), 1–9.

Dawes, S. (2010). Governance in the digital age: A research and action framework for an uncertain future. *Government Information Quarterly, 26*(2), 257–264.

Dawes, S., & Helbig, N. (2015). The value and limits of government information resources for policy informatics. In E. W. Johnston (Ed.), *Governance in the information era: Theory and practice of policy informatics* (Chap. 2) (pp. 25–34). Routledge.

Dominguez, L. R., Sanchez, I. M. G., & Alvarez, I. G. (2015). Determining factors of E-government development: A worldwide national approach. *International Public Management Journal, 14*(2), 218–248.

Dorf, M. C., & Sabel, C. F. (1998). *A constitution of democratic experimentalism, Cornell Law Faculty Publications* (p. 120). https://scholarship.law.cornell.edu/facpub/120.

Dunleavy, P., Margetts, H., Bastow, S., & Tinkler, J. (2006). *Digital era governance: IT Corporations, the State, and E-Government*. Oxford University Press.

Dunleavy, P. (2009). Governance and state organization in the digital era. In C. Avgerou, R. Mansell, D. Quah, & R. Silverstone (Eds.), *The Oxford Handbook of Information and Communication Technologies*. Oxford University Press.

Dunleavy, P., & Carrera, L. (2013). *Growing the productivity of government services*. Edward Elgar Publishing.

Dunleavy, P., & Margetts H. (2015). The second wave of digital-era governance: A quasi-paradigm for government on the web. *Philosophical Transactions of the Royal Society A, 371*, 20120382. http://oxis.oii.ox.ac.uk/wp-content/uploads/sites/5/2015/07/Margetts-Dunleavy.pdf.

Farrell, H. (2015). The consequences of the internet for politics. *Annual Review of Political Science, 15,* 35–52.

Fung, A., Gilman, H. R., & Shkabatur, J. (2012). Six models for the internet & politics. *International Studies Review, 15,* 30–47.

Gil-Garcia, J. R., & Luna-Reyes, L. F. (2006). Integrating conceptual approaches to E-Government. In M. Khosrow-Pour (Ed.), *Encyclopedia of E-Commerce, EGovernment and Mobile Commerce*. Hershey, PA: Idea Group Inc.

Gil-Garcia, J. R., Dawes S. S., & Pardo, T. A. (2017). *Digital government and public management research: Finding the crossroads* (pp. 633–646). Public Management Review. Routledge.

Hardin, G. (December 13, 1968). The tragedy of the commons. *Science, New Series, 162*(3859), 1243–1248. https://www.jstor.org/stable/1724745?origin=JSTOR-pdf&seq=1#page_scan_tab_contents.

Hood, C., & Margetts, H. (2007). *The tools of government in the digital age* (2nd edn.). Red Globe Press.

Janowski, T. (2007). Digital government evolution: From transformation to contextualization. *Government Information Quarterly, 32,* 221–236.

Lember, V., Kattell, R., & Tonurist, P. (2018). Technological capacity in the public sector: The case of Estonia. *International Review of Administrative Sciences,* 1–26.

Maniam, A. (2016). From scarcity to generativity: New approaches to governing resource. In *Ethos issue* (p. 16). Accessed online at https://www.csc.gov.sg/articles/from-scarcity-to-generativity-new-approaches-to-governing-resources.

Margetts, H. (2008). Public management change and e-government: The emergence of digital era governance. In A. Chadwick, & P. N. Howard (Eds.), *Routledge handbook of internet politics*. Routledge.

Margetts, H. (2008). The internet and public policy. *Policy & Internet, 1*(1), 1–21.

Mauss, M. (1966). *The gift; forms and functions of exchange in archaic societies*. Cohen and West.

Moon, M. J., Lee, J., & Roh, C.-Y. (1966). The evolution of internal IT applications and e-government studies in public administration: Research themes and methods. *Administration & Society, 46*(1), 3–36. https://doi.org/10.1177/0095399712459723.

Merton, R. K. (2014). *Social theory and social structure*. Collier Macmillan Publishers: The Free Press.

Kettl, D. F. (2008). *The next government of the United States: Why our institutions fail us and how to fix them*. W. W. Norton & Company Inc.

O' Reilly, T. (2011). Government as a platform. *Innovations, 6*(1), 13–40.

Ostrom, V., Tiebout, C. M., & Warren, R. (1961). The organization of government in metropolitan areas: A theoretical inquiry. *American Political Science Review*, 831–842.

Peters, B. G. (2011). *Steering from the centre*. University of Toronto Press.

Pollitt, C. (2011). Technological change: A central yet neglected feature of public administration. *The NISPAcee Journal of Public Administration and Policy III, 2*, 31–53.

Putnam, R. D. (2011). Bowling alone: America's declining social capital. *Journal of Democracy, 6*(1), 65–78.

Rosenbloom, D. H., & Rosenbloom, D. D. (1995). *Public administration: Understanding management, politics, and law in the public sector*. New York: McGraw-Hill Companies.

Samuelson, P. A. (November, 1954). The pure theory of public expenditure. *The Review of Economics and Statistics, 36*(4).

Media

Hvistendahl, M. (2017). *Inside China's vast new experiment in social ranking*. Wired Business, 14 December 17. https://www.wired.com/story/age-of-social-credit/. Accessed on 1 May 2018.

Heller, N. (2017). *Estonia, the digital republic*. December 18 & 25, 2017. https://www.newyorker.com/magazine/2017/12/18/estonia-the-digital-republic. Accessed on May 1, 2018.

Mistreanu, S. *Life inside China's social credit laboratory*, April 3, 2018. https://foreignpolicy.com/2018/04/03/life-inside-chinas-social-credit-laboratory/. Accessed on May 1, 2018.

Aaron Maniam is currently working on doctoral research at the Blavatnik School of Government, University of Oxford. His work involves a comparative study of the digital transformation of public sector agencies in Estonia, New Zealand and Singapore. A policymaker by profession, he was previously Senior Director (Industry) at Singapore's Ministry of Trade and Industry, responsible for coordinating economic policies and regulating the manufacturing, services and tourism sectors, as well overseeing long-term economic transformation. He joined the Singapore government in 2004, serving on the North America Desk of the Foreign Service (2004–2006) and at Singapore's Embassy in Washington, DC (2006–2008), where he was the principal coordinator for Congressional liaison and issues relating to the Middle East. He was posted to the Strategic Policy Office (SPO) at the Public Service Division in 2008, where he worked on scenario planning and analysis of long-term trends relevant to Singapore. He was appointed the first Head of the Singapore Government's newly-formed Centre for Strategic Futures (CSF) in January 2010, while retaining his SPO portfolio. In July 2011, Aaron was appointed Director of the Institute of Policy Development (later renamed the Institute of Public Sector Leadership) at the Civil Service College (CSC), which organizes leadership training program for public sector talent (the top 1% of the public sector workforce). In 2012, he started the CSC Applied Simulation Training (CAST) Laboratory, an experiment to apply principles of "serious play" to training public officers to deal with complex environments. He led efforts to develop the College's curriculum on complexity science and convened the College's multi-sector interest groups on Complexity and Governance. He has taught courses on leadership and governance at the National University of Singapore's Scholars Programme and Singapore Management University. He is a World Economic Forum Young Global Leader, a member of the Forum's Global Future Council on Agile Governance, and a Fellow of the Royal Society for the encouragement of Arts, Manufactures and Commerce (RSA).

Chapter 15
The Challenges of Digitalization for the (German) State

Valentin Gauß

Introduction

A public discussion of the opportunities and risks of digitalization is mostly concerned with technical, economic or social aspects. The area of "governance" (i.e., political control) is largely ignored by the public. But the political control of digital transformation is a central issue, not only for politics and administration, but also for society. It is highly relevant how the transformation process of digitalization is accompanied and controlled by the state. It is becoming increasingly clear that the digital space must not develop into a "legal vacuum" and that states must adapt to the "way of thinking" of digitalization within the framework of their legislation. Looking at the political decision-making process, however, we get the impression that the state and public administration can no longer fully implement their own requirements for shaping and controlling policy in the course of increasing digitalization (Schallbruch 2018).

The complex network of responsibilities on the political levels (federal, state and local authorities) within Germany is developing into the driving force behind a growing inability to act politically. In addition to vertical and horizontal responsibilities, the lack of technical and process knowledge with regard to digitalization deprives the state of its "strength to act" and presents new challenges for legislation and creative powers (Schallbruch 2018).

In this article, examples of the fundamental challenges of the transformation process are presented, and selected practical approaches are outlined.

V. Gauß (✉)
Ministry of Transport of the Federal State of Baden-Württemberg, Stuttgart, Germany
e-mail: valentin.gauss@vm.bwl.de

© Springer Nature Switzerland AG 2020   207
D. Feldner (ed.), *Redesigning Organizations*,
https://doi.org/10.1007/978-3-030-27957-8_15

The Challenges of a Complex Distribution of Responsibilities

German industry, as the "workbench of the world" (Martini et al. 2016), is already in a good position for the digital transformation process thanks to excellently trained experts and the cooperation with universities, technical colleges and research institutes that is taking place on various levels (ibid.). Citizens are also taking advantage of the opportunities offered by digitalization. More and more areas of daily life—such as shopping, the use of financial services or travel planning (hotel and flight bookings, arrival and departure)—are being optimized using software and apps. However, current insights regarding public administration paint a different picture: The conversion of internal administrative processes and work structures, as well as outward-directed citizen services, have fallen short of our own expectations. Taking the European Commission's Digital Index as a benchmark, Germany ranked 14th out of 28 last year, but only 21st for government services (European Commission 2018).

One obstacle seems to be the structure of German bureaucracy. In Germany, the tasks involved in digitalization are not coordinated by a central authority; rather, several state institutions are working on leading the country into the digital age. At the federal level alone, several ministries are working on designing a concept for digitalization. Digitalization is conceived and carried out in the respective area of responsibility of a department. The same is happening at the state level. There is a department dealing with digitalization in the Ministry of Transport in Baden-Württemberg, and also in the state's Ministry of Economics and the Ministry of the Interior.

In addition, there is a State Minister for Digitalization in the Chancellor's Office, as well as various federal authorities, such as the Federal Office for Information Security. At the state level, there are 16 state governments, each pursuing their own strategies. This shows that German federalism also diversifies responsibilities in the area of digitalization. At first glance, this seems to be a sensible division, since digitalization affects all areas of life and, thus, in keeping with federalism, regional interests are given weight. A closer look, however, reveals a patchwork approach that reduces efficiency.

For example, the responsibility of the Integrated Traffic Control Centre (IVLZ)—a joint control center of the city of Stuttgart, the Stuttgarter Straßenbahn AG (SSB), the fire department and the police—ends at the city limits. Measuring facilities in the superordinate non-municipal road network are maintained by the Road Traffic Control Center (SVZ) of Baden-Württemberg. The automatic exchange of measurement data is not always possible, even though a lot has happened in the recent past. If there is a larger volume of traffic in the surrounding area, which is moving toward the city of Stuttgart, the intervention in road traffic can usually only take place when the traffic is already jammed on inner-city roads (expert discussion with IVLZ in 2018).

The different responsibilities and the lack of coordination mean that the tools provided by digitalization are not used consistently. A comprehensive exchange of information does not take place in many areas—essentially, many political institutions only act within their respective field of work. For this reason, so-called pilot and model projects are being funded at various points, without the findings of these projects having been consolidated nationwide. Instead, it may happen that different actors promote and implement similar projects.

As a result, it becomes clear that the existing structures fall short of the opportunities of networked and, thus, efficient working. But this is precisely what is needed to understand and positively exploit the changes associated with digitalization.

Challenges for Legislation and the Creative Power of the State

At this time, digitalization is clearly a challenge for the state. Some authors, such as Schallbruch (2018), even go so far as to speak of excessive demands. Digitalization and the associated transformation of society and the economy are questioning the way in which society as a whole and all actors involved (civil society, economy and science) are controlled and organized.

Up to now, state action has followed a recurring pattern. In short, there is a social problem at the beginning of every state action. Particular interests and state interests are being balanced, and then a bill is proposed. This proposed bill is then questioned, revised and ratified. If the challenges continue after the law has been passed, or new problems arise, the process starts again. The opinion-forming and weighing process that takes place takes time. It can take years from legislative initiative to adoption.

The rapid emergence of new innovations and technical possibilities in the context of digitalization poses a major challenge to the political decision-making process to date. Solutions to a social problem must be found much faster, and political decision-makers must respond more flexibly to technical innovations. Digital structures are subject to permanent change and often no longer correspond to the original object of regulation until a new law or regulation is implemented. In Germany, the process is currently often slowed down by a culture of risk minimization that prevails within the administration. Avoiding risks is in itself nothing that the actors can be criticized for. In the context of the fast pace of economic and technical development, however, it becomes apparent that the administration loses important flexibility.

An example from the field of platform economy reveals how important it is that the state understands the social relevance of emerging digital applications and responds appropriately to them. The original idea of Airbnb was to provide temporarily unused private rooms to people who wanted to spend their holidays in direct contact with locals. However, an alternative business model developed relatively quickly. Increasingly, entire apartments are being rented out permanently to holiday guests and withdrawn from the regular housing market. Especially in popular tourist destinations, this leads to further pressure on the housing market. The local authorities have been powerless for a long time and have only recently begun to look for solutions to make the "holiday homes" accessible to the housing market again. This example clearly shows that state actors have not recognized—or at least underestimated—the social relevance of "rental platforms" for a long time. As a result, they responded rather late to the housing market situation in many places.

The fact that the state has not yet completely lost its ability to act with regard to digital platforms is exemplified by the travel service provider Uber. The driving service (UberPop) offered by the American company Uber has managed to change the taxi business in many countries within a short time. In Germany, this service was available in Berlin, Hamburg, Munich and Frankfurt am Main, among others. Unlike conventional taxi companies, which are subject to many government requirements, Uber waived these requirements. Uber requires a driver's license and information about the score at the Federal Motor Transport Authority (KBA) as proof of suitability. Calibrated odometers (taximeters) do not exist, and the medical health of the drivers (e.g., eyesight test) is not checked either. Because of these irregularities and the protests of licensed taxi operators, the government felt it had a responsibility to act as quickly as possible. Uber's offer was declared unlawful and was then prohibited (Linke 2015). From a state perspective, the protection of service providers and passengers had to be maintained. In addition, examples such as Airbnb and Uber show, however, that many innovations are created in the context of digitalization, and demands from citizens (e.g., for cheap holiday apartments, cheap taxi rides) can be met. The extent to which the state should make more active use of such innovations for itself is only raised here as a further-reaching question.

State regulations are generally based on sanctions for non-compliance (Stemmer 2016). In the real (analogue) world, the state can enforce compliance with standards and laws through its administrative authorities on the federal, state and local level and the subordinate enforcement bodies (e.g., public prosecutor's office, police or revenue office). It becomes apparent that the political decision-making process and democratic institutions in the digital space are increasingly reaching their limits in this area. In the virtual (digital) world, laws are more difficult to enforce and abuse and violations more difficult to sanction (Schallbruch 2018). For example, the identification of individual persons is difficult (keyword: Real name regulation).

State control also fails because services are globally oriented and are often hosted on non-European servers and databases. This is particularly obvious in the de facto monopoly position of individual—mainly U.S.—economic players (Google, Facebook, Amazon, Apple and Microsoft). The digital platforms and their services are located on U.S.-American servers and are, therefore, initially subject to U.S.-American law and are thus protected from interference by the German state.

The German state is increasingly realizing that different actors in the digital space are evading its control and sanction capacity more and more. For this reason, it has passed the Network Enforcement Act (NetzDG). In this context, however, we have to ask: to what extent can the state reaffirm its power through the new law?

Basically, the aim of the NetzDG is that no agitation or unconstitutional statements may be spread on social networks. The NetzDG is intended to help enforce a kind of jurisdiction on digital platforms, such as Facebook or Twitter. In fact, however, the state does not "enforce its own democratically legitimized right on the platforms, rather it accepts the normative power of these platforms" (Schallbruch 2018). The state is thus relinquishing power and responsibility to companies and disempowering itself. Schallbruch (2018) calls this phenomenon "digital enforcement deficit". The state no longer plays the role of an acting protagonist; rather, it relies on companies to enforce existing laws.

A further example of a decline in the creative power of the state can be seen in traffic control and traffic diversion in the event of a breakdown. Before the advent of digital navigation systems, government agencies—such as the Road Traffic Control Center Baden-Württemberg (SVZ-BW) or the Integrated Traffic Control Center (IVLZ) of the city of Stuttgart—were solely responsible for controlling traffic. If a malfunction occurs, the responsible authorities weigh the individual interests of residents against state control interests and then set up a diversion. Restricted areas or residential areas are excluded in most cases. So-called "back ways" are thus only used by locals.

Digital traffic platforms, on the other hand, focus on a route optimized for each individual case. If the road traffic regulations permit using a certain road, the navigation system will recommend using this route, disregarding the interests of local residents. While state actors, for example, take into consideration the impact of a traffic backlog on the higher-level road network (federal and state roads) and leave out the catchment areas of kindergartens, private providers such as Google (Google Maps) or Apple (Apple Maps) do not take such social concerns into account. In this context, the question arises as to whether the state is already losing its regulatory power, and whether private-sector companies are abolishing previously applicable "standards".

This is confirmed in a study by the Institute for Transportation Studies (ITS) at UC Berkeley. It turned out that private navigation services behave "selfishly". They always suggest the fastest route for each individual user. If only 20% of motorists follow a route proposed by private navigation service providers, this leads to congestion problems on the downstream road network, as a model calculation on American roads has shown (Madrigal 2018). In Germany, too, people are increasingly following the on-board units or smartphone recommendations, thus causing congestion in the downstream road network (expert discussion with IVLZ Stuttgart in 2018).

It is evident that digitalization poses far-reaching challenges, especially for legislation and the creative power of the state. Political actors must become aware of these challenges to make sure their actions will be part of the current change.

Selected Approaches

As already outlined above, due to the large number of political actors involved in the field of digitalization, an overall cross-departmental strategy is not yet discernible in many places. Instead, individual projects and model projects determine day-to-day business. A unique project-based approach will rarely exploit the potential of digital transformation (Wegener et al. 2016).

This must be overcome. Stemmer (2016) points out, for example, that the first important step is to establish a comprehensive, systematic and strategic IT control within the public administration. State digitalization strategies, IT summits and the digital agendas of the federal government point in the right direction; however, "in their current form they are not yet sufficient to really do justice to the importance of digitalization" (Stemmer 2016).

Initial examples in the administration show that networked and comprehensive action is gaining in importance. For example, the green-black state government of Baden-Württemberg has adopted a comprehensive digitalization strategy (digital@bw) and anchored it in the coalition agreement. An interministerial working group—led by the Ministries of the Interior, Digitalization and Migration—has been set up to ensure that the transformation of society as a whole is a success. Regular meetings and data exchange strengthen cooperation between the individual ministries.

Literature also suggests that politicians and administrators should reconsider the promotion of individual pilot and model projects, many of which are technology-based. This does not mean abandoning pilot or so-called lighthouse projects altogether. New technologies must be tested in a lab-like environment, as this is the only way to identify possible risks at an early stage. However, it is no longer appropriate to initiate one pilot project after the other without rolling out the results nationwide or state-wide. Moreover, it is important to think in a networked way.

Focusing on model projects carries the risk of losing oneself in the small scale. According to Wegener et al. (2016), the goal is for the state to focus on developing overall societal frameworks and norms within which economic innovation and social development are possible. Political goals and agile iterative procedures in administration replace static and linear planning (Wegener et al. 2016).

This cultural shift to evidence-based and impact-oriented management requires more digitalization professionals who not only can implement IT processes, but also understand how digital business models work. It is not enough for individual units or official units to control digitalization. Since the transformation is comprehensive, all units must also understand how decisions affect society and administration internally (Schallbruch 2018; Stemmer 2016; Wegener et al. 2016).

In addition, it is proposed that a certain willingness to take risks and to tolerate errors be developed within the administration and the policy in order to promote speed and efficiency. Up to now, the administration has acted in such a way that errors are systematically excluded. However, the proposal does not aim to ignore all risks and make ill-considered decisions. Schallbruch (2018) instead means that risks should be weighed appropriately, but not every contingency should be excluded.

It is becoming increasingly clear that digital processes are characterized by a high degree of complexity and require sound technical and process knowledge. It's precisely these requirements that the political system must adapt to. Within individual digital structures, there are significant interactions and dependencies, which must also be taken into account in legislation.

In contrast to the hierarchically structured control of a state, with its top-down decision-making structures and "departmental thinking", so-called "digital governance" must make use of soft systems methodology and network-like processes. Possible forms of action include bottom-up decision-making structures, open government and public private partnerships. New ways of describing rules and standards must also be found, particularly in the digitalization debate and the associated "finding of roles" for the state. With regard to politics and administration, this means, according to Stemmer (2016), a digital transformation of the system in which the state must maximize the "value contribution" of digital technologies, structures and processes throughout society while, at the same time, keeping the associated risks at bay.

Summary and Outlook

We have tried to show that digital transformation poses great challenges for politics and administration, because established structures repeatedly reach their limits. These challenges include the complex distribution of responsibilities between the political actors. For the legislation and the creative power of the state, in particular, it becomes more and more relevant to understand the changes that are triggered by digitalization and to use them for themselves.

The "digital world" is characterized by rapid change and networked structures. For the state to make future decisions in line with technological developments, rather than being overtaken by them, politicians and administrators need to understand how digital network societies work. The resulting logic of action must then be transferred to the public sector and its working methods. It is important that political actors also comprehend and understand digitalization, so that an effective state can set the guidelines within which social change takes place. In this context, it is essential that comprehensive governance of digitalization is established.

If such a reorientation in the field of digitalization is consistently pursued, the further question can even be posed as to whether a German solution is still the right regulatory framework at all. Digitalization is not a phenomenon that occurs within national borders and can be regulated by national governments. Rather, it is a global development that also requires global or, in the first step, at least European solutions. At the present time, however, it seems more appropriate to initially implement a governance of digitalization on a national level, so that supranational networking can then be pursued.

References

Europäische Kommission. (2018). *Digital Economy and Society Index (DESI) 2018 - Country Report Germany*. Online: http://ec.europa.eu/newsroom/dae/document.cfm?doc_id=52214. August 30, 2019.

Linke, B. (2015). Gewerbefrei oder „Uber"reguliert? – Die Vermittlung von Personenbeförderungsdiensten auf dem Prüfstand. *Neue Zeitschrift für Verwaltungsrecht (NVwZ), 2015 Heft 8*. 476–479.

Madrigal, A. C. (2018). *The Perfekt Selfishness of Mapping Apps*. Online: https://www.citylab.com/transportation/2018/03/the-perfect-selfishness-of-mapping-apps/555683/. August 20, 2018.

Martini, M., Fritzsche, S., & Kolain, M. (2016). *Digitalisierung als Herausforderung und Chance für Staat und Verwaltung: Forschungskonzept des Programmbereichs „Transformation des Staates in Zeiten der Digitalisierung*. Speyer: Deutsches Forschungsinstitut für öffentliche Verwaltung.

Schallbruch, M. (2018). *Schwacher Staat im Netz: Wie die Digitalisierung den Staat in Frage stellt*. Wiesbaden: Springer Fachmedien Wiesbaden GmbH.

Stemmer, M. (2016). *Digitale Governance - Ein Diskussionspapier*. Hrg: Kompetenzzentrum Öffentliche IT. Fraunhofer Institut für Offene Kommunikationssysteme FOKUS, Berlin.

Wegener, N., Schoellhammer, R. G., Köhl, S., Wolf, P., Klessmann, J., & Parycek, P. (2016). *Digitale Vernetzung von Staat mit Wirtschaft und Gesellschaft –Akteursorientierte Handlungsempfehlungen für Politik und Verwaltung*. Hrg: Fraunhofer Institut für Offene Kommunikationssysteme FOKUS, Berlin.

Valentin Gauß is currently working as Personal Assistant to the Minister for Transport of the Federal State of Baden-Württemberg. Valentin joined the ministry in 2016. That time he served as a desk officer at the department of digitalization with special reference to rural areas. Valentin was working on scenario planning and analyzed long-term trends with relevance for traffic planning of the Federal State of Baden-Württemberg. A geographer and political scientist by profession, Valentin started his professional career at the Rhein-Neckar-Verkehrs GmbH (RNV), one of the largest major transportation alliances in Germany. He holds a master's degree (M.Sc.) in geography from the Ruprecht-Karls University in Heidelberg. In his studies he focused also on political science (University of Heidelberg), transport and regional planning (University of Stuttgart).

Chapter 16
E-Estonia—"Europe's Silicon Valley" or a New "1984"?

Florian Hartleb

Introduction

Are we living in a "brave new world", as the futurist novelist Aldous Huxley described? Can we talk about a revolutionary stage for our societies and economies politicians have not realized yet (Precht 2018)? Or are we living "smart"? Tax declaration with just a few clicks, having our daily life completely without papers, founding a company within a few minutes, and the possibility for e-voting? A country where the administration is fully digitalized, and the society is fully integrated? This is far from being a utopia, or a scenario for the future or within academic debates—a country already did that years ago: the small Baltic (or Nordic, as it internally prefers) country called Estonia—an area as big as the Netherlands, but with only 1.3 million inhabitants, like the city of Munich. In the past, it was a part of the Eastern Bloc and occupied for half a century by the Soviet Union (with dark times especially during Stalinism and as part of the Russification policy, which included settlement policies and restrictions on daily life).

These days, Estonia gets a lot of attention as a laboratory for the possible future of our states and societies. The immediate difference between e-stonia and other countries.

The *New Yorker* wrote in December 2017 about "Estonia, the digital republic: Its government is virtual, borderless, blockchained, and secure. Has this tiny post-Soviet nation found the way of the future?" (Heller 2017). The article said, "E-Estonia is the most ambitious project in technological statecraft today" (ibi). However, it wasn't made as an ambitious project but as a pragmatic solution that was the most efficient

Political Consultant for the Estonian state in the field of human rights, gender equality and against working discrimination.

F. Hartleb (✉)
Tallinn, Estonia
e-mail: florian_hartleb@web.de

© Springer Nature Switzerland AG 2020   215
D. Feldner (ed.), *Redesigning Organizations*,
https://doi.org/10.1007/978-3-030-27957-8_16

one for a country that had to start from "zero" after gaining re-independence in 1991. Young decision-makers, such as historian Mart Laar, saw a light at the end of the tunnel.

Other countries seem to be far behind; for example, the economic superpower Germany, where the process of digitalization is just starting—almost two decades later than in Estonia, at least in terms of e-government but also related to the digital mindset and infrastructure. Access to the Internet became a basic human right in Estonia back in 2000, prompting an intensification of efforts to expand connectivity to rural areas, which would enable the government to develop and offer its online services more widely and equitably among the citizens.

Concerning the languages, Estonia seems to be quite isolated (an extra tree with Finnish and Hungarian as counterparts of the Finno-Ugric language family); but concerning the digital world, it seems to speak in the common language.

In terms of innovation and digitalization, the country seems to be in the center of the tree—at least in Europe—and also for NATO since the NATO Cooperative Cyber Defence Centre of Excellence is based in Tallinn. Estonia portrays itself as a "startup" ("state-up") and digital wonder. The e-Estonia showroom, close to the Tallinn airport, is advertising this, in the capital—the former industrial park Telliskivi—reflects this spirit. Indeed, the potential for innovation seems to be high in a country where Skype was intellectually engineered in 2003 and a competition of ideas is promoted.

Observations in this article are based on self-experiences, on four years as an e-resident of Tallinn, and on self-organized trips for business people, politicians and students. A Bertelsmann-Stiftung study surveyed the 50+ generation between 1st March till 30th April 2016, in Estonia (in Estonian language and anonymous). 143 out of 212 (67% quota) surveyed persons answered (Hartleb 2016, 2018).

The following analysis will serve as an in-depth analysis of the digitalization process of a country, including discussing the potential risks. The German magazine *Der Spiegel* (online version) labelled it the "European Silicon Valley" (Kaminski 2015), whereas the magazine's print version fully rejects the model in a sarcastic tone, calling it "Cyberblabla in Laptopia" and identifying a gap between rhetoric and reality (Schmundt 2016). Bloomberg.com regards Estonia as an "overhyped Silicon Valley" (Bershidsky 2015). In other words: Is e-Estonia just a marketing label, a role model and trendsetter, the mirror of a digital revolution, or a new "1984", a totalitarian approach in the sense of George Orwell? Can the member state of the European Union be regarded as a European answer to the USA, as well as to the emerging Asian market: China, South Korea, Singapore and India? The debate itself is astonishing since in 1991, the country started rebuilding itself on the ground of a collapsed planning economy based on collectivism, without getting huge support and financial aid such as happened with Eastern Germany, for example (the so-called Solidaritätszuschlag).

The E-Estonia Movement as Top-Down Process

The e-Estonia movement has emerged as a top-down process—with strategy and vision—and is sometimes even described as a "digital ideology" (Vaarik 2015). Young Estonian politicians, until now often under 40 years old (including the current and previous Prime Ministers), after the generational cut in a young post-communist society, decided that the country would compensate for a relative lack of raw materials through technology. The country's digital transformation began at the very top. Since 1999, the Estonian cabinet has worked entirely digitally—starting with desktop computers and later incorporating laptops and tablets (see Fig. 16.1). On the one hand, there was no bottom-up approach with interest groups, street movements and revolutionary anti-establishment parties—such as the Pirates, a political movement, which was successful for a couple of years in Sweden, Iceland and Germany. On the other hand, there was also no conservative resistance against the technological process.

The country started the digitalization process at the end of the 1990s with online-banking, which was the first e-service that most people quickly used and are most confident in (see Fig. 16.2). The largest online banks have been subsidiaries of existing Estonian banks (formerly Hansapank, etc.) and affiliates of Scandinavian banking groups (Swedbank, SEB, Nordea or Danske Bank). But there is also a historical linkage, specifically regarding the Estonian Personal Identification Code, or *isikukood*. The isikukood was created in 1990, even before the liberation of Estonia from the Soviet Union. It was conceived as an instrument against occupation, because it enabled dissident movements to register and organize the native Estonian population—an act that was explicitly forbidden by the Soviets. Later, the isikukood

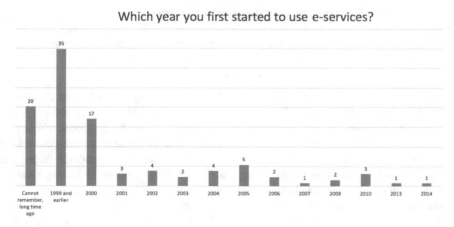

Fig. 16.1   Poll. *Own source* Hartleb (2016)

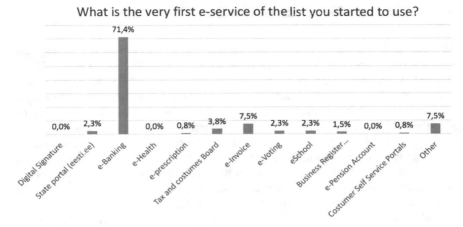

Fig. 16.2 Poll. *Own source* Hartleb (2016)

was integrated into the new "eID" system described below and extended to physically resident non-citizens. This is, perhaps, one reason for its acceptance within the population.

At the beginning of the twenty-first century, the state started systematically, with a step-by-step approach:

- 2000—World's first e-government with paperless e-cabinet sessions
- 2000—World's first mobile parking system
- 2001—Introduction of X-Road, the decentralized backbone of all public e-services
- 2002—Introduction of mobile payments
- 2005—World's first internet voting in official elections
- 2009—Electronic registration of companies (up and running in 15 min)
- 2013—Estonia becomes a pilot country in computer-based math education
- 2014—99% of banking transactions are done electronically in Estonia
- 2014—Estonia launches the world's first e-residency.

In 2018, almost all Estonians are part of the digital society, including the Russian minority, which is less educated and not fully integrated (some don't even have any citizenship, the so-called grey passport). Estonian citizens use the electronic ID cards introduced in 2002. They not only serve as proof of identity, but also provide access to over 200 online government services. These include a virtual health database, through which doctors deliver and renew prescriptions. Nothing is handwritten anymore; instead, you can go online to book your appointment at doctors and specialists. The same services have been available via mobile ID since 2007. Visiting a post office is becoming a thing of the past as the digital signature is now as common as the handwritten. Indeed, Estonians can sign all manner of forms digitally—as easily as converting a Word document to pdf. By 2012, 95% of the country's citizens had filed their tax returns online—more than anywhere else in the world (according to the state officials). Estonians don't know copy shops or the job of being a tax adviser.

From December 2014 on, Estonia has opened its e-services to the world by offering the ability to become a "digital citizen" (without having the analog citizen rights or distributing a visa for digital nomads). Anybody can be an Estonian e-resident, getting an Estonian government digital ID and receiving access to digital signing, online banking and digital services for establishing and running their companies (Kotka et al. 2015; Sikkut 2017: 95). Especially for freelancers, this sounds interesting (33.438 applicants till 29th May 2018). Countries such as South Korea want to follow this business-driven pilot project.

The Role of the X-Road

In Estonia, a smart chip national ID card or a special SIM card in a mobile phone offers the opportunity for electronic authentication and digital signing. Both are issued by the state. The setup is based on a two-factor authentication: a combination of a physical key (the ID card or mobile ID) and electronic key of a PIN code that only the individual knows. The backbone of Estonia's digital security is a blockchain technology called K.S.I. The basic key to understand how it functions: Estonia relies on a government-run technology infrastructure, called X-Road, that links public and private databases into the country's digital services (see Fig. 16.3). X-Road is a platform, an environment, for efficient data exchange, but at the same time, it has no

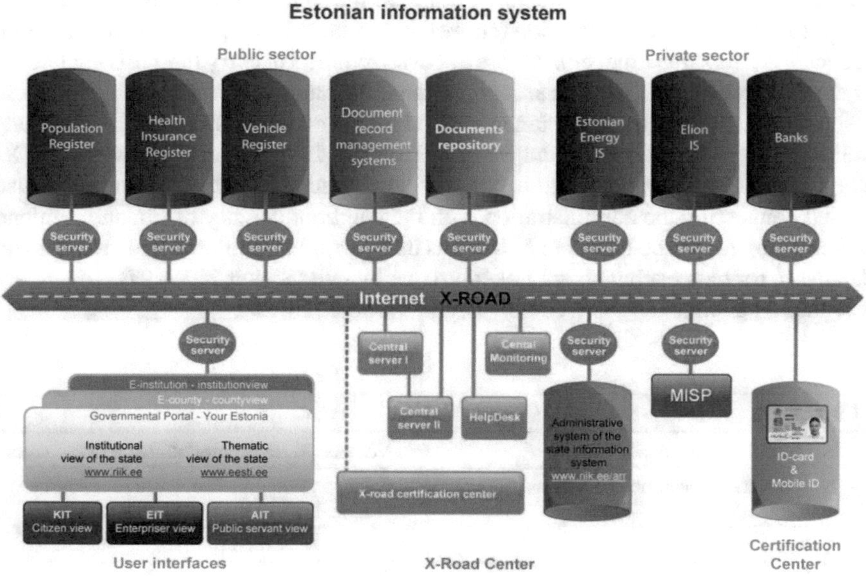

Fig. 16.3 X-road system. *Source* Estonian World (2013)

monopoly over individual data repositories that belong to the www.eesti.ee, the Governmental Portal. The software allows e-service providers and databases to exchange data and members of the public to retrieve their official documents. However, this needn't conjure up images of an Orwellian future, a new *1984*; authorities only have access to information that concerns them directly. What's more, users can see who has viewed details and when—and unauthorized access is severely punished. Data do not get lost or stolen. All personal information is kept on separate servers and behind distinct security walls of government agencies, but the system allows the state and businesses, like banks, to share data when individuals give consent.

Estonian state institutions are structurally incentivized to join the X-Road, simply because they can design services that would not be as efficient or convenient to develop and maintain individually.

– Decentralization
– Interoperability
– Open platform
– Open-ended process.

Estonia has adopted the "once only rule": Under Estonian law, government agencies should not ask people for data that any other agency holds. X-Road is particularly suitable for queries involving multiple agencies and information sources. For example, checking vehicle registration data requires data retrieval from the population registry and vehicle registry (i.e., two otherwise unconnected data repositories). According to the State Information Authority, the conventional offline approach would require three police officers working on the request for about 20 min. With the X-Road, the entire information retrieval is conducted by one police officer within seconds. At the same time, citizens are not even required to carry their driving license or the car's registry documents around, as the information system that the police use displays the status of these documents based on the driver's ID card or license plate number in real time (Solvak and Vassil 2016: 28). E-ambulance is keyed onto X-Road and allows paramedics to access patients' medical records. Estonia has also started connecting the administration with the neighbor country of Finland. Finland is then going to use the Estonian X-Road as its data exchange platform. Both national databases for data exchange are linked up one by one (Sikkut 2017: 97).

Mutual benefit example (see Solvak and Vassil 2016: 29).

Citizens	Civil servants
– No administrative paperwork	– No need to revise mountains of documents
– No personal appearance (time wasted by waiting, etc.)	– Focus on outcome (no waste of resources, practical use of once-only principle)
	– Access to X-road

Pillars

- Population registry
- Social insurance board
- Health insurance
- Tax and customs office
- etc.

The *New York Times* stated in 2014, "Estonia's willingness to use digital products sets it apart from France and Germany, where people have objected to keeping data online. Estonians have embraced the concept. (…) While Europe and the United States debate the role of technology in people's daily lives, Estonia has welcomed it as a fact of life, largely shooing away concerns about data privacy that have become hot-button issues elsewhere" (Scott 2014). In other words: The country obviously implemented the so-called Nordic values (Vaarik 2015):

- Transparency (fewer "dark" secrets, more open data)
- Freedom (protection of human rights and fundamental freedoms, net neutrality, etc.)
- Responsibility (civil duties, footprint, protecting the weak)
- Trustworthiness (predictability, keeping promises)
- Egalitarianism (equal opportunities)
- Inclusion and cooperation (access, inclusive politics, communities)
- Innovation (striving for the better, economic development).

The Estonian system is based on trust—a key word the digital pioneers are using (Fig. 16.4). It seems that the population has the same approach. This is an astonishing aspect due to the history of the country being occupied from totalitarian regimes— Stalinism and National Socialism—in the twentieth century.

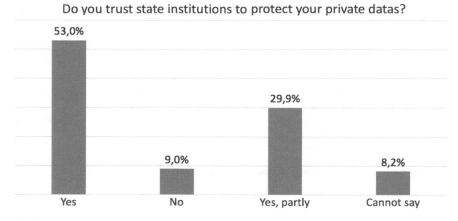

Do you trust state institutions to protect your private datas?

53,0%	9,0%	29,9%	8,2%
Yes	No	Yes, partly	Cannot say

Fig. 16.4 Own Poll. *Own source* Hartleb (2016)

The Life-Long Change of Mindset

e-Estonia is a mirror of a digital (r)evolution we are all facing. As the Estonian Margus Simson, who has worked for a long time as a consultant in the IT sector (including banking) charted the transformation (see Fig. 16.5).

Is the *homo technicus* the next stage in human evolution? Homo Technicus could be defined as a human living in symbiosis with technology and machines. A human constantly connected to the digital world, such as the Internet or the cloud. Estonia could be regarded as a role model with regards to the entire life cycle of its citizens (see Fig. 16.6). A newborn baby is already part of the system. The data can automatically go from the birth hospital directly to the state registration and to the doctors. Only the name must be chosen.

Fig. 16.5 Digital revolution. *Source* Simson (2018)

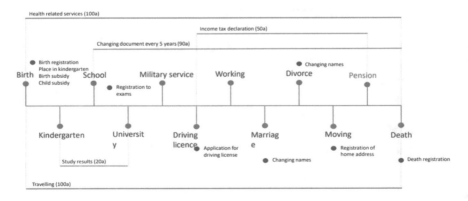

Fig. 16.6 Life-long model of a digital citizen. *Source* Simson (2018)

Aspects of Democracy and Participation

Would the term "technocracy" be adequate to describe the philosophy of e-Estonia? The term would refer to a "management of society by technical experts" (Kalb 1999)—here, the creators of the platforms and the interest groups for a neoliberal state based on IT. This statement would neglect the advantages of a digital implementation. In principle, throughout the country, access to the Internet is guaranteed, even on the small islands (4G standard). This means that the often referred to gap between the center and the periphery is diminished in the digital society. Does the system bring more democracy? The e-voting system, which one third of the voters are using (according to the last elections), didn't fulfill some optimistic assumptions when it was first introduced (see Fig. 16.7). At least, it hasn't increased the low turnout (Solvak and Vassil 2016) but it is still the only country in the world providing remote electronic means to its citizens (Vinkel and Krimmer 2017: 179). A critical mass exists that also currently exists in most western democracies, the typical populist party voter:

– disappointed by politics;
– anti-establishment attitude;
– open to conspiracy theories.

Unlike in Germany or the U.K., senior citizens in Estonia are fully on board with the digital revolution (see Fig. 16.8). In particular, the subject of a clash between the generations is nonexistent because the process started two decades ago.

And it is not only the older generations coming to grips with digitization. Children are learning programming in primary school—solving technical problems, as well

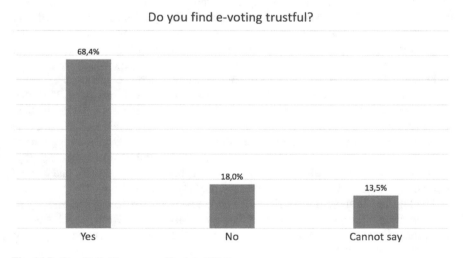

Fig. 16.7 Own Poll. *Own source* Hartleb (2016)

Do you find the digital progress has increased a gap (divide) between the generations?

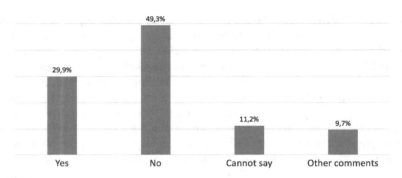

Fig. 16.8 Own Poll. *Own source* Hartleb (2016)

as building and controlling robots. If Europe aims to compete digitally with the U.S., Israel, India, China and Singapore, it will have to follow Estonia's lead.

A Fragile, Vulnerable System?

In 2007, Estonia was the target of a massive cyberattack. The online services of Estonian banks, media outlets and government bodies were taken down by unprecedented levels of Internet traffic. Massive waves of spam were sent by botnets, and huge numbers of automated online requests swamped servers. The result for Estonians citizens was that cash machines and online banking services were sporadically out of action; government employees were unable to communicate with each other via email; and newspapers and broadcasters suddenly found they couldn't deliver the news. Estonian government officials said that evidence suggested the attack was orchestrated by the Kremlin, but they did not have any concrete evidence. 2007 was a wake-up call, helping Estonians to become experts in cyber defense today. That type of situation has never happened again (BBC news 2017). Estonia's online medical portal routinely crashed after digital prescriptions were introduced in 2010, because retirees—the main users of regular prescriptions—kept signing into the system to renew their medication on the day they all received monthly social security payments. In 2017, one decade after the cyberattack, the national security had to deal with another potential threat: the theoretical hacking of the system.

The bug lies in the chipset's firmware code that generates key pairs, and it was discovered by a team of researchers at Masaryk University in Brno, Czech Republic. Infineon security chips manufactured from 2012 onwards, including the latest versions, are all vulnerable (Leyden 2017). The Estonian government revealed the original flaw in September 2017, but gave no details. At the time, it said the flaw affected 750,000 ID cards, and it closed its public key database as a precautionary measure. The citizens and e-residents had to update their certifications (PIN codes).

Estonia is the only country until now to have fully realized the vulnerability of its critical governmental data and acted to ensure its protection. The state authorities decided to have a server resource that is 100% under the control of the Estonian government but is located outside of Estonian territory. Currently, there are specific procedures that are followed to back up necessary data and applications, so that the service availability can be restored using the backup copy if necessary. The first data embassy is now established in Luxembourg as a start (others should follow). The goal is to procure resources under bilateral agreements from the "Government Clouds" of states that are "friendly to Estonia". The Estonian state would sign a bilateral treaty, under which Estonia will rent special floor space or an enclosed room in an existing data center that has been constructed and operates according to necessary standards.

Estonia became the first European country to openly discuss the prospect of a digital currency managed by the government and offered to the country's more than 20,000 e-residents. Kaspar Korjus, born in 1987, was till February 2019 the head of the 15-person team for the e-residency program, and he proposed that the nation have its own cryptocurrency, the so-called Estcoin. Does this go too far? European Central Bank (ECB) President Mario Draghi rejected this idea straightaway with the argument that "a rise in popularity in so-called cryptocurrencies, which are normally issued by private companies and exist only in electronic form, has been worrying the ECB, which has said they could in theory erode its control over the supply of money" (Reuters.com 2017). In other words, Estonia's only currency should be the Euro. Another point for debate is the abuse of the e-residency for people who seek to easily obtain entrance to the European market. The question arises of how the Estonian banks can overview the transfer of money, including the payment of taxes, etc. The skepticism is as growing from the bank side (Merten 2018). Even the digital mastermind Taavi Kotka expressed his concerns recently on TV.

How to Transform or Export the Model?

The Estonian government calculates that e-government achieves a cost savings of 2% of GDP each year. The boldest savings claims include hospital waiting times cut by one-third and digital elections that cost less than half their analog equivalent (Heath 2017). Saving time and money—is this not an incentive to export this model, which is taken for granted? The former President Toomas Hendrik Ilves, who grew up and studied in the USA, represents a hands-on mentality and a cosmopolitan approach. As the representative of the state between 2006 and 2016, he can be regarded as the father of the digital strategy. In 2017, he was honored to receive the Reinhard-Mohn-Prize from the Bertelsmann-Stiftung, dedicated to "smart countries". Estonia competed with countries such as Sweden, Austria and Israel and won the competition. In his award speech, Ilves pointed out the basics for creating a digital state:

1. You need a strong digital identity, guaranteed by the government. (…)
2. To get the benefits of digitization, you need to give this digital identity legal status (i.e., to make a digital signature equivalent to a physical signature). (…)
3. The identity must be mandatory and universal. Why? If getting it is optional, only about 15–20% of the population will take it. (…) In four months, Estonia went from individual users of digital prescriptions to having 98% as users. In four months. No one uses paper prescriptions, except for tourists.
4. Use the power of the ID to transform bureaucracy. Bureaucracy is some 5000 years old. (…) With a digital ID, all the necessary searches are done in parallel. This is why in Estonia we have a "once only" regulation (…)
5. You need the proper architecture for the back end. Indeed the back end is the backbone of the system. We use a distributed data exchange layer, which means every interaction is directly between the user and the server, and it is authenticated each time. (…)

All these solutions are technological and digital, but all these solutions require analogs: policies, laws and regulations (Ilves 2017: 6–7). Even after the hacking issue in autumn 2017, the population trusts the system, a core value for the digital society.

Germany is a very different case in terms of having a completely different system design in comparison. Some people are sarcastic or even jealous, believing that paper-based processes are fully secure. Returning to Germany, what is missing is sustainability. Only companies that seek concrete B2B relationships have different experiences. In general, the "German angst" (debate about the risks, as opposed to the opportunities) is laden with the following arguments:

– Data protection (with some irrational fears and the existing trust in the analog system)
– Federalism
– Older generation (two-class society, digital divide)
– Only for nerds (experience with the Pirate parties—the party focused on never-ending debates; in a party congress, they discussed the possibility of time travel (see Hartleb 2013).

But more importantly, there is a resistance among civil servants in the public sector who are afraid of any changes while being part of a secure subsystem that follows the status quo (Besitzstandsdenken). In Estonia, a distinction exists between the public and private sector. Being a civil servant is not in the focus of Estonian university alumni—a big difference, for example, to Germany in terms of state philosophy and basics. The Estonian state has the primary goal to be more and more efficient and providing services within a close cooperation with the business, whereas the German state is based on hierarchies and loyalties, inner-systems based on a strong civil servant sector with tendencies of keeping the status quo (Drechsler 2018). Between 1st March and 15th May in 2017, in a poll in the public administration in Bavaria about current attitudes toward digitalization it became visible that for them "this topic is something for the IT person". What also became obvious was that the process towards digitalization is seen as relevant but only realistic for implementation if IT and cyber security will be guaranteed. The widespread lack of skills, and very little attention of public authorities to the citizen's needs may be main reason to that (Hartleb 2017).

The German business sector must consider that, even when it comes from the powerful voice of Industry 4.0. The difference to Estonia is already huge. Also the refugee crisis has shown the need for a professional data exchange; in addition to this, the revolutionary digitalization of the labor markets requires another mindset. A European boost is needed for that—otherwise the winning trophy will go to U.S. companies like Amazon, Facebook, Google, etc., or to China, Singapore or India. It is not a question of money, but of an open mindset diminishing the power of bureaucracy and lobbyist resistance to the "winds of change".

References

BBC news. (2017). How a cyber attack transformed Estonia. 27 April 2018. http://www.bbc.com/news/39655415. Accessed on May 27, 2018.

Bershidsky, L. (2015). Estonia's Overhyped Silicon Valley. In Bloomberg.com, 5 March 2018. https://www.bloomberg.com/view/articles/2015-03-05/estonia-s-overhyped-silicon-valley. Accessed on May 27, 2018.

Drechsler, W. (2018). Software—das Ende des Staates? Was Europa von Estland lernen kann. In Rat für Forschung und Technologieentwicklung Re:thinking Europe (pp. 288–305). Positionen zur Gestaltung einer Idee, Wien.

Estonian world. (2013). Estonia to export its Data Exchange Layer X-Road to Finland. http://estonianworld.com/technology/estonia-export-data-exchange-layer-x-road-finland/. Accessed on May 27, 2018.

Hartleb, F. (2013). Digital campaigning and the growing anti-elitism: The Pirates and Beppe Grillo. *European View, 12*(1), 135–142. (Springer Press).

Hartleb, F. (2016). E-government. Von Estland lernen? Innovative Verwaltung. *Fachzeitschrift für modernes Verwaltungsmanagement, 7–8*, 38–41.

Hartleb, F. (2017). Montgelas 4.0. Der Freistaat Bayern auf dem Weg zu einem modernen e-government. Ergebnisse einer repräsentativen Umfrage im Auftrag von Adobe Systems. Munich. https://blogs.adobe.com/digitaleurope/files/2017/06/Montgelas-4.0.pdf. Accessed on May 27, 2018.

Hartleb, F. (2018). Die Mär von der Zweiklassengesellschaft durch Digitalisierung. Empirische Befunde aus dem IT-Land Estland. Verwaltung & Management. *Zeitschrift für moderne Verwaltung, 24*(2), 100–106.

Heath, R. (2017). Test driving the ultimate connected society: (E-)stonia. In Politico, 23 May 2018. https://www.politico.eu/article/test-driving-the-ultimate-connected-society-e-stonia/. Accessed on May 27, 2018.

Heller, N. (2017) Estonia, the digital republic. In New Yorker, December 2017, https://www.newyorker.com/magazine/2017/12/18/estonia-the-digital-republic. Accessed on May 27, 2018.

Ilves, T. H. (2017). Dankesworte for the win of the Reinhard Mohn-Prize 2017 'Smart country'. Speech manuscript, Bertelsmann-Stiftung, Gütersloh, 29 June 2017. http://www.bertelsmann-stiftung.de/fileadmin/files/Projekte/72_Reinhard_Mohn_Preis/Rede_Reinhard-Mohn-Preis-2017_Dankesrede-Toomas-Hendrik-Ilves-englisch_20170629.pdf. Accessed on May 27, 2018.

Kalb, J. (1999). After technocracy and postmodernism. *Modern Age, 41*(2), 168–172.

Kaminski, K. (2015). Estland, das Silicon Valley Europas? In Spiegel online, 14 March 2015. http://www.spiegel.de/netzwelt/web/estland-ein-einblick-in-die-start-up-szene-von-tallinn-a-1022184.html. Accessed on May 27, 2018.

Kotka, T., Alvarez del Castillo, C. I. V., & Korjus, K. (2015). Estonian e-Residency: Redefining the National-State in the Digital Era. Working Paper, University of Oxford. http://www.politics.ox.ac.uk/materials/centres/cyber-studies/Working_Paper_No.3_Kotka_Vargas_Korjus.pdf. Accessed on May 27, 2018.

Leyden, J. (2017). Never mind the WPA2 drama… Details emerge of TPM key cockup that hits tonnes of devices. The Register, 16 October 2017. https://www.theregister.co.uk/2017/10/16/roca_crypto_vuln_infineon_chips/. Accessed on May 27, 2018.

Merten, M. (2018). E-residency. Taugt Estland als Vorbild für andere Staaten? In WirtschaftsWoche, 23 March 2018. https://www.wiwo.de/lifestyle/digitale-verwaltung-taugt-estland-als-vorbild-fuer-andere-staaten/22574632-all.html. Accessed on May 27, 2018.

Precht, D. (2018). *Jäger, Hirten, Kritiker. Eine Utopie für die digitale Gesellschaft.* München: Goldmann.

Reuters.com. (2017). ECB's Draghi rejects Estonia's virtual currency idea. 7 September 2017. https://www.reuters.com/article/us-ecb-bitcoin-estonia/ecbs-draghi-rejects-estonias-virtual-currency-idea-idUSKCN1BI2BI. Accessed on May 27, 2018.

Schmundt, H. (2016). Estland. Cyblabla in Laptopia. In Der Spiegel, 9 January 2016 (pp. 102–103).

Scott, M. (2014). Estonians embrace life in a digital world. In New York Times, 8 October 2014. https://www.nytimes.com/2014/10/09/business/international/estonians-embrace-life-in-a-digital-world.html. Accessed on May 27, 2018.

Sikkut, S. (June, 2017). Creating the super smart society. *IM io. Das Magazin für Innovation, Organisation und Management,* (2), 94–97.

Simson, M. (2018). Presentation, E-Estonia, E-Estonia-showroom, Tallinn.

Solvak, M., & Vassil, K. (2016) E-voting in Estonia. Tartu. https://skytte.ut.ee/sites/default/files/skytte/e_voting_in_estonia_vassil_solvak_a5_web.pdf. Accessed on May 27, 2018.

Sundberg, M. (2014). Old world language families. http://www.sssscomic.com/comic.php?page=196. Accessed on May 27, 2018.

Vaarik, D. (2015). White Paper on Estonian's digital ideology. Tallinn. https://www.mkm.ee/sites/default/files/digitalideology_final.pdf. Accessed on May 27, 2018.

Vinkel, P., & Krimmer, R. (2017). The how and why to internet voting an attempt to explain E-Stonia. In R. Krimmer et al. (Eds.), *Electronic Voting* (pp. 178–191) Heidelberg: Springer.

Dr. Florian Hartleb is currently working in Tallinn, Estonia as a political consultant dealing with the digital society, including a project for the Bertelsmann Stiftung on "smart countries". He also works for the Estonian state in the field of human rights, gender equality and against working discrimination (equality body). Next to it, he is currently lecturing at Catholic Universität Eichstätt in Germany. He also contributes to international media and events on current political topics such as populism, euroscepticism and terrorism. He studied political science, law and psychology at Eastern Illinois University (US) and the University of Passau and gained his doctorate from Chemnitz University of Technology in 2004. Florian Hartleb subsequently worked as a consultant in the German parliament (Bundestag), as a research associate at Chemnitz University of Technology and as a professor for political management at a private university in Berlin. He has co-authored high-school textbooks and published a book on lone wolf terrorism (in German, Einsame Wölfe, Hoffmann und Campe: Hamburg, 2018; in English as revised and updated version for Springer in 2020).

Chapter 17
Governance and Digital Transformation in Hong Kong

Stephen Thomson

Digital transformation is changing the way in which we live. In Hong Kong, there is evidence that some businesses are using digital transformation to offer new products and services, and to change the way in which they offer existing products and services. This is seen in such diverse areas as online retail banking, ubiquitous digital payment cards, car-sharing services, pay-as-you-go bicycle booking apps, accommodation booking services, instant voice messaging as an alternative or supplement to face-to-face communication and e-channels at immigration checkpoints, with smart ID card and thumbprint recognition technology. Nevertheless, there are many examples of commercial enterprises in Hong Kong that have not readily embraced digital transformation or even modernization, such as taxis (most of which do not accept card payments; taxi operators also fiercely resist the expansion of Uber in Hong Kong), workplaces (which often use clocking-in devices and fax machines rather than more modern alternatives) and a lukewarm approach to flexible working arrangements (Siu 2017).

The government of the Hong Kong Special Administrative Region (HKSAR) has launched a number of initiatives aimed at facilitating and harnessing digital transformation. Since 1998, the government has operated a Digital 21 Strategy aimed at developing digital initiatives in Hong Kong. In 2015, the Innovation and Technology Bureau was established as a government department, with responsibility for formulating innovation and technology policy, and which oversees the Innovation and Technology Commission and the Office of the Government Chief Information Officer. In 2017, the government published a Smart City Blueprint (CIO HK 2017), which is divided into six major areas, namely Smart Mobility, Smart Living, Smart Environment, Smart People, Smart Government and Smart Economy. This includes some fairly modest proposals, such as providing real-time information on franchised

S. Thomson (✉)
School of Law, City University of Hong Kong, Hong Kong SAR, China
e-mail: sthomson@cityu.edu.hk

Ombudsman, City of Hong Kong, Hong Kong SAR, China

© Springer Nature Switzerland AG 2020
D. Feldner (ed.), *Redesigning Organizations*,
https://doi.org/10.1007/978-3-030-27957-8_17

buses through mobile devices and at bus stops, facilitating a QR code standard to promote the wider use of mobile retail payments, digitization of public and private sector data and the phasing down of coal-fired electricity generation (which is important, but long overdue).

However, it also includes some more ambitious proposals, such as setting up a Big Data Analytics Platform for facilitating healthcare-related research, introducing smart lamppost technology and using remote sensor devices to monitor air pollution, cleanliness of public places and usage of litter and recycling bins. Funding and operational support has also been made available to promote and advance the digital transformation agenda, including the Innovation and Technology Fund, the Innovation and Technology Fund for Better Living, the Research and Development Cash Rebate Scheme and three incubation schemes run by the Hong Kong Science and Technology Parks Corporation. The government is also increasingly using digital solutions to execute service delivery as part of an e-governance strategy, including in relation to government procurement (e-Procurement), management of tax affairs (eTAX), local university admissions (JUPAS), immigration services, license and permit applications, processing of port formalities (eBS) and bill payments.

Digitization and digitalization can also support values of transparency and open access in government (Collin 2015). This is particularly important in Hong Kong, which lacks important democratic mechanisms found in many Western jurisdictions. The government completed the first part of its electronic information management study in 2010, recommending a government-wide strategy on transparency and open access which develops electronic records management, shared among government bureaux and departments, to reduce implementation costs, time and risk and, thereby, improve operational efficiency (CIO HK 2010). The government is also rolling out an information technology infrastructure for human resources management, known as Government Human Resources Management Services. This was initially implemented in four government bodies, namely the Efficiency Unit, the Rating and Valuation Department, the Office of the Government Chief Information Officer and the Civil Service Bureau. The infrastructure is supported by a Government Cloud Platform, which was proposed to achieve cost savings through economies of scale and resource sharing; time savings through streamlined procurement and system implementation and on-demand service provision; enhanced agility in meeting the demand of information technology services; and fostering development of the information technology industry (CIO HK 2012a). The "GovCloud" was launched in 2013 (CIO HK 2012b) with an initial financial commitment of 242 million HKD (approximately 30.8 million USD).

The government also relaunched its Public Sector Information portal in 2015 to facilitate the dissemination of datasets provided by governmental and public bodies. This makes available a diverse range of datasets, including those relating to climate, the environment, education, employment, health, housing, social welfare and transport. Hong Kong was ranked in 24th place (joint with Germany and Romania) of 94 jurisdictions in the most recent Global Open Data Index (2016/2017), which measured the state of open government data publication. Among the lowest ranked aspects of Hong Kong's data were administrative boundaries, company registers, government spending and land ownership, while the highest ranked aspects were national statistics, procurement and air quality. Hong Kong ranked behind a smaller number of other Asian jurisdictions, such as Taiwan (1st), Japan (16th) and Singapore (17th), but ahead of others such as India (32nd), Thailand (51st), Indonesia (61st) and Myanmar (Burma) (94th). It is worth noting that there are questions over the methodology used. For example, Hong Kong is given the lowest possible rating for "locations", which is a rating for a database of postcodes and corresponding spatial locations in terms of latitude and longitude, or similar coordinates in an openly published coordinate system. However, Hong Kong does not use a system of postcodes, and there has been no comprehensive address standardization. It is also questionable how useful a coordinate system would be in a territory with such densely populated urban areas. In addition, conspicuously absent from the 2016 index were Mainland China and South Korea. Rankings for individual aspects of data were measured according to whether the data was openly licensed, in an open and machine-readable format, downloadable at once, up-to-date, publicly available and available free of charge.

It is nevertheless important to note that the use, or increased use, of information technology does not constitute "digital transformation" in the full sense of the term, for it does not necessarily involve deep process changes. Thus many aspects of e-governance comprise digitization rather than digitalization, though even digitization can raise important issues of approach to governance (Reitz 2006). For example, the more government-related information that is made available in digital rather than analog format, the greater the importance that persons are given the necessary skills and tools to access and utilize that information. This may be particularly relevant to the elderly or those who are financially unable to afford personal internet access. Accordingly, there may be implications for public provision of internet access and training opportunities in the use of information technology. It may be noted in this regard that Hong Kong public libraries provide computer facilities through which members of the public may access the Internet free of charge, and the government's Wi-Fi Program makes free wireless Internet access available at over six hundred locations in Hong Kong. There may, however, be data privacy concerns when using public Internet facilities, notwithstanding the protections of the Personal Data (Privacy) Ordinance (cap. 486). Nevertheless, Hong Kong has yet to embrace e-governance across the board. For example, persons interested in serving on an advisory or statutory body are invited—on a difficult-to-locate page of the Home Affairs Bureau website—to download and complete a curriculum vitae form for subsequent mailing to the government for inclusion in a centralized database known as the Central Personality Index.

The Hong Kong Monetary Authority has also launched a number of initiatives related to smart banking, such as a Faster Payments System, an Enhanced Fintech Supervisory Sandbox 2.0, a policy framework on Open Application Programming Interface (Open API) and the promotion of virtual banking (HK MA 2017). Hong Kong's ability to successfully adapt to and utilize fintech may affect its ability to maintain its role as one of the world's leading financial centers. Notably, the development of the fintech environment in Hong Kong has been said to be behind that of China and Singapore (Chen and Woodhouse 2016; John 2017).

With digitization and digitalization come privacy challenges. For example, the HKSAR Government has, since 2003, issued smart identification cards which store digital information. These can be used for a range of purposes, including immigration control, personal digital certificates for executing secure electronic transactions, library services and health services. Nevertheless, the digitization and centralization of personal data must comply with privacy standards. A review of the Immigration Department's Smart Identity Card System was undertaken by the Privacy Commissioner for Personal Data in 2010, which found that, overall, the Immigration Department had "appropriate policies, practices and guidance in place in handling and processing personal data system[s]", but that further improvements were necessary (HK PC 2010). The Privacy Commissioner for Personal Data has also produced various guidance notes that explain the legal requirements relevant to data privacy in such areas as collection and use of biometric data (HK PC 2015), the Electronic Health Record Sharing System (HK PC 2016) and physical tracking and monitoring through electronic devices (HK PC 2017). These are grounded in the legal regime provided by the Personal Data (Privacy) Ordinance, enforcement of which has reportedly improved in recent years (Hogan Lovells 2014).

Furthermore, while there has been some degree of regulation of emerging technologies, there are concerns that regulation is not keeping pace with technological innovation. For example, regulation of the Internet of Things is largely left to the general law of Hong Kong (Mo and Mok 2016), though a Wireless Internet of Things Licence was introduced in 2017 (HKSAR 2017). Regulation of artificial intelligence (AI) is also broadly left to the general law: it was recently pointed out that, under the law of Hong Kong, if an employer used an AI system to recruit employees, but the system discriminated against candidates (such as pregnant women who would require maternity leave), it is not clear whether and to what extent the employer would be liable under the existing regulatory framework (Singh 2017). In fact, according to the Asian Index of Artificial Intelligence, which analyzed eight major Asian economies for preparedness for and resilience to AI-led changes, Hong Kong ranked in penultimate position. China led the rankings in first position, followed by Singapore (2nd), India (3rd), Japan (4th), Taiwan (5th), South Korea (6th), Hong Kong (7th) and Indonesia (8th). Hong Kong fell significantly behind in AI preparedness, measured by the prevalence of overall startup activity; venture capital raised by top AI startups; students enrolled in science, technology, engineering and mathematics (STEM) subjects at top-ranked universities; and AI publication volume (ABC 2017, p. 3). Meanwhile, it fared just above average in terms of the resilience of the economy to broader structural changes brought about by AI, determined by references to indicative government policies on AI and the employment structure of the economy, particularly the share of middle-skilled work that is vulnerable to AI disruption (ABC 2017, p. 5). Hong Kong also had the second lowest absolute proportion of AI resilience in terms of government AI policies, only slightly ahead of Indonesia in that regard (ABC 2017, p. 5). It was concluded that:

Hong Kong punches notably below its weight relative to its Asian Tiger peers on the [Asian Index of Artificial Intelligence] measure of resilience and preparedness, although it is home to a relatively high number of top computer scientists and it ranks just after Singapore in terms of the number of academic papers published (ABC 2017, p. 15).

The overall picture of digital transformation in Hong Kong is therefore mixed. While the HKSAR Government has made several commitments to, and launched a number of innovations promoting, digital transformation, there remain significant doubts over its approach to regulation of the digital sphere.

First, the authorities have displayed a reactive rather than a proactive approach to regulation. This is by no means unique to governance in Hong Kong, but it is part of a broader concern about the extent to which public authorities have sufficient foresight in their modernization strategy. For example, the Civil Aviation Department recently conducted a public consultation on the regulation of unmanned aircraft systems (UAS), more commonly known as drones (CAD 2018). The current legislative regime provides limited and fragmented regulation of UAS operation. This includes a provision against the reckless or negligent use of UAS, which are considered to be aircraft (Air Navigation (Hong Kong) Order (cap. 448C), Art. 48), the requirement that operators of UAS for hire or reward must obtain a permit from the Civil Aviation Department (Air Transport (Licensing of Air Services) Regulations (cap. 448A), Reg. 22), and the requirement for compliance with telecommunications (Telecommunications Ordinance (cap. 106)) and data privacy laws (Personal Data (Privacy) Ordinance (cap. 486)). However, there is no general registration requirement for UAS or their operators, with a Certificate of Registration and a Certificate of Airworthiness required only in relation to aircraft (including UAS) weighing over seven kilograms without fuel (Air Navigation (Hong Kong) Order (cap. 448C), Arts. 3, 7 and 100). One of the proposals under consultation was that an online registration system be established for owners of UAS weighing over 250 grams, though it has been noted that such a requirement is already in place in Mainland China and the United States (CAD 2018), suggesting that Hong Kong is lagging behind other developed jurisdictions in appropriately regulating smaller UAS. Though the Civil Aviation Department has already published limited information ("Safety Tips for Operating Unmanned Aircraft Systems") on areas in which UAS must not be operated (such as the vicinity of Hong Kong International Airport, Victoria Harbor and Shek Kong—all of which affect aircraft approach and departure paths), it has been noted that other congested areas in Hong Kong could be designated as no-fly zones (CAD 2018). While this would alleviate some of the safety and privacy challenges that UAS pose in congested areas, much of Hong Kong's urban areas are densely populated and could thus potentially be off-limits to UAS.

Another challenge in Hong Kong is regulatory enforcement. An insightful example relates to the existence of legislation that prohibits the idling of motor vehicle engines (Motor Vehicle Idling (Fixed Penalty) Ordinance (cap. 611)), yet there is widespread disregard for that legislation and insufficient enforcement thereof. There are also serious doubts as to whether the legislation constitutes any substantial form of deterrent. The driver of a motor vehicle who violates the prohibition on the idling of engines does not commit an offence (Motor Vehicle Idling (Fixed Penalty) Ordinance (cap. 611), s. 7) but is liable to pay a fixed penalty of only 320 HKD (around 41 USD). This is itself notable considering the tendency of Hong Kong law to impose relatively heavy criminal penalties as a means of deterrent. Contrast this provision with, for example, the sanctions for unlawfully parking a vehicle (including a bicycle – Road Traffic Ordinance (cap. 374), s. 2), which constitutes an offence carrying a fine of 2000 HKD (around 255 USD—Road Traffic (Parking) Regulations (cap. 374C), s. 4(5). In short, the financial penalty for unlawfully parking a bicycle is over six times greater than that for unlawfully idling a motor vehicle engine. The former is also an offence, whereas the latter is not. This contrast is particularly astounding considering that the idling of motor vehicle engines has such patent deleterious effects on human health and the natural environment.

The HKSAR Government's statement of support for digital transformation is also contradicted by some of its actions, which show resistance to digital disruption of traditional industries and sectors. A police crackdown on Uber in May 2017 (Lo and Yau 2017) signaled that ride-sharing was not being encouraged in Hong Kong, even though it could result in greater competition, superior service quality, improved consumer choice and value for money, and enhanced environmental standards given Uber's emphasis on digitized consumer interaction. In November 2017, the Hong Kong Consumer Council recommended gradual regulatory reform, whereby the taxi market would be opened up to e-hailing services. It described the Government's position—in particular, its policy on the introduction of franchised taxis (THB 2017)—as "not appear[ing] to fully satisfy the call from consumers" and "not well suited to the emergence of existing E-hailing services that are popular in many countries" (CC 2017). The Government's reluctance to innovate in the ride-sharing market appears largely to be driven by political considerations; in particular, resistance from traditional taxi operators and monopolization of the taxi license market. Indeed, no new taxi licenses have been issued by the Government since 1994, with the exception of a small number of new licenses issued on a largely rural and sparsely populated island, namely 25 licenses issued in 2016 for Lantau Island (THB 2017). This is despite the increase in the population of Hong Kong since 1994 from just over six million to over seven million people. The lack of (digital) innovation in the Hong Kong taxi industry is a sign of conservatism and defense of the status quo in the face of political pressure, and casts doubt on the Government's commitment to embrace digital transformation, in addition to commercial innovation more broadly. The potential of the Government's digital transformation strategy will be limited if it simply panders to existing commercial interests.

This feeds into a broader issue of economic, social and political culture. Despite Hong Kong being one of the world's leading financial centers and commercial ports, and offering a low taxation regime that can attract investment and stimulate economic growth, Hong Kong tends in many respects towards passivity, and lags behind other Asian (not to mention Western) economies in terms of innovation and modernization. There is in Hong Kong a seemingly insatiable commercialism and a culture of wastefulness, coupled with a poor record on recycling and a general lack of concern for the environment. To take just one example, 35% of municipal solid waste (MSW) sent to landfill sites in 2016 (the most recent year for which statistics were available) was food waste, and this figure was an increase on previous statistics on food waste in commercial and industrial sectors (EPD 2016). Waste paper accounted for 22% of landfilled MSW, and waste plastics accounted for 21% of landfilled MSW (EPD 2016, p. 6). Construction waste disposed of at landfill sites had increased by 5.3% on the previous year (EPD 2016, p. 6), and the average daily quantity of total solid waste disposed of at landfill sites had increased by 1.5% on the previous year, continuing an annual trend of growth (EPD 2016, p. 5). In the context of failing policies on waste and pollution, there is concern that some of the Government's announced Smart City measures relating to pollution and environmental sustainability are at best inadequate and at worst gimmicks. However, the fact that the general population appears to acquiesce (not to mention participate) in the overall culture of wastefulness signifies that there may be insufficient political incentive for the Government to genuinely change course. There is no merit in digital transformation for its own sake: it must be part of a broader effort to improve living conditions. In the context of failing policies on waste and environmental sustainability, resistance to innovation in particular industries and sectors, and a generally reactive rather than proactive approach to regulation of the digital sphere, the Government's stated support for digital transformation may be less encouraging than it at first appears.

However, there is significant pressure on Hong Kong to remain competitive. It is one of the world's leading financial centers, having recently moved ahead of Singapore into third place in the Global Financial Centres Index, trailing only New York and London (Z/Yen 2018). Hong Kong International Airport has the largest cargo volume of the world's airports, handling over 4.6 million metric tonnes in 2016 (the most recent year for which figures are available), representing an increase of 3.48% on the previous year (ACI 2017). Hong Kong also has the fifth largest cargo volume of the world's container ports, handling 20.8 million TEU in 2017, representing a 5.9% increase on the previous year (JOC 2018). Nevertheless, Hong Kong faces increasing competition from other financial and commercial centers, both in China and in the Asia-Pacific region. Though factors such as regulation, the fiscal environment and economic and political stability will continue to play a major role in determining the economic competitiveness of Hong Kong, future investment, attraction of foreign capital and the routing of trade through Hong Kong will be materially affected by the extent to which the economy and infrastructure are sufficiently modernized. This includes the development and harnessing of an effective digital transformation strategy to maximize efficiency, optimize supply and distribution chains, reduce pollution and environmental footprint, improve sustainability, minimize risk and promote economic growth.

It is actions, rather than words, by which the HKSAR Government's commitment to digital transformation will be measured, in the context of a broader strategy to improve quality of life and care for the environment in which we live. The Government has signaled that it supports the digital transformation agenda, but the pace of change is slow and the direction has yet to be convincingly innovative or proactive. There also needs to be a change in culture and attitude towards regulation and the environment if the benefits of digital transformation are to be fully realized. Otherwise, the risk is not only that digital transformation is a short-lived fad, but that a failure to fully realize its potential represents a missed opportunity to improve living conditions.

References

Articles

Agawu, E. (2017). What's next for e-government? Innovations in e-government through a cybersecurity lens. *Nebraska Law Review, 96*, 364.

JOC. (2018). Top 50 world container ports 2017. *Journal of Commerce* (August 2018) https://www.joc.com/port-news/growth-accelerates-top-global-ports_20180816.html?destination=node/3443561.

Mo, C., & Mok, J. (2016). The internet of things: Legal issues from a Hong Kong law perspective. *Hong Kong Lawyer* (October 2016) http://www.hk-lawyer.org/content/internet-things-legal-issues-hong-kong-law-perspective.

Reitz, J. C. (2006). E-government. *American Journal of Comparative Law, 54*, 733.

Singh, H. (2017). Artificial intelligence could discriminate and companies would get away with it—Experts explain why Hong Kong laws need to catch up. *South China Morning Post*, 26 September 2017. http://www.scmp.com/news/hong-kong/economy/article/2112786/updated-laws-needed-hong-kong-can-embrace-artificial.

Books

Collin, J. (2015). Digitalization and dualistic IT. In J. Collin, et al. (Eds)., *IT leadership in transition: The impact of digitalization on finnish organizations* (Aalto University Publication Series, Science + Technology, July 2015) (pp. 29–34).

Laws

Air Navigation (Hong Kong) Order (cap. 448C).

Air Transport (Licensing of Air Services) Regulations (cap. 448A).

Motor Vehicle Idling (Fixed Penalty) Ordinance (cap. 611).

Personal Data (Privacy) Ordinance (cap. 486).

Road Traffic Ordinance (cap. 374).

Road Traffic (Parking) Regulations (cap. 374C).

Telecommunications Ordinance (cap. 106).

Media

Chen, C., & Woodhouse, A. Hong Kong falling behind on fintech, experts warn. *South China Morning Post*, 4 May 2016. http://www.scmp.com/business/markets/article/1941384/hong-kong-falling-behind-fintech-experts-warn.

HK MA. (2017). Hong Kong Monetary Authority. *Press Release: A New Era of Smart Banking* (29 September 2017) http://www.hkma.gov.hk/eng/key-information/press-releases/2017/20170929-3.shtml.

HKSAR Government. (2017). HKSAR Government. *Press Release: Communications Authority creates new Wireless Internet of Things Licence* (1 December 2017) http://www.info.gov.hk/gia/general/201712/01/P2017120100271.htm.

John, A. Plenty of chatter, and even some action, as Hong Kong makes progress on fintech. *South China Morning Post*, 28 June 2017. http://www.scmp.com/tech/start-ups/article/2100298/plenty-chatter-and-even-some-action-hong-kong-makes-progress-fintech.

Lo, C., & Yau, C. *22 Uber drivers arrested in undercover Hong Kong police operation. South China Morning Post*, 23 May 2017. http://www.scmp.com/news/hong-kong/law-crime/article/2095336/21-uber-drivers-arrested-hong-kong-undercover-police.

Siu, P. Hong Kong employers open to flexible work arrangements but fear system abuse: study. *South China Morning Post*, 24 January 2017. https://www.scmp.com/news/hong-kong/economy/article/2064990/hong-kong-employers-open-flexible-work-arrangements-concerned.

Yiu, E. Flexible working options not just wanted by Asia's mothers, but by fathers too. *South China Morning Post*, 29 October 2017 https://www.scmp.com/business/companies/article/2117496/flexible-working-options-not-just-wanted-asias-mothers-fathers.

Studies and Guidance

ABC. (2017). Asia Business Council. *Artificial Intelligence in Asia: Preparedness and Resilience* (September 2017) http://www.asiabusinesscouncil.org/docs/AI_briefing.pdf, pp. 3, 5, 15.

ACI. (2017). Airports Council International. *Cargo Traffic 2016 (Annual)* (2017) http://www.aci.aero/Data-Centre/Annual-Traffic-Data/Cargo/2016-final-summary.

CAD. (2018). Civil Aviation Department. *Regulation of Unmanned Aircraft Systems in Hong Kong: Consultancy Study and Way Forward* (3 April 2018) https://www.cad.gov.hk/reports/UAS_public_consultation_Eng.pdf. pp. 3, 17.

CC. (2017). Consumer Council. *More Choices, Better Service: A Study of the Competition in the Personalised Point-to-Point Car Transport Service Market* (November 2017) https://www.consumer.org.hk/sites/consumer/files/competition_issues/CarHailing/report.pdf, p. 60.

Civil Aviation Department, *Safety Tips for Operating Unmanned Aircraft Systems*. https://www.cad.gov.hk/reports/CAD%20Leaflet-UAS.pdf.

EPD. (2016). Environmental Protection Department. *Monitoring of Solid Waste in Hong Kong: Waste Statistics for 2016* (December 2017) https://www.wastereduction.gov.hk/sites/default/files/msw2016.pdf, p. 6.

HK PC. (2010). Office of the Privacy Commissioner for Personal Data, Hong Kong, *Privacy Compliance Assessment Report on Smart Identity Card System (SMARTICS)* (July 2010) (pp. 81–82).

HK PC. (2015). Office of the Privacy Commissioner for Personal Data, Hong Kong, *Guidance on Collection and Use of Biometric Data*. https://www.pcpd.org.hk//english/resources_centre/publications/files/GN_biometric_e.pdf) (July 2015).

HK PC. (2016). Office of the Privacy Commissioner for Personal Data, Hong Kong, *Personal Data (Privacy) Ordinance and Electronic Health Record Sharing System (Points to Note for Healthcare Providers and Healthcare Professionals)* https://www.pcpd.org.hk//english/resources_centre/publications/files/eHRSS_Points_to_Notes_ENG.pdf (February 2016).

HK PC. (2017). Office of the Privacy Commissioner for Personal Data, Hong Kong, *Physical Tracking and Monitoring Through Electronic Devices*. https://www.pcpd.org.hk//english/resources_centre/publications/files/physical_tracking_e.pdf, (May 2017).

Hogan Lovells, *An Overview of Hong Kong's Personal Data (Privacy) Ordinance: Key Questions for Businesses* (March 2014) http://www.hoganlovells.com/files/Uploads/Documents/14.03_PDPO%20Key%20Questions%20for%20Businesses.pdf, pp. 1 and 4.

Open Knowledge International, *Global Open Data Index 2016/2017*. https://index.okfn.org/place/.

THB. (2017). Transport and Housing Bureau, *Public Transport Strategy Study—Role and Positioning Review: Personalised and Point-to-Point Transport Services* (CB(4)666/16–17(05)) (March 2017) https://www.legco.gov.hk/yr16-17/english/panels/tp/papers/tp20170317cb4-666-5-e.pdf, p. 2.

Z/Yen. (2018). Z/Yen and China Development Institute, *The Global Financial Centres Index 24* (September 2018).

Web References

CIO. (2010). Office of the Government Chief Information Officer, *Strategies & Government IT Initiatives: Electronic Information Management* 2010. https://www.ogcio.gov.hk/en/our_work/strategies/initiatives/eim/.

CIO. (2012a). Office of the Government Chief Information Officer, *Legislative Council Panel on Information Technology and Broadcasting: Implementation of a Government Cloud Platform* (LC Paper No. CB(1)1783/11–12(06)) (May 2012) http://www.legco.gov.hk/yr11-12/english/panels/itb/papers/itb0514cb1-1783-6-e.pdf.

CIO. (2012b). Office of the Government Chief Information Officer, *"Item for Finance Committee: New Subhead 'Implementation of a Government Cloud Platform'"* (FCR(2012-13)39) (May 2012) http://www.legco.gov.hk/yr11-12/english/fc/fc/papers/f12-39e.pdf.

CIO. (2017). Office of the Government Chief Innovation Officer, Hong Kong Smart City Blueprint for Hong Kong (15 December 2017) https://www.smartcity.gov.hk/.

Office of the Government Chief Information Officer, *Strategies & Government IT Initiatives: Government Human Resources Management Services*. https://www.ogcio.gov.hk/en/our_work/strategies/initiatives/shrms/.

Dr. Stephen Thomson is an Associate Professor in the School of Law, City University of Hong Kong, where he is Director of the JSD and Ph.D. programmes. He is a Legal Adviser to the Ombudsman of Hong Kong, a member of the Constitutional Affairs and Human Rights Committee of the Law Society of Hong Kong, and an examiner on the Law Society of Hong Kong's Overseas Lawyers Qualification Examination. He was recently a Herbert Smith Freehills Visitor at the Faculty of Law, University of Cambridge. Dr. Thomson is the author of 'Administrative Law in Hong Kong' (Cambridge University Press, 2018, with a foreword by Hon. Andrew Li), and 'The Nobile Officium: The Extraordinary Equitable Jurisdiction of the Supreme Courts of Scotland' (Avizandum, 2015, with a foreword by Rt. Hon. The Lord Hope of Craighead). Dr. Thomson has published in a number of leading law journals in the UK, US, Australia and Hong Kong. He holds the degrees of LL.B. (Hons.) (First Class), LL.M. (Res.) (by Thesis), Ph.D. and Dip.L.P. from the University of Edinburgh, and has acted in a consultancy and advisory capacity to public bodies and law firms in both Hong Kong and the UK.

Chapter 18
Blockchain—The Savior of Democracy?

Alexander Braun

Introduction

The last couple of years have seen an unparalleled rise of populism in the political arena. From Hungary to Poland, Italy to the United Kingdom, the United States to Brazil, the Philippines and many fractional movements all around the world, political strongmen have exploited these tendencies successfully for their rise to power and to undermine basic democratic principles and institutions.

It is no accident that this development coincides with trust in politicians reaching an all time low, paving the way for simple answers in an increasingly complex world. The financial crisis of 2008 brought the economic system to the brink of collapse and resulted in the worst global recession since 1929. This was accompanied by a massive loss of trust in banks, bankers and financial markets, as the bundling and sale of sub-prime mortgages by investment banks had shaken the global financial system to its core. It is therefore no coincidence that just one month after the collapse of the investment bank Lehman Brothers in September 2008, a mysterious Satoshi Nakamoto published a paper entitled *Bitcoin: A Peer-to-Peer Electronic Cash System*, introducing the concept of a decentralized currency outside the realms of the traditional banking system.

While *Bitcoin's* potential for substituting fiat currencies issued by national banks and disintermediating these institutions in the process still remains to be seen, the technology behind it—the *Blockchain*—enables parties to interact with each other without requiring a third party as an intermediary to build the trust required for a transaction. It has therefore triggered a tidal wave of research projects across all industries where trust is required to enable a market to flourish.

A. Braun (✉)
Creative Construction Heroes, Berlin, Germany
e-mail: alexander@creativeconstruction.de

© Springer Nature Switzerland AG 2020
D. Feldner (ed.), *Redesigning Organizations*,
https://doi.org/10.1007/978-3-030-27957-8_18

With the erosion of trust in politicians and democratic institutions at the center of the recent rise in populism, could the *Blockchain* tackle these trust-issues and usher in the next phase of evolution for democracy?

Loss of Trust in Politics

Although the population's trust in bankers against the backdrop of the financial crisis can only be described as desolate and has provided the initial spark for *Bitcoin* and the *Blockchain*, there is one group that enjoys even less trust among the population: politicians (see Fig. 18.1). At the global level, they are experiencing new lows every year—with negative consequences for the democratic process and the democratic system itself.

Public trust in government is close to historic lows under the administration of Donald Trump in the U.S. (see Fig. 18.2), but international comparisons demonstrate, that this problem is by no means an isolated one (see Fig. 18.3).

However, as the increase in protest movements, petitions and NGOs worldwide shows, this is not due to a lack of will to participate in the democratic process (Firth 2015).

Although there are many potential reasons for the loss of trust, there are a number of key causes that are beyond dispute:

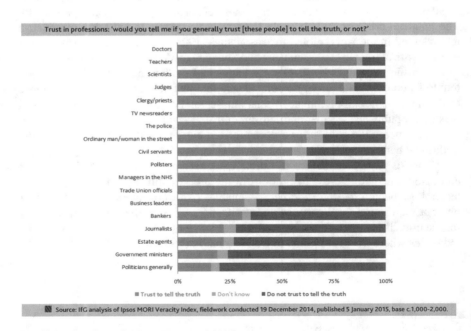

Fig. 18.1 Trust in professions (Freeguard 2015)

Public trust in government near historic lows

% who trust the govt in Washington always or most of the time

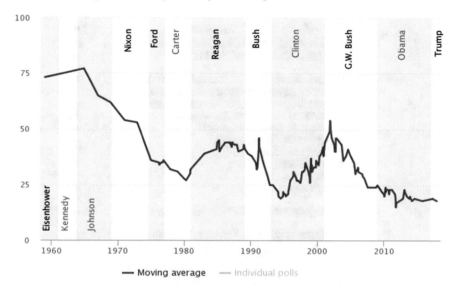

Fig. 18.2 Public trust in government (Pew Research Center 2017/12)

1. **Lack of trust in democratic elections**: Fairness of elections is in doubt.
2. **Lack of trust in the democratic representatives**: Corruption is rampant and fulfillment of election promises is in doubt.
3. **Lack of a sense of influence**: Elections are the only events to influence the process, with no participation in actual decisions made within a cycle.

Lack of Trust in Democratic Elections

Electoral fraud is commonplace in many countries. This can be done either directly—by manipulating ballots and ballot boxes, or by voters not being allowed to vote—or indirectly, by only selectively allowing opponents to vote. The structure of the attribution of votes also strengthens the perception that the democratic process can only be unfair (Stewart 2010). Political representatives—such as U.S. President, Donald Trump—are also actively working to further damage trust in the democratic process by accusing political opponents of manipulating elections (Cillizza 2018).

Few worldwide have a lot of trust in their government

How much do you trust the national government to do what is right for our country?

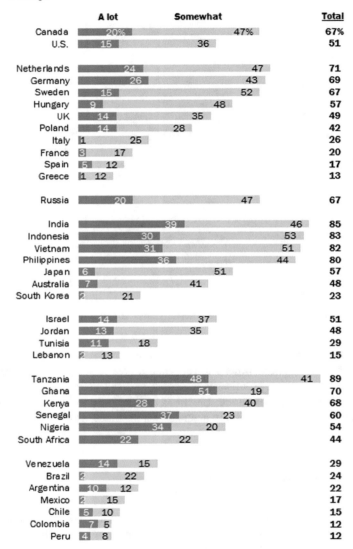

Note: Question not asked in Turkey.
Source: Spring 2017 Global Attitudes Survey. Q4.

PEW RESEARCH CENTER

Fig. 18.3 Trust in government worldwide (Pew Research Center 2017/10)

Lack of Trust in the Democratic Representatives

Corruption is the order of the day in many countries. Even if there are clear rules in many places regarding party donations, political decisions in favor of a stakeholder group and lobbying repeatedly cause headlines and erode the population's trust in their political representatives. Non-compliance with election promises is the norm and further contributes to electoral disenchantment.

Lack of a Sense of Influence

The political representatives are always elected for a period of several years, during which the population is largely excluded from participation in decisions that are made during this period. While the population's trust in the democratic process and the independence of its representatives is low anyway, the power exercised during this period—with only limited chances for influence, which is exercised only occasionally in the elections, but without direct democracy in central decisions—continues to contribute to the feeling of impotence with regard to the political decisions taken. This feeling is amplified, for example, by outcomes of referenda such as the recent one in Germany concerning the operation of Tegel Airport in Berlin, which the current government openly does not intend to honor (Mallwitz 2018).

To grasp the potential of the *Blockchain* to address these problems, a basic understanding of the technology and architecture is required.

The Basic Functionality of a Blockchain

The technology enabling the decentralized digital currency *Bitcoin* is called *Blockchain*. Its use cases reach far beyond establishing a digital currency, however. It promises enormous efficiency gains in all industries that have hitherto relied on intermediaries to establish trust between the contracting parties and to monitor the execution of codified agreements: notaries, attorneys, rights-holders, land registries, banks—there is hardly an industry in which key players currently needed for it to function could not be substituted.

A *Blockchain* system consists of two central components:

1. **Peer-to-Peer Network (P2P)**: A peer-to-peer network consists of a large number of computers (called nodes) that are randomly connected to each other. This enables a decentralized network in which no single node is a single point of failure and in which no single node has the ability to censor information and withhold it from the other nodes in the network, since all other nodes can just bypass that node to redistribute the information. These nodes manage the *Blockchain* database, communicating changes (called transactions) between each other.

2. **Blockchain Database**: The *Blockchain* database stores the complete history of all transactions and the order in which they took place. It consists of blocks containing transactions stacked on top of each other, forming a chain (hence the name *Blockchain*). From the first block containing the *Empty State* (so-called *Genesis Block*) nodes involved in the Peer-to-Peer (P2P) network begin to transmit transactions (=changes of state). Due to the P2P structure, it is difficult to determine who sent a transaction, and censorship of a transaction is not possible. These transactions are collected in a pool of transactions (a transaction block). When a new block is added to the network, a consensus is formed about the order of the grouped transactions, and a cryptographic signature is added to the end of the block. This cryptographic signature links the new block to the previous block and contains some validation information. If any part of the information contained in this block were modified, the signature would no longer be valid, and any node on the P2P network would clearly see that this block had been tampered with. Any block following a manipulated block would also be declared invalid. Since this database is available as a directory of all transactions on all nodes, *Blockchains* are also referred to as *Distributed Ledger Technology (DLT)*. In contrast to a traditional centralized database, which represents a single-point-of-failure and one single target for hackers seeking to change the state of the database, the distributed nature of the *Blockchain* database would require a malicious party to access and manipulate all databases on all nodes at the same time. This and the cryptographic security measures ensure that the *Blockchain* lists the immutable, permanent history of all database changes.

Smart Contracts and Ethereum

In addition to the *Bitcoin Blockchain*, in the meantime, other *Blockchains* have been created that extend the range of functionality and attempt to eliminate identified shortcomings. The most prominent representative is the *Ethereum Blockchain*, whose currency is the so-called *Ether*. This is a programmable *Blockchain* that enables the development of *Smart Contracts*. *Smart Contracts* can be used to map contracts that are automatically executed as soon as the stored condition is fulfilled, using the if-then conditions familiar from programming.

For example, rights management organizations and collecting societies, which are responsible for compensating musicians using a complex and inaccurate tracking and distribution system with a large administrative overhead, can be replaced entirely. In a *Smart Contract*, all artists involved in a song—from composers and songwriters to each individual performer—can be registered, along with their respective shares in the revenue generated. The payout is automatically triggered by the occurrence of the condition defined in the smart contract—such as the streaming of the associated song on a platform such as *Spotify*. There is no longer a need for an organization to monitor, and also no more risk that an intermediary enriches himself and doesn't distribute the funds to the beneficiary parties in accordance with the contract.

Political Trust on the Basis of Blockchain

The *Blockchain* establishes trust without intermediaries,

- as its decentralized architecture eliminates single-points-of-failure,
- the entries of transactions are confirmed by a consensus of nodes and are immutable, and
- the open nature of these transactions creates transparency for everyone.

Could these characteristics provide a solution for the problems of trust in the democratic institutions identified above?

Lack of Trust in Democratic Elections

The registration and identification of voters involves a great deal of administrative effort and offers a multitude of opportunities for those in power to influence the process in their favor. Whether certain groups are not admitted, or the process of admission is subject to such formal and logistical hurdles that they are difficult for certain groups to overcome, has an impact on participation in the election and, thus, also on the outcome of the election (Brock 2017).

There are also many possibilities for manipulation in the course of the election itself: from voting districts that are specifically grouped in such a way that the results are skewed in a certain direction, or persons who vote in several places due to (intentionally) incorrect identification, or votes cast for deceased persons, or ballot boxes that disappear or are exchanged—the list of manipulation possibilities is long (Oliver 2017).

While there have already been attempts with electronic voting systems, according to the nonprofit organization *Black Box Voting*, these have always had problems with compliance with all central integrity requirements (Finley 2014):

- Who can vote?
- Who has voted?
- Counting the votes
- Ensuring the integrity of the votes cast

Holding a secret ballot was previously impossible with online voting systems, since any system that guaranteed the correct counting of votes inevitably also made it possible to identify each person's selection in the vote.

In all these dimensions, the *Blockchain* offers the potential for a remedy: while maintaining the anonymity of each individual vote, the entitlement of each voter to vote and the status of his vote can be clearly identified. He can make his decision encrypted from anywhere that he has Internet access with his smartphone and is, therefore, not subject to logistical hurdles. Each vote is fixed in the *Blockchain* without the possibility of subsequent manipulation.

Lack of Trust in the Democratic Representatives

Bribes in the public sector account for USD 2 trillion, or two percent of GDP worldwide (IMF 2016, p. 5). The introduction of cryptocurrency, which is transferred via the *Blockchain* and thus documents the payment flows to project partners in a comprehensible and forgery-proof manner—reduces the leeway for corruption. In contrast to anonymous *Blockchains*—in which only the IDs of the wallets appear and, thus, do not allow any conclusions to be drawn about the owner of the wallet—non-anonymous *Blockchains* containing personal information of real interactors can be constructed for this purpose using *Ethereum*. *Smart Contracts* can also be used to store the conditions of each transaction, whereby each payment is linked to the fulfillment of clearly defined criteria (Aldaz-Caroll 2018). Apart from making it more difficult to make unjustified payments, it is also possible to significantly speed up corruption investigations, which can now be carried out immediately instead of over the 15 months currently required on average.

Even further into the future, election promises could also be made in the form of *Smart Contracts* before the election. Not only would politicians have to be measured against these thanks to transparency, but they could also be bound by sanctions, which would be automatically implemented.

However, the limits of such a solution are reached where the *Blockchain* is bypassed and payments are not made in cryptocurrency, or where there is a benefit beyond direct payments. Other starting points for manipulation are the elements in the system that have to confirm that the condition defined in the *Smart Contract* has been reached. The assumption that the *Blockchain* would ensure trust here is therefore incorrect. It does not guarantee trust, but merely pushes it to the periphery to the confirming party in the physical world, which in turn provides entry points for manipulation (such as a sensor that measures the emergence of external conditions or about the achievement of a specific phase of construction). These problems have been identified, however, and potential solutions are in the works (Orcutt 2018). The limits of smart contracts are also shown where no simple "yes–no" or "if–then" execution of a contract can take place, since, for example, evaluative criteria about a condition are relevant (Braun 2018).

Therefore, only that which is processed within the *Blockchain* is transparently traceable, which is not a panacea against fraud and corruption. A *Blockchain*-based structure can, however, make illegal payments considerably more difficult and increase transparency, thereby also increasing the acceptance of the work of the public sector. Significant flows of payments and benefits that bypass the *Blockchain* would also reveal the unusual contractual terms and conditions that are opposed to these transparent *Smart Contracts* and, thus, lead to a close examination. Against the background of the scale of the problem, even a gradual improvement, which is possible with the use of the *Blockchain*, has great potential.

However, the *Blockchain* and the *Smart Contracts* based on it are still in a rather early stage of development, which is often compared to the early days of the *Internet*. This is why errors in the program code (bugs)—which exist in every piece of software and are only gradually eliminated—always attract attention. This is the normal evolutionary path of maturation and has made the *Internet* what it is today, via trial-and-error. In the context of *Blockchain* and *Smart Contracts*, however, this circumstance poses a not insignificant problem: The identification and subsequent correction of errors runs counter to the core of the technology, which is supposed to rule out subsequent changes (immutability). So if the software code is the law, aren't the bugs included in the code also the law? Contrary to this basic view, the *DAO (Decentralized Autonomous Organization)* enforced a *Blockchain* change and reimbursed USD 200 million worth of *Ether* to its investors after hackers had exploited a bug in the underlying *Smart Contract* and had, in the meantime, stolen USD 50 million worth of *Ether* (The Economist 2016).

Lack of a Sense of Influence

After the people's representatives are elected, the citizens—with the exception of direct democracies—have little influence on their decisions over a longer period of time, and real accountability rarely exists. Here, the *Blockchain* provides a remedy in the sense that decisions to be made can be put into a digital referendum without the need for a great deal of administration. Each person entitled to vote must be clearly identified and, at the same time, anonymous, and the result cannot be subsequently manipulated. On a trial basis, a *Blockchain* technology optimized for polling has already been successfully used in several countries (Houser 2018).

However, activists are already going much further in their own efforts. They want to replace the existing system of representative democracy with a *Blockchain*-based *liquid democracy*: every decision would be made directly by the people instead of by politicians. These decisions would no longer have to be bound to national borders. *Pia Mancini*, co-founder of *Sovereign*—an open source application with a decentralized governance—deplores the misrepresentation of citizens in the current system. The Argentinian raises the question of why she should be represented in international negotiations on climate protection by the Argentine government, whose stance on the subject she does not support. The *Internet* has created an environment in which geographical boundaries play far less of a role than ideological ones. She is convinced that, as citizens of the twenty-first century, we are currently doing our best to interact with institutions designed for the nineteenth century based on fifteenth century information technology. A balancing act that cannot go well and, therefore, urgently requires a fundamental overhaul of the system (Nave 2017).

The *Blockchain* offers the technical opportunities for this, regardless of how one perceives such a form or participation and whether—beyond smaller countries such as Switzerland—enough citizens can be continuously activated for such a participation process in order to achieve a sufficiently perceived legitimacy for it on a broad basis. Since it also seems unrealistic to *Mancini* to replace the existing system overnight, she wants to establish the direct influence in the existing system first. Therefore, she has founded a party and put it up for election. Elected representatives should be bound to the votes made via the *Blockchain* with regard to their votes in parliament. If citizens consider themselves to be incompetent in a subject area to be voted on, they can transfer—via *Blockchain*—the right to vote to another person, who can thus cast more than one vote. The susceptibility of such a structure to corruption is to be limited by the upper limits on the number of votes (Jacomet 2017).

Blockchain Projects in the Political Space

A number of organizations are involved in the field of *Blockchain*-based democracy and are developing systems to make this possible. The following summary provides an overview:

- DemocracyOS: http://democracyos.org/
- Democracy Earth Foundation: https://www.democracy.earth/
- Flux: https://voteflux.org/
- Agora: https://agora.vote/
- Innovote: http://inno.vote/
- Ballotchain: http://www.reply.com/en/content/ballotchain
- Voatz: https://voatz.com/
- Polys: https://polys.me/
- United Vote: https://united.vote/
- DAO (Decentralized Autonomous Organization): https://www.ethereum.org/dao
- Follow My Vote: https://followmyvote.com/.

Conclusion

As described above, the Blockchain offers considerable potential for making a large number of structures much more democratic—and not just much more efficient than the established structures, but also for promoting citizens' self-determination and counteracting disenchantment with politics. In addition to the technological challenges described above, however, there are hurdles here that are primarily owed to the motivations of the decision-makers. Many politicians are therefore still hesitant in their support of the *Blockchain* for the following reasons:

- **Decentralization of the Systems**: The *Blockchain* shifts power structures from a centralized hierarchy to decentralized and flat structures. This is not necessarily in the interest of the respective leaders, since—in a functioning democracy—we assume that the elected representatives represent and assert the interests of the population in the best possible way. The will to surrender power is thus limited, as this calls into question—in addition to one's own influence—why the representatives of the people are elected and entrusted with governing at all. What is the raison d'être of a representative democracy if we do not trust the elected representatives?
- **Technical Complexity**: The *Blockchain*—with P2P networks, *Smart Contracts* and cryptography—has a technical complexity that is difficult for most to comprehend. Many politicians lack a sufficient understanding of the technology and, thus, the competence to make a qualified assessment of its potential.
- **Tendency to Avoid Risk**: Just as most politicians lack the competence to assess the potential, they also lack the expertise to assess the risks associated with the *Blockchain*. The manager of the International Policy Lab at MIT, Daniel Pomeroy, sees this as the central hurdle for translating scientific solutions into political goals. While researchers can comfortably deal with uncertainties and probabilities, politicians want to deal with absolutes. Although it is highly unlikely that the *Blockchain* would be hacked, and no one has succeeded yet, it is theoretically possible.

It's certainly not foreseeable today whether models like liquid democracy will actually prevail and, if necessary, fundamentally change democracy and make politicians superfluous. However, this is less a question owed to technical feasibility than to the motivation, organization and incentivization of all the actors involved.

Irrespective of the degree of democratic participation—up to and including liquid democracy—the *Blockchain* can already provide a far-reaching spectrum of advantages for the political process that can increase political confidence and make the system much more efficient and transparent. In the comparison of possible solutions, the focus should therefore not be on *Blockchain* versus a perfect system, but on the advantages that *Blockchain* can provide in comparison to current problematic structures. There will be no perfect system that provides no opportunities for manipulation. However, although still in its infant state of technological maturity, the *Blockchain* shows a clear path towards becoming far superior to the current system in all of the three dimensions that are responsible for the population's loss of confidence in politics.

References

Aldaz-Caroll, E. (2018/2). *Can cryptocurrencies and blockchain help fight corruption?* https://www.brookings.edu/blog/future-development/2018/02/01/can-cryptocurrencies-and-blockchain-help-fight-corruption/. Accessed November 26, 2018.

250   A. Braun

Braun, A. (2018). *Trends 2018: Logistics—The age of enlightenment* https://www.creativeconstruction.de/lp/trends2018/logistics-the-age-of-enlightenment/. Accessed November 26, 2018.

Brock, T. (2017/11). *Blockchain, voting & elections: What's the problem?* https://youtu.be/L4zR6mTUPlk/. Accessed November 26, 2018.

Cillizza, C. (2018/10). *Donald Trump warns people to beware of non-existent voter fraud.* https://edition.cnn.com/2018/10/22/politics/donald-trump-voter-fraud/index.html. Accessed November 26, 2018.

Finley, K. (2014/5). *Out in the open: An open source website that gives voters a platform to influence politicians.* https://www.wired.com/2014/05/democracy-os/. Accessed November 26, 2018.

Firth, N. (2015/4). *Better than a ballot box: Could digital democracy win your vote?* https://www.newscientist.com/article/mg22630180-400-better-than-a-ballot-box-could-digital-democracy-win-your-vote/. Accessed November 26, 2018.

Freeguard, G. (2015/2). *Public trust in public servants—In six graphs.* https://www.instituteforgovernment.org.uk/blog/public-trust-public-servants-%E2%80%93-six-graphs. Accessed November 26, 2018.

Houser, K. (2018/3). *Sierra Leone just held the world's first blockchain-powered election.* https://futurism.com/sierra-leone-worlds-first-blockchain-powered-election/. Accessed November 26, 2018.

IMF. (2016/5). *Corruption: Costs and mitigating strategies.* http://www.imf.org/external/pubs/ft/sdn/2016/sdn1605.pdf. Accessed November 26, 2018.

Jacomet, N. (2017/1). *Democracy earth, the promise of a safe and independent online voting system.* https://medium.com/open-source-politics/democracy-earth-the-promise-of-a-safe-independant-online-voting-system-37366935db5e. Accessed November 26, 2018.

Mallwitz, G. (2018/3). *Berliner Senat will Tegel-Volksentscheid nicht umsetzen.* https://www.morgenpost.de/berlin/article213858169/Berliner-Senat-will-Tegel-Volksentscheid-nicht-umsetzen.html. Accessed November 26, 2018.

Nave, K. (2017/12). *Democracy 2.0: How blockchain technology is unveiling a new type of democracy.* https://www.delltechnologies.com/en-us/perspectives/democracy-2-0-how-blockchain-technology-is-unveiling-a-new-type-of-democracy/. Accessed November 26, 2018.

Oliver, J. (2017/4). *Gerrymandering: Last week tonight with John Oliver.* https://youtu.be/A-4dIImaodQ. Accessed November 26, 2018.

Orkutt, M. (2018/11). *Blockchain smart contracts are finally good for something in the real world.* https://www.technologyreview.com/s/612443/blockchain-smart-contracts-can-finally-have-a-real-world-impact/. Accessed November 26, 2018.

Pew Research Center. (2017/12). *Public trust in government: 1958-201.7* http://www.people-press.org/2017/12/14/public-trust-in-government-1958-2017/. Accessed November 26, 2018.

Pew Research Center. (2017/10). *Many unhappy with current political system.* http://www.pewglobal.org/2017/10/16/many-unhappy-with-current-political-system/. Accessed November 26, 2018.

Stewart, I. (2010/4). *Electoral dysfunction: Why democracy is always unfair.* https://www.newscientist.com/article/mg20627581-400-electoral-dysfunction-why-democracy-is-always-unfair/. Accessed November 26, 2018.

The Economist. (2016/7). *Not-so-clever contracts.* https://www.economist.com/business/2016/07/28/not-so-clever-contracts. Accessed November 26, 2018.

Alexander Braun Online with the first AOL CDs since 1995, Alexander has been active in the development of Internet projects and startups since 1999. From 2003 to 2007 he worked in management positions at Bertelsmann in London, Shanghai and Toronto and developed Internet-based business models in the media industry. As the founder and CEO of the digital strategy consultancy CREATIVE CONSTRUCTION HEROES, Alexander advises corporate clients from a wide range of industries on the digital transformation of their business model. He has authored and co-authored books on artificial intelligence and chatbots, Web 2.0 and the digital transformation of companies and business models. Alexander graduated with a master's degree in Business and Economics from University of St. Gallen and attended Executive Education at INSEAD and the Massachusetts Institute of Technology (MIT).

In Tech We Trust?

Chapter 19
Digital Propaganda—Russia or the Kid Next Door?

Sarah Lohmann

Introduction: Digital Propaganda and the Global Political Balance of Power

Eighteen years ago, 10 countries gathered with U.S. officials in Bratislava to discuss the road forward for membership in NATO. At the meeting, each participant received a paper memo to their personal attention from the Russian Embassy, stating that it regarded "NATO's enlargement plans as a grave mistake provoking negative changes of military-strategic landscape and division lines in Europe" and Slovakia's aims at security and foreign policy as "so many fabrications" (Means 2001a, b).

While the non-veiled threat was meant to bring fear, it brought the opposite from the participants, serving as a rallying cry for democratic change among the former Soviet bloc nations and their NATO supporters.

Russian propaganda has changed significantly since those days almost two decades ago. Through digital espionage, it has gotten to know its audience better, personalizing the messaging, cloaking it in the voice of a fellow national, delivering it to the receivers' social media accounts, and playing on fears and divisions in the receivers' community.

Jowett and O'Donnell define "propaganda" as "the deliberate and systematic attempt to shape perceptions, manipulate cognitions, and direct behavior to achieve a response that furthers the desired intent of the propagandist" (Jowett and O'Donnell 2006). Indeed, influencing a nation's perception of security has long been one of Russia's objectives in its information operations.

In the early years of NATO enlargement into Eastern Europe, for the countries on the cusp of freedom, long after any military threat from Russia was gone, perception was everything. At another NATO expansion meeting in Riga half a year after the

S. Lohmann (✉)
American Institute for Contemporary German Studies, Johns Hopkins University, Washington DC, USA
e-mail: slohmann@aicgs.org

© Springer Nature Switzerland AG 2020
D. Feldner (ed.), *Redesigning Organizations*,
https://doi.org/10.1007/978-3-030-27957-8_19

Bratislava memo, the emotional perceptions linked to the past played a role in the will to be a part of the military alliance.

"A society is secure if it feels secure", Vaidotas Urbelis, then Lithuania's international relations director for the Ministry of Defense, said at the meeting (Means 2001a, b). Without NATO membership, "There wouldn't be that kind of insurance. The things we are doing probably may disappear. It's not about fact. It's about perception", former Latvian State Secretary of Defense Edgars Rinkēvičs told the author.

Today, the changes to the military-strategic landscape the Russians warned of almost two decades ago is being framed through digital influence in a realm not bound by territory. This is happening even as the impact of that digital propaganda has vast potential to change the global political balance of power and superpowers' claims on national sovereignty.

Digital propaganda used by Russia during the 2016 U.S. election and the German 2017 election, among others, have caused the current intelligence and government investigations to focus on Russia. But during the U.S. midterm elections in November 2018 and the European Parliament elections in 2019, should governments have been primarily fixated on a single, foreign "enemy"? How should each country overcome these disparate challenges? Can democratic nations create effective defense efforts that transcend domestic borders? To answer these questions, this piece will first examine what is meant by digital propaganda, who is using it during elections with what tools, what is currently being done to counter the challenges and what remains to be done.

Defining Digital Propaganda

Shaping perception plays a vital role in the success of the use of digital propaganda. For the purposes of this publication, Jowett and O'Donnell's definition of propaganda above can be expanded to include the use of digital methodologies to influence perception, cognition and behavior.

The Oxford Internet Institute further defines computational propaganda as "the use of algorithms, automation, and human curation to purposefully distribute misleading information over social media" (Woolley and Howard 2017). It later expanded the definition to include specific types of propaganda, including "the automated dissemination of fake news, misinformation, propaganda and other forms of junk news" (Neudert et al. 2017).

Digital propaganda is sometimes used interchangeably with the term computational propaganda by Oxford University's Computational Propaganda Research Project (Sanovich 2017). This article will discuss only the digital aspects of propaganda—that is, those algorithms, bots and automation associated with distributing different forms of propaganda on social media—rather than computational aspects, such as those focused on hardware.

Actors

Special Counsel Mueller's office said in a court filing 12 May 2018 that Russian intelligence agencies were continuing to attempt to meddle in U.S. midterm elections. This was corroborated in February by then-CIA director Mike Pompeo in his testimony before the Senate Intelligence Committee (Gerstein 2018).

Former U.S. Director for National Intelligence Daniel Coats likewise warned, "there should be no doubt that Russia perceives its past efforts as successful and views the 2018 U.S. midterm elections as a potential target" (Rosenberg et al. 2018).

Ads placed 18 months before the election through 2017 targeted everyone from "people who like" Bernie Sanders, Second Amendment rights, Martin Luther King Jr., Hillary Clinton, Black Lives Matter, and Muslims living in the United States. The messages were targeted to inflame tensions around race, religion and cultural pride (U.S. House of Representatives Permanent Select Committee on Intelligence 2018a, b). The April findings of the House Intelligence Committee on Russian interference on U.S. elections provided evidence regarding the vulnerability of U.S. voters for information operation campaigns. According to statements made by Sens. Burr and Warner on 16 May 2018, following the release of the Senate Intelligence Committee report on the 8th of May, the American election system remains at risk. Sen. Warner of Virginia said that "The Russian effort was extensive, sophisticated and ordered by President Putin himself" (Warner 2018).[1]

The Russian information operation used targeted ads created by, among others, the Internet Research Agency (IRA), a Russian troll farm that also used fake U.S. personas to spread propaganda on divisive issues. The ads reached 11.4 million Americans on Facebook alone. Worse, at least 126 million Americans were exposed to the content created by 470 IRA Facebook pages (U.S. House of Representatives Permanent Select Committee on Intelligence 2018a, b).

Facebook confirmed that a new campaign to meddle with the U.S. 2018 midterms has been underway since May 2017. With similar targeting techniques to the previous digital propaganda campaign, the events organized by the fake pages had the potential to inflame violence. The 32 fraudulent accounts—one of which was co-administrated by the indicted Internet Research Agency—on Facebook and Instagram have been removed, and users informed that fake accounts invited them to come to counter-protest rallies, such as "Unite the Right 2" and "#AbolishICE", which was against the actions of the U.S. Immigration and Customs Enforcement agency. In total, 30 events were organized by the fake accounts, and one account alone had 290,000 followers, with $11,000 being spent on 150 ads (Menn and Paresh 2018).

Across the Atlantic, Europe had concerns about meddling ahead of the European Parliament elections in 2019, after information campaigns targeted the French and German elections and the Catalonian independence referendum in 2017. Stratcom, an EU-foreign service counter-propaganda unit, documented more than 3500 cases of pro-Kremlin disinformation in the European media in the two years before the

---

[1]This paragraph and the three following were published in similar form by the author in: Lohmann (2018).

election (Rettman 2018). NATO Secretary General Jens Stoltenberg said ahead of the elections that NATO was countering disinformation campaigns "online, on paper, in the air, in different platforms", (Stoltenberg 2019). During the German elections, Russia promoted the far-right AfD in its state-controlled networks and planted disinformation about the refugee-friendly Angela Merkel. One such campaign, known as the "Lisa affair", widely covered a false story about a girl who was supposedly raped by refugees. The misleading coverage of the German election season helped prompt Merkel to travel to visit Putin in Sochi in May 2017 to warn about election meddling (Shuster 2017).

The Russian botnet IRA also meddled in the French elections, when 30,000 fake IRA accounts were detected and destroyed by Facebook (Bünte 2018). Russian-language faceless accounts thought to be botnets were also notably used the day before the German election to bolster the turn out of the AfD through fake news stories spread through fake accounts that claimed that AfD voters could be disenfranchised of their voting rights (Czuperski 2017).

With all the focus on Russian intervention, China's more subtle methods often get overlooked. With ten times the number of troll farms, and more personnel, financial and infrastructure resources, China is playing a long-term game in its democracy interference efforts (Lohmann 2017). The Chinese Communist Party is investing $10 billion a year in influence abroad under General Secretary Xi Jinping (Parello-Plesner 2018). Rather than focus on the next upcoming election, China's goal in its digital propaganda efforts is a global rebalance of power with China as the leading superpower. Yet democracy defenders should be aware of the digital propaganda tools that China's Communist Party is using.

In October 2018, for example, the Dengfeng City Network Information Office sent propaganda directives for all personnel to share stories on US relations via WeChat and Weibo which would recharacterize facts in the US-China trade war. Employees were to negate that "Chinese growth as 'driven by U.S. investment in China'" and promote the story that "The General Downhill Trend of America Has Not Changed" (China Digital Times 2018). Beyond such overt "influence" campaigns, China also focuses its efforts on influence operations (IO) that are "covert, corrupt, or coercive" (Parello-Plesner 2018). Here, several examples of the target countries, including the United States, due to its superpower status, and Australia and New Zealand, due to their influence on the South China Sea, should be mentioned.

In Australia, Prime Minister Malcolm Turnbull introduced legislation in late 2017 to fight foreign interference after Australian Senator Sam Dastyari had to resign in a scandal around CCP influence. The senator had allowed his office to accept legal and travel payments from the company of Chinese billionaire Huang Xiangmo, who held the press conference with him jointly. Using CCP talking points at a press conference, the Senator had answered a question about China's activity in the South China Sea in direct contrast to the Labor Party's policy, claiming that China's artificial island building and military activity there were none of Australia's business and that "the Chinese integrity of its borders is a matter for China" (McDermott 2017).

New Zealand was likewise asked to consider passing similar legislation after a Chinese-born member of parliament in New Zealand was investigated by the Security Intelligence Service due to his advocating CCP policies while serving on the "five eyes" intelligence alliance and getting Chinese funding for his National party (Anderlini 2017).

In the United States, China has continued its policy of culling influence with the media, think tanks, politicians and business through funding coming from the CCP and the "United Front", the Work Department of the CCP which works on influence operations through financing everything from student exchanges to business ventures (Parello-Plesner 2018).

Tools

With all this evidence of Russian interference and more subtle Chinese operations, why shouldn't the United States and Europe focus their energy singularly on defeating these foreign meddlers? It's more complicated than that. In *New America*'s policy paper "Digital Deceit", authors Ghosh and Scott explain that while Facebook and Twitter are getting the heat for being taken advantage of by Russia, the whole social media industry is built to grab users' attention and sell them products or policies. According to the authors, this includes everything from behavioral data tracking, which aggregates data on purchases and places visited by a user, to search engine optimization, which allows search engine algorithms to be compromised so that results are skewed. Even artificial intelligence, allowing message targeting and system-operated campaign management, can be used for digital propaganda purposes (Ghosh and Scott 2018).

That means that while the Russians may be tracking users' online profiles and behavior in order to influence who they vote for, it is equally possible that domestic political parties, fringe nationalist extremist groups and special interest lobbies are doing the same. That does not mean that all of those groups have access to the same amount of sophisticated information, the same level of finesse in targeting users, or a grand, long-term strategy for how voters should be influenced. It does mean that influencing elections is a much more complex operation, with many more actors tugging on users' perception than can be defined by one enemy or methodology.

The use of bots—robots that forward content on the Internet—and fake news to influence perception has been highly successful in the last two years. For example, during the 2016 U.S. election, 400,000 bots tweeted for and against Clinton and Trump. After the third presidential debate, *Wired* reported that seven times as many messages were sent by bots from pro-Trump accounts than pro-Clinton accounts (Zaleski 2016).

But calculating whether a bot is being used depends on "likelihood", and those who analyze bot activity have different cut-off percentages for likelihood. Using a 60% likelihood, rather than the typical 47% likelihood used by the bot recognition computer program BotOrNot, only 3.4 percent of pro-Trump Tweets came from bots—as opposed to 3.2 percent for Clinton, and 1.7 percent for Sanders—over Memorial Day weekend 2016, when there were still three main contenders (McGill 2016).

Just as important as who or what is amplifying a message is what they are amplifying. In Michigan, in the days around the U.S. presidential election Nov. 1–11, 2017, in a sampling of 22 million tweets, professional news and "junk" news ("junk" news being news that is at least partially not based on fact) were shared equally (Howard et al. 2017). While Michigan certainly isn't representative of all of the United States, the Oxford Internet Institute's research on digital propaganda in both Germany and the United States is worth analyzing, due to its large amount of sample data during pre-election peak tweeting periods.

In Germany during the 2017 presidential elections, professional news was shared four times as much as junk news, but the majority of junk news shared was based on far-right sources (Neudert et al. 2017). In addition, ahead of the German parliamentary elections, from 1 to 10 September 2017, 15% of the far-right Alternative für Deutschland-related tweets were automated (44,533 botted tweets of 905,465 tweets sampled). By comparison, the next highest percentage of bots used was for Die Linke, at 12.3% (1819 bots). The governing Christian Democratic Union/Christian Socialist Union coalition had 13,099 botted tweets (7.3%) (Neudert et al. 2017).

With all that perception influencing going on from different actors, be it political players, foreign enemies or extremist groups, what are the remedies? The challenge, Ghosh and Scott argue, is that a foreign actor's information operation using a digital propaganda campaign operates with the same tools as paid advertising (Ghosh and Scott 2018). This means that the methods are difficult to distinguish, and thus difficult to stop. If countries begin regulating the use of bots and online ads to keep foreign actors at bay, they must be prepared for the fact that the same regulation will have an impact on e-trade.

Yet while the methods used by the different actors may be the same, the impact on civil society is vastly different. If foreign powers successfully use targeted digital propaganda to interfere in a campaign, trust in democratic institutions flounder. When that foreign power programs bots to amplify the voices of targeted religious, political, cultural or special interest groups to fan the flames of division, countryman can turn against countryman, violence can be incited and civil society damaged. But when bots are used to spread information about public emergencies, civil society can be helped. When they are used to inform a user of their choice of products, the user may be annoyed, but the negative impact on democracy is minimal.

The Way Forward and the Transatlantic Relationship

The key to limiting the negative impact of disinformation campaigns lies in informing users of the online actors conducting them, their target groups, the factualness of their posting and whether the posting is paid advertising. This kind of regulation will need to come from the legislature: self-screening among different social media platforms will not work equally well without accountability mechanisms from the law. A couple of laws are already working toward this on each side of the Atlantic, but more is needed.

Germany's Network Enforcement Act, which went into effect on 1 October 2017, punishes the platforms more than it does the bad actors. The law requires social media platforms to delete propaganda, hate speech and fake news within 24 h of notification of a complaint, or face fines of 50 million Euros (Dwyer 2017).[2] This causes the social media platforms to need to hire thousands of employees at high costs to the company, while the terrorists and extremists who post violent content often go unpunished. While the law theoretically creates an incentive for hate speech and propaganda distributors to post less, by the time the 24-h notification window is over, they have often already received the viewership and multiplication of the message they needed for their cause.

A similar draft law in the United States, called the Honest Ads Act, would track foreign financing of ads costing over $500 during an election, while not tracking unpaid speech on social media platforms (U.S. House of Representatives 2017). This means that interest groups or foreign powers that track user behavior and use botnets to send out organic messaging to influence that behavior would still fly under the radar screen. It also means that the digital propaganda campaign to affect the US mid-term elections, with $11,000 spent on 150 ads, would also normally not be mandated to be trackable. At the same time, outlawing botnets is not realistic, as these can be used to produce a users' desired results in Internet searches for both research and product purchase.

Several initiatives that defend against digital propaganda campaigns can be found across the transatlantic community. The big tech companies have introduced voluntary programs to help fight or identify information campaigns. To ensure that readers get the facts straight, the Google News Initiative in the U.S., called the Digital News Initiative in Europe, is retraining its algorithms to prioritize breaking news with "authoritative" sources rather than the most recent posts, to cut down on disinformation during breaking news events (de Looper 2018). Facebook also launched an initiative this year to train readers how to identify fake news in their news feed (Jefferson 2018).

There are dozens of other initiatives by third parties to apply ratings systems to filter bias and fake news. The Global Investigative Journalism Network compares 24 of these fake news tracking initiatives across the world (Bell 2017). For example, the University of Michigan uses a patented system called an "All Sides Bias Ratings Page" to classify news according to political bias from left-leaning to right-leaning (University of Michigan Library 2018). The challenge is how to get the general public to use such systems, as most of the third-party initiatives are opt-in programs that are time consuming to set up and, in some cases, costly.

Election commissions in the United States and Europe can also ensure their voting publics are equipped with objective information about elections, candidates and

---

[2]The previous two sentences and paragraph to follow were published in similar form by the author in: Lohmann (2018).

parties. This will increase trust in democratic institutions. The role of nonpartisan civil society actors will be increasingly important in this process. Scott has proposed the vital importance of transparency around political elections. In addition to being informed about actors and who they are targeting, users should be informed when they have been exposed to disinformation and how their personal information is being used to manipulate them (Ghosh and Scott 2018).

Much of this was the aim of the General Data Protection Regulation (GDPR), which is supposed to protect the personal data of residents of the European Union, but similar measures are lacking in the United States. The problem is, GDPR still allows users' Internet use to be tracked through cookies and for them to be manipulated. It just puts the onus on the user to wade through pages of legalese on how to opt out every time they pull up a website (Dixon 2018).

Conclusions

Professor Corneliu Bjola of Oxford University and Professor James Pamment of Lund University argue in their book, *Countering Online Propaganda and Extremism*, that "in the digital age, information is now weaponized". They claim that the solutions, including strategic communications and counter-disinformation campaigns, "must be tailored to the broader context of the diplomatic relations of the relevant parties" (Bjola and Pamment 2018).

At the time of this publication, diplomatic relations between the United States and Russia are tense, to say the least, while Berlin edges toward rapprochement with Moscow due to German economic interests. Yet Chancellor Merkel and President Trump have each certainly sent a differentiated strategic message to the Kremlin, despite Russia's digital propaganda directed at both countries.

President Trump initially did not acknowledge the extent of Russia's interference, while Merkel drew a red line on election meddling during her May 2017 Sochi visit. The U.S. has indicted 13 Russian nationals and three Russian entities for their information warfare during the 2016 presidential elections, and the investigations continue.

Germany's new Bundeswehr Cyber and Information Space Command (CIR) in Bonn aims to come up with defensive strategies against everything from attacks on infrastructure to Russia's new digital propaganda (Schimmeck 2017). But with less spectacular results from the Russian disinformation campaign in Germany, the focus on countering Russia's propaganda is not as streamlined as in the United States.

So are these efforts enough, and should Washington and Berlin keep their sites set on Russia and China? Not unlike the alliance and security decisions that were being discussed almost two decades ago in Bratislava, the question could perhaps be formulated differently: "Do we choose democracy or division?" The challenge today is that this threat to national security for both old and new NATO countries does not just come from a single foreign actor. The expansion of democracy, as well as the defense of our nations, is being compromised by the subtle dance of multiple actors beyond our borders, who can easily capitalize on the political, cultural and religious divisions within. But then, as now, the citizens of Europe and the United States have a choice to remain united and a choice in how much power to give those actors.

Yes, together, Germany and the United States need to remain watchful of China and Russia's digital propaganda and to inform their publics of its influence in the public debate around elections. At the same time, citizens in both countries can diffuse hate speech and encourage their governments to strengthen transparency mechanisms online, so that social media users are aware of the many actors influencing their perceptions, be it political interest groups, Russia or the loner activist kid next door.

References

Anderlini, J. (2017, September 13). China-born New Zealand MP probed by spy agency. *Financial Times*.
Bell, F. (2017, May 8). A global guide to initiatives tackling 'Fake News'. *Global Investigative Journalism Network*.
Bjola, C., & Pamment, J. (Eds.). (2018). *Countering online propaganda and extremism*. Routledge.
Bünte, O. (2018, April 4). Facebook löscht erneut Seiten und Accounts der russischen Troll Fabrik IRA. *Heise Online*. https://www.heise.de/newsticker/meldung/Facebook-loescht-erneut-Seiten-und-Accounts-der-russischen-Trollfabrik-IRA-4010575.html. Accessed June 22, 2018.
Czuperski, M. (2017, September 23). #Election Watch: Final Hours Fake News Hype in Germany: Bots and trolls push vote-rigging claim ahead of German election. Atlantic Counsel, @DFRLab.
de Looper, C. (2018. March 20). The Google News initiative aims to step up the fight against Fake News. *Digital Trends*. https://www.digitaltrends.com/mobile/google-news-initiative/. Accessed June 21, 2018.
Dixon, M. (2018, May 24). GDPR should have made cookies toast. *Fortune*. http://fortune.com/2018/05/24/gdpr-data-privacy-cookies/. Accessed June 22, 2018.
Dwyer, C. (2017, May 3). Facebook plans to add 3,000 workers to monitor, remove violent content. *NPR*.
Gerstein, J. (2018, June 12). Mueller sees Russian effort to influence 2018 elections. *Politico*.
Ghosh, D., & Scott, B. (2018, January 23). Digital deceit: The technologies behind precision propaganda on the internet (pp. 5–32). *New America*. https://www.newamerica.org/documents/2068/digital-deceit-final.pdf. Accessed June 22, 2018.
Howard, P., Bolsover, G., Kollanyi, B., Bradshaw, S., & Neudert, L.-M. (2017, March 26). Junk News and Bots during the U.S. Election: What were Michigan voters sharing over Twitter? Oxford University. www.oii.ox.ac.uk/.../What-Were-Michigan-Voters-Sharing-Over-Twitter-v2.pdf. Accessed June 22, 2018.
Jefferson, M. (2018, May 24). Fighting Fake News, Facebook unveils new initiatives. *Mediatel*. https://mediatel.co.uk/newsline/2018/05/24/fighting-fake-news-facebook-unveils-new-initiatives/. Accessed June 21, 2018.
Jowett, G., & O'Donnell, V. (2006) *Propaganda and persuasion* (pp. 7, 269). Thousand Oaks, CA: Sage Publications.
Lohmann S. (2017, February 9). *In a world of cyber threats, isolationism will never win*. AICGS Online, Johns Hopkins University. https://www.aicgs.org/publication/in-a-world-of-cyber-threats-isolationism-will-never-win/. Accessed November 30, 2018.
Lohmann, S. (2018, May 22). *Election meddlers and transatlantic remedies*. AICGS Online, Johns Hopkins University. https://democrats-intelligence.house.gov/facebook-ads/social-media-advertisements.htm. Last accessed June 22, 2018.

McDermott, Q. (2017, November 29). Sam Dastyari defended China's policy in South China Sea in defiance of Labor policy, secret recording reveals. *ABC News.* https://www.abc.net.au/news/2017-11-29/sam-dastyari-secret-south-china-sea-recordings/9198044. Accessed November 30, 2018.

McGill, A. (2016, June 2). Have Twitter Bots infiltrated the 2016 election? *The Atlantic.* https://www.theatlantic.com/politics/archive/2016/06/have-twitter-bots-infiltrated-the-2016-election/484964/. Accessed June 22, 2018.

Means, S. (2001a, December 11). The Russian face in Riga: NATO as a salve for post-Soviet wounds. *The Washington Times.*

Means, S. (2001b, May 18). Tomorrow's NATO: expanding democracy to Russia's back door. *The Washington Times.*

Menn, J., & Paresh, D. (2018, July 31). Facebook says it identifies campaign to meddle in US elections. *Reuters.* https://www.reuters.com/article/us-usa-election-facebook/facebook-says-it-identifies-campaign-to-meddle-in-2018-u-s-elections-idUSKBN1KL2FG. Accessed August 4, 2018.

Neudert, L., Kollanyi, B., & Howard, P. (2017, September 19). Junk News and Bots during the German Parliamentary Election: What are German voters sharing over Twitter? COMPROP DATA MEMO 2017.7. Oxford University, pp. 1, 4.

Parello-Plesner, J. (2018, June 20). The Chinese Communist Party's Foreign Interference Operations: How the U.S. and Other Democracies Should Respond. The Hudson Institute.

Rettman, A. (2018, January 18). Next Year's EU Election at risk of Russian meddling. *EU Observer.* https://euobserver.com/foreign/140598. Accessed June 22, 2018.

Rosenberg, M., Savage, C., & Wines, M. (2018, February 13). Russia sees midterm elections as chance to sow fresh discord, intelligence chiefs warn. *New York Times.* https://www.nytimes.com/2018/02/13/us/politics/russia-sees-midterm-elections-as-chance-to-sow-fresh-discord-intelligence-chiefs-warn.html. Accessed June 19, 2018.

Sanovich, S. (2017). *Computational propaganda in Russia: The origins of digital misinformation* (pp. 1–5). Working Paper No. 2017.3, Oxford University. http://comprop.oii.ox.ac.uk/wp-content/uploads/sites/89/2017/06/Comprop-Russia.pdf. Accessed June 22, 2018.

Schimmeck, T. (2017, March 31). Militärs mit Computermaus und Laptop. *Deutschlandfunk.* http://www.deutschlandfunk.de/das-neue-cyber-kommando-der-bundeswehr-militaers-mit.724.de.html?dram:article_id=382767. Accessed June, 22 2018.

Shuster, S. (2017, September 25). How Russian Voters Fueled the Rise of Germany's Far Right. *Time.*

Stoltenberg, Gen. J. (2019, February 28). Address by NATO Secretary General Gen. Jens Stoltenberg at GLOBSEC public event in Kosice, Slovakia, 28 Feb. 2019. https://www.nato.int/cps/en/natohq/opinions_164133.htm?selectedLocale=en. Accessed July 8, 2019.

University of Michigan Library. (2018, April 28). *Where do news sources fall on the political bias spectrum?* http://guides.lib.umich.edu/c.php?g=637508&p=4462444. Accessed June 22, 2018.

U.S. House of Representatives. (2017). *HR 4077: Honest Ads Act.* https://www.govtrack.us/congress/bills/115/hr4077/summary. Accessed June 22, 2018.

U.S. House of Representatives Permanent Select Committee on Intelligence (2018a). *Facebook Ads.* https://democrats-intelligence.house.gov/facebook-ads/. Accessed May 18, 2018.

U.S. House of Representatives Permanent Select Committee on Intelligence. (2018b). *Social media advertisements.* https://democrats-intelligence.house.gov/facebook-ads/social-media-advertisements.htm. Accessed June 22, 2018.

Warner. (2018, May 16). *Senate Intel Completes Review of Intelligence Community Assessment on Russian Activities in the 2016 U.S. Elections.* https://www.warner.senate.gov/public/index.cfm/pressreleases. Last accessed June 22, 2018.

Woolley, S., & Howard, P. (Eds.). (2017). *Computational propaganda worldwide: Executive summary* (p. 3). Working Paper 2017.11. Oxford, UK: Project on Computational Propaganda. http://comprop.oii.ox.ac.uk/wp-content/uploads/sites/89/2017/06/Casestudies-ExecutiveSummary.pdf. Accessed June 22, 2018.

Zaleski, A. (2016, October 11). How Bots, Twitter, and Hackers pushed Trump to the Finish Line. *Wired.* https://www.wired.com/2016/11/how-bots-twitter-and-hackers-pushed-trump-to-the-finish-line/. Accessed June 21, 2018.

Dr. Sarah Lohmann is a Senior Cyber Fellow at the American Institute for Contemporary German Studies, Johns Hopkins University, where she works on strengthening agreement between German and American policy makers on issues of cyber defense and digital propaganda. Dr. Lohmann has also served as a university instructor at the Universität der Bundeswehr since 2010. She achieved her Doctorate in Political Science there in 2013, when she became a senior researcher on the faculty of State and Social Sciences. Previously, she was a press spokeswoman for the U.S. Department of State. She has been published in peer-reviewed journals and books, written and edited several books for internal government circulation, and published over a thousand articles in international press outlets. She received her Masters from American University, Washington, DC and her Bachelors from Wheaton College, Illinois.

Chapter 20
Cyberspace as Military Domain: Monitoring Cyberweapons

Thomas Reinhold

Introduction

Over the last several years, a growing number of military forces worldwide have started to recognize cyberspace as the next military domain whereas the questions of how to regulate this development with measures of arms control and if this works at all for this domain have yet to answered. The strategies that military forces have been prepared (UNIDIR 2013) often involve the establishment of offensive capabilities, sometimes for deterrence reasons or seen as the appropriate measure to react to cyberattacks by actively disturbing or even destroying the attackers IT systems which is described in terms like "active defense" or "hack back" (see exemplary NATO 2010). The necessary "cyberweapons capabilities" of software or hardware with disruptive or destructive effects are actively developed (see exemplary DARPA 2012) and had already been used (US-ICS-CERT 2016; US-DOD 2016), although the cases of cyber incidents so far all happened outside of officially declared wars, and the attribution of cyberattacks to state actors is hard to prove. Nevertheless, many incidents are supposed to be performed by state actors like the so called "BlackEnergy" malware that affected the Ukrainian electric power industry (US-ICS-CERT 2016). A few cases exist where military strategies explicitly include cyber warfare capabilities, such as in the U.S. fight against the ISIS terror group (see US-DOD 2016). On the other hand, the international community currently struggles to come to an agreement on binding norms of state behavior and how established rules of international law can apply to this new domain (Tikk and Kerttunen 2017). The debates include the challenge of determining an appropriate response to the ongoing militarization of cyberspace, the question of how to slow down the armament and the prevention of an arms race in this domain. Furthermore, the attempt to apply established measures of arms control or non-proliferation, as well as lessons learned from

T. Reinhold (✉)
Institute for Peace Research and Security Policy (IFSH), Hamburg, Germany
e-mail: info@cyber-peace.org

other military technological developments, quickly comes to a stop due to specific technical features in cyberspace. Against this background, the following article will look at the core principles of arms control and the problems when applying these to the cyberspace domain. It will use as examples the lessons learned from nuclear disarmament as the most assessed arms control and arms monitoring area from the recent decades. The comparison will be used to develop concepts and approaches for applicable cyber arms control measures and to formulate the outlook for necessary treaties and international institutions.

The Roots and Core Principles of Arms Monitoring

The concept of arms monitoring is a general term that is often used in the context of arms control and non-proliferation. The overall function of arms control is the prevention of conflicts and the stabilization of international state relations by reducing the motivation of adversaries for preventative or pre-emptive military operations to destroy military capacities, as well as for the reduction of the probability of the application of specific military weapon systems (Müller and Schörning 2006). These goals are tackled on different levels and by different measures. Neuneck and his fellow authors give an overview (Mölling and Neuneck 2001) that differentiates using the following categories and correlating measures:

- Geographic measures: demilitarized regions, security zones
- Structural measures: defensive orientation of force structures
- Operative measures: limitation of maneuvers, omission of provocative actions
- Verification measures: data exchange, inspections
- Declaratory measures: abandon the first use of nuclear weapons
- Technology-related measures: limitation, reduction or destruction of certain weapons or technologies
- Proliferative measures: prohibition or restriction on the export of militarily relevant technologies
- Selective measures: prohibition or restriction of the use of certain weapons and methods of war
- Actor-related measures: prohibition, restriction or permitting of specific groups of actors
- Goal-related measures: safeguard clauses, prohibition of attack on particular, especially civil, targets.

These specific measures are embedded in treaties or agreements where parties bilaterally or multilaterally declare their intent for specific actions or their omission and the dedicated procedures and actions. A popular example is the "Convention on the Prohibition of the Development, Production, Stockpiling and Use of Chemical Weapons and on their Destruction", often abbreviated as CWC, that had been negotiated by the UN, entered into force in 1997 and established the Organization for the Prohibition of Chemical Weapons (OPCW) to control the implementation of the treaty (United Nations 1992). Such treaties and binding agreements as well as the customary international law create the international law that defines the rules for state behavior and interactions. One of the main principles of these rules is the convention "pacta sunt servanda" (Wehberg 1959). This centuries-old principle, that translates to "agreements must be kept", had been explicitly formulated 1969 in the "Vienna Convention on the Law of Treaties" (United Nations 1969) and entered into force in 1980. The convention describes that "every treaty in force is binding upon the parties to it and must be performed by them in good faith". This general rule brought to light the question of how treaty members are able to surveil and control the mutual compliance of agreed terms and how this should be performed. This task, which is described as verification, is an important measure for international security politics and mostly integrated in verification regimes, a concept that is based on the regime theory of Robert O. Keohane (Robert and Martin 2009). Verification regimes are either integrated to existing treaties or stand for themselves and consist of the following different parts:

- The treaty agreement itself.
- The rules that the treaty members agree to follow in combination with specific thresholds, binding instructions or forbidden activities.
- The practical measures that treaty members or specifically entrusted authorities are allowed to perform in order to control the compliance of the other treaty members.
- The definition of the authority that is allowed to make decisions regarding the compliance and consequences that states agree to perform and bear when the agreed rules are not followed.

In other terms, verification and the task of controlling and monitoring weapons is always a very context specific definition of what is getting controlled, how, by whom and for what purpose.

Principles of Nuclear Weapons Monitoring

One of the most intense verification debates of the last fifty decades concerns the risks and threats of nuclear armament. The most commonly known institution in the context of these debates is the International Atomic Energy Agency (IAEA), an international independent organization that had originally been founded in 1957 for the promotion and development of the peaceful usage of nuclear energy (IAEA 1961). It directly reports to the United Nations General Assembly and the United Nations Security Council. Since its foundation, its tasks have fundamentally changed. With the international adoption of the Treaty on the Non-Proliferation of Nuclear Weapons (NPT 1968), the IAEA had been put in charge of different treaties (Neuneck 2017) *"to establish and administer safeguards designed to ensure that special fissionable and other materials, services, equipment, facilities, and information made available by the Agency or at its request or under its supervision or control are not used in such a way as to further any military purpose; and to apply safeguards, at the request of the parties, to any bilateral or multilateral arrangement, or at the request of a State, to any of that State's activities in the field of atomic energy"* (IAEA 2018a; IAEA 1961). These safeguards (IAEA 1968) are practical measures that reflect the core of nuclear weapons monitoring and address two different dimensions: "horizontal" and "vertical" non-proliferation. Horizontal non-proliferation is the challenge of preventing and regulating the spread of nuclear weapons to new state and non-state actors by banning the trade of nuclear arms, as well as stopping capabilities for the production of nuclear weapons or feasible material. The term vertical non-proliferation, on the other hand, describes measures to control the technological advancement and stockpiling of nuclear weapons by nuclear powers (Goldansky 1988). One of the most recent tasks of the IAEA, which should be used as a demonstrative example for the different levels of nuclear arms control, is the supervision of the JCPOA nuclear treaty agreement (Joint Comprehensive Plan of Action) (IAEA 2016), which had been negotiated with the Islamic Republic of Iran by the five permanent members of the United Nations Security Council, Germany (P5+1) and the European Union over thirteen years, came into force in January 2016 and is still active despite the one-sided termination of the agreement by the United States under US President.

Iran's compliance is controlled by verification measures that are integrated into this treaty as safeguards. They enable IAEA staff members to get access to nuclear and research facilities, shut down and seal critical industrial hardware, install surveillance cameras, control industrial plants, count the equipment in nuclear facilities, take samples from nuclear material, as well as measure the radiation level of devices and places. As already pointed out, these verification measures are always practical steps that tightly concentrate on specific aspects of the controlled technology, the outcomes of which can be compared against threshold values, "do's and don'ts" or lists of forbidden technological procedures. Such monitoring measures always need to be very specifically tailored to the controlled technology and the monitoring context and can therefore strongly differ for different kinds of situations. From a broader and more generalized perspective, they can be categorized into four areas of restrictions that directly relate to applicable monitoring principles (Neuneck 2012):

- Geographical restrictions that regulate the allowed or prohibited location of specific goods, which are controlled by locating and visually monitoring (like ultra violet and x-ray imaging or aerial and satellite photography) these goods. An example for such monitoring measures is the Treaty on Open Skies (OSCE 1992), which came into force in 2002 and is currently ratified by 34 states. It allows unarmed aerial surveillance flights over the entire territory of the treaty members.
- Limitations in terms of the amount or even the complete prohibition of the possession of goods are controlled by counting and cataloging the goods. This can include the reduction of existing capacities. An example is the "Strategic Arms Reduction Treaty—New START" (NTI 2010) as the successor to former treaties (START I from 1991 and START II from 1993) between the United States and the Russian Federation. The treaty entered into force in 2011 and is valid until 2020, and it regulates the further nuclear arms reduction of both countries. The treaty

establishes a commission and dedicated rules and deadlines for inspections and its bilateral organization.

- Definitions of threshold values for specific properties of physical, chemical or biological states of goods are controlled by measuring or scientifically estimating these properties. An example is the already mentioned JCPOA treaty with the Islamic Republic of Iran (IAEA 2016). Among other things, the treaty contains agreements to reduce the enrichment level of uranium to a degree that enables medical treatments and research but prevents the fast weaponization of the uranium for nuclear bombs (IAEA 2018b).
- Restricting the proliferation of goods is controlled by tracing the goods, regulating or prohibiting their trade. An example of a non-proliferation treaty is the Treaty on the Non-Proliferation of Nuclear Weapons (NPT 1968), which 191 states currently adhere to. This treaty directly shaped the role and responsibilities of the IAEA that, among other things, enables the organization to inspect nuclear facilities. An additional protocol of the treaty extends these rights to include unannounced inspections and is currently signed by 139 states.

Established Measures and Their Applicability in Cyberspace?

This chapter will assess the questions about how these measures, the experiences and lessons learned can be applied to cyberspace and the challenges of an ongoing cyber armament:

In contrast to all other domains, cyberspace has some specific technical features that differ strongly from all other domains and have an important impact on the application of monitoring approaches. Often these technical features render established measures useless, because they are designed for physical domains like sea, air, land or space and rely on features of these domains that cyberspace does not provide. Therefore, the technical specifics of cyberspace have to be taken into account when thinking about monitoring and arms control in this domain.

Virtuality

First of all, cyberspace is by design a "virtual" domain. In theory, data is stored and processed by a specific IT system that has a geographical location and falls under a legislatively responsible jurisdiction. On the other hand, data can be seamlessly copied and—especially in the cloud computing age—is often transferred and stored on other IT systems for availability issues or split up into multiple parts to be processed on different and sometimes even geographically distributed IT systems. This means that even if hardware itself always has a physical representation, in practical terms, the data itself, its storage and processing cannot be reasonably attributed to a specific geographical location and a specific nation states sovereignty.

Distribution

Another relevant aspect of software, like any other digital information, is that every piece of such data is stored physically in different ways, such as magnetic fields on classic hard drives or electromagnetic states on solid state drives, but that this storage takes place distributed within other data fragments. The handling of data as logical entities, like files, is a mere abstraction of operating systems and the physical storage most likely isn't carried out in a cohesive manner. This means that data itself has no specific coherent physical representation, and digital information cannot be handled as a unique and autonomous self-contained entity like a missile, a tank or a test cube. Furthermore, it also does not produce any kind of reliable "traces" when moved or copied, traces that could be used for monitoring. Any way of "counting" and limiting software is rendered meaningless by these aspects.

Attribution

A third technical feature of cyberspace is commonly known as the attribution problem. This term describes the problem and the ambiguity of assigning any kind of activity within cyberspace to its origin and the presumed actor that intentionally performed this activity. The necessity for attributing an attack to its origin and therefore identifying the attacking party is a key element to the states right for self-defense under the UN Charta. Attribution of cyberattacks is currently considered to be the main problem when applying international law and its rules of state behavior to cyberspace (see example Guerrero-Saade and Raiu 2017) because digital data transfer happens over multiple steps of involved IT system and cyberattackers use this feature to create a complex path from the system that controls an attack to its target. Recreating this path potentially involves the necessary cooperation of each of these "hubs". This technical feature provides multiple possibilities for adversaries to cover their tracks and use IT systems of uninvolved third parties. It also means that even if the source system of any data access is identified, it is unclear if the system itself had been hacked and misused. This principle also affects the question of how goods can be assigned to their owners, as well as the task of regulating their proliferation.

Dual Use

The last feature of cyberspace specifically concerns the technical equipment that is necessary for its infrastructure—the networking and computing devices, from servers to home electronics, or even embedded controlling devices and the software they are running—the 'Internet of Things'. All of this technology can be used for military as well as civilian purposes without being able to draw a distinct line between these usage scenarios. Therefore, it cannot be generically prohibited or allowed for arms control reasons. Furthermore, the dual use character of goods means that it's not the good itself but its precise usage that determines whether or not it falls under the negotiated agreements of arms control and disarmament. The task of defining lists of such goods and the necessary special control and monitoring has been performed for several decades for nuclear, chemical and biological goods. Its most popular example is the Wassenaar Arrangement (Wassenaar 1996), a treaty between 42 currently participating states that have agreed upon dedicated arms and export control, as well as sharing trade data for such sensitive goods as a measure of trust and confidence building. The treaty had been broadened in 2013 to include "intrusion software" (Wassenaar 2017) that can be used either for surveillance or to break and undermine IT security measures or otherwise manipulate IT systems.

In comparison with former dual-use approaches—where a relevant factor for the regulation of chemical, biological or nuclear goods was either the sheer amount of specific materials, the necessary equipment or specific military delivery systems that can be monitored and verified—the dual-use character of IT hardware and software is even more distinct. This means that, for cyberspace and its necessary technological infrastructure, it is not possible to differentiate between goods, because both the hard- and software are the same for civil, economic and military purposes. This also affects any approach towards differentiation between legitimate goods that distinctively serve military defensive measures and those whose primary purpose is for offensive measures. Even malware or software exploits that can be used offensively are also necessary to test and increase the cyber security of IT systems. A popular example for this case are penetration testing tools: software that is specifically designed to attack and penetrate IT systems and networks to detect flaws, weaknesses and security problems. These tools are an important instrument for IT security practitioners and its regulation can affect the protection of IT systems. On the other hand, its detection during theoretical inspections doesn't necessarily prove any non-compliance to a treaty. Therefore, only the usage of tools is decisive regarding the offensive or defensive application of goods. Any verification regime rules that declare certain behaviors forbidden need to implement measures for controlling the specific application of IT goods, which is not practically implementable as argued before.

Concepts for Cyber Arms Monitoring and Control

An important step for arms control and monitoring is the definition of the subject that needs to be regulated. Aside from the mentioned Wassenaar Agreement, this step has not yet been performed in internationally binding treaties. The presented specifics of cyberspace showed that such a definition has to consider more than the aspects of the usage, the intention of use and the effects of a tool over the specific technical features. A fitting definition that comprehensively reflects these is given by Stefano Mele:

> [a cyberweapon is] A part of equipment, a device or any set of computer instructions used in a conflict among actors, both National and non-National, with the purpose of causing, even indirectly, a physical damage to equipment or people, or rather of sabotaging or damaging in a direct way the information systems of a sensitive target of the attacked subject (Mele 2013).

Based on this definition, and in the light of the technical specifics of cyberspace, the core questions of monitoring and arms control—*"what to control, how to control it, by whom and for what purpose"*—raise the concerns about what aspects can actually be monitored in this domain. An assessment of suitable and measurable parameters also needs to evaluate the degree of explanatory power that a specific parameter can provide, as well as the question how the measurement can be performed. On the other hand, the extent of necessary alteration of hardware or software for monitoring purposes will affect the applicability and the political acceptance of possible monitoring regimes. With regard to this consideration, the paper takes the establishment of any first steps for cyber arms monitoring as a starting point and concentrates on parameters and measures that "look from the outside" on IT systems and the networks and do not require a modification of existing IT hardware or software infrastructures.

Physically Obvious Parameters

The first set of measurable parameters can be defined as these parameters that are physically obvious, hard to disguise or manipulate and obvious to monitor. They are applicable to monitor the tendency of technological developments, the establishment of new cyber capacities and will reveal significant changes. The drawback of these parameters is that they will not be applicable to monitoring the real time activities of actors like clandestine cyber operations. The parameters are:

- The overall power supply and the current power consumption of IT infrastructures
- The available cooling power and current thermal power production of IT infrastructures
- The network bandwidth and transmission capacities and current flow rates of data transmissions
- The number of interconnections to other external civil or commercial networks and their maximum and current transmission performance

- The required maintenance staff for the IT infrastructure
- The available computing processing and network processing power, as well as storage capacities. Measurement of these parameters requires direct access to the controlled systems.

Parameters of the Extent of Usage and Adaptation of Existing Tools

The other set of parameters applies to the usage of IT systems and aims to measure or monitor their specific application. They qualify for the real time control of cyber operations and activities but can still be gathered "from the outside". The drawback of these parameters is that they are capable of monitoring cyber activities in such detail that they can potentially reveal unwanted or even secret information. Their application will therefore be limited to situations that justify such intrusiveness. This could be either high risk contexts with a strong potential for military misconceptions and escalations or as a strong political signal of transparency and trustworthiness by unilateral declarations of a state. The applicable parameters are:

- The meta data of incoming and outbound network connections like senders and receivers, as well as the type and amount of transferred data
- The amount of usage of anonymization services or network encryption services
- The acquisition, possession and stock piling as well as the usage of software and hardware vulnerabilities like exploits for known security problems. Such vulnerabilities and code that uses these flaws are the "weapons material" for intrusive cyber tools ("cyber weapons") and necessary to overbear IT security measures, get access to IT systems, transfer the payload and perform the intended operations.

The above differentiation demonstrates that the question of the purpose of each monitoring measure needs to address specific situations and political agreements, either to provide oversight for the technological advancements or to restrict and control the deployment of specific offensive cyber operations. With regard to the task of applying established verification principles in cyberspace, the principle that seems to be most applicable is the definition of any kind of thresholds. It paradigmatically reflects that not the presence but the extent of the usage of goods in cyberspace defines compliance or noncompliance with an agreement. Approaches like restricting possession and/or proliferation of goods currently fail, as shown, due to the technical nature of the domain. On the other hand, the analysis of the necessary monitoring procedures reveals that there are already existing methods in computer science that have been developed for comparable protection and control claims, but that have not yet been used in the context of arms control and disarmament. For example, the question of how to control and restrict the usage of IT goods to allowed clients has been a challenge for the IT economy since the early days of software development and marketing. Over the last decades, a lot of effort has been put into digital rights and intellectual property protection systems and digital usage restrictions like the digital rights management technologies (DRM) with hardware dongles or online software authentication. A similar situation exists for the question of uniquely identifying IT systems in networks. The new internet addressing system—IPv6—provides technologies and capacities to provide unique addresses for all IT devices, which can help to overcome the attribution problem when applied to relevant networks like those that are used by military forces or intelligence services. Such mechanisms can, for example, provide a way of marking military cyber forces and their activities. The examples show that arms control and disarmament are merely new ways of looking at the challenges of interconnected global IT systems from a political and international security standpoint that don't necessarily require the development of new technologies, but rather apply and adapt existing tools and concepts in the light of different goals. It's not the perfect solution to technological problems, but it raises the question of how current systems can be shaped for a technical restriction of states and military forces to apply military pressure over cyberspace, as well as the question of how to control these restrictions.

Conclusion and Outlook

The previous explanations showed the necessity of—as well as the different problems with—the task of arms monitoring in cyberspace. They also demonstrated that many of the lessons learned from former technological developments steps cannot be applied or projected on this new artificial domain, which fundamentally differs in important technical aspects. In comparison to nuclear arms control and disarmament, the challenge of cyber armament monitoring has one strong advantage. The relevant domain is—in contrast to air, space, sea and land—completely man made, and all its rules are based code (see exemplary the "Code is Law" argumentation, Lessig 2006). Every functional principle is defined and created by people or, rather, international committees like the standardization-focused Internet Engineering Task Force (Bradner 1999) or the more research-focused Internet Research Task Force (IRTF 2018). These committees develop new technologies for cyberspace and decide about their deployment. This provides a strong point for legislation and means that principles can be further established to support the peaceful development of this domain, to create transparency where it's necessary and to support measures for international political stability. On a national level, recent political debates on the implementation and institutionalization of processes—such as a vulnerabilities equities process that makes decisions about the disclosure of computer security vulnerabilities that are used or held secret by state institutions—will provide important experience for how the assessment of hazardousness and the possible impact of malicious cyber tools can be used for future arms control institutions.

Bibliography

Bradner, S. (1999). *Internet Engineering Task Force (IETF)* (p. 1). Open Sources: Voices from the Open Source Revolution.

DARPA. (2012). Broad Agency Announcement—Foundational Cyberwarfare (Plan X), DARPA-BAA-13–02. Arlington, USA. Retrieved from https://govtribe.com/project/darpa-baa-13-02-foundational-cyberwarfare-plan-x.

Guerrero-Saade, J. A., & Raiu, C. (2017). Walking in your enemy's shadow: When fourth-party collection becomes attribution hell. In *Virus Bulletin Conference*.

Goldansky, V. (1988). Connection between horizontal and vertical proliferation of nuclear weapons. In J. Rotblat & L. Valki (Eds.), *Coexistence, cooperation and common security*. London: Palgrave Macmillan.

IAEA. (1961). *The agency's safeguards*. The International Atomic Energy Agency, Geneva, Switzerland. Retrieved from https://www.iaea.org/sites/default/files/publications/documents/infcircs/1961/infcirc26.pdf.

IAEA. (1968). *The agency's safeguard systems*. The International Atomic Energy Agency, Geneva, Switzerland. Retrieved from https://www.iaea.org/sites/default/files/publications/documents/infcircs/1965/infcirc66r2.pdf.

IAEA. (2016). *Iran and the IAEA: Verification and monitoring under the JCPOA*. The International Atomic Energy Agency, Geneva, Switzerland. Retrieved from https://www.iaea.org/sites/default/files/5722627.pdf.

IAEA. (2018a). *The statute of the IAEA*. The International Atomic Energy Agency, Geneva, Switzerland. Retrieved from https://www.iaea.org/about/statute.

IAEA. (2018b). *Statement on Iran by the IAEA Spokesperson on May 1, 2018*, Geneva, Switzerland. Retrieved from https://www.iaea.org/newscenter/pressreleases/statement-on-iran-by-the-iaea-spokesperson.

IRTF. (2018). Internet Research Task Force. Retrieved from https://irtf.org/.

Lessig, L. (2006). *Code: And other laws of cyberspace, Version 2.0*. Center for Internet and Society Standford.

NPT. (1968). *Treaty on the non-proliferation of nuclear weapons*. Retrieved from https://www.state.gov/documents/organization/141503.pdf.

Mele, S. (2013). *Cyber-weapons: Legal and strategic aspects*.

Mölling, C., & Neuneck, G. (2001). Präventive Rüstungskontrolle und Information Warfare. In Rüstungskontrolle im Cyberspace. Perspektiven der Friedenspolitik im Zeitalter von Computerattacken, in: Dokumentation einer Internationalen Konferenz der Heinrich-Böll-Stiftung am 29./30. Juni 2001 in Berlin, S. 47–53.

Müller, H., & Schörning, N. (2006). Rüstungsdynamik und Rüstungskontrolle: Eine exemplarische Einführung in die internationalen Beziehungen Nomos, 2006, Aussenpolitik und Internationale Ordnung.

NATO. (2010). *Cyber war and cyber power. Issues for NATO doctrine*. Rome: NATO Defense College.

Neuneck, G. (2012). Confidence building measures—Application to the cyber domain. In *Cyber Security Conference*, Berlin.

Neuneck, G. (2017). 60 Jahre nuklearer - Prometheus oder Sisyphos? Vereinte Nationen Magazin, 2017.

NTI. (2010). Treaty between the United States of America and the Russian Federation on measures for the further reduction and limitation of strategic offensive arms (New START). *Nuclear Threat Initiative*. Retrieved from http://www.nti.org/media/documents/new_start.pdf.

OSCE. (1992). *Treaty on Open Skies. Organization for Security and Cooperation in Europe*, Vienna, Austria. Retrieved from https://www.osce.org/library/14127.

Robert, K., & Martin, L. (2009). The promise of institutionalist theory. *International Security*, *20*(1), 39–51. http://www.jstor.org/stable/2539214. (Published by : The MIT Press Stable Robert O. Keohane and Lisa L. Martin).

Tikk, E., & Kerttunen, M. (2017). *The Alleged Demise of the UN GGE: An Autopsy and Eulogy.* New York, The Hague, Tartu, Jyvaskyla: Published by the Cyber Policy Institute.

United Nations. (1969). *Vienna Convention on the law of treaties.* United Nations. Geneva, Switzerland. Retrieved from https://treaties.un.org/doc/publication/unts/volume%201155/volume-1155-i-18232-english.pdf.

United Nations. (1992). *Convention on the prohibition of the development, production, stockpiling and use of chemical weapons and on their destruction.* Retrieved from https://treaties.un.org/doc/Treaties/1997/04/19970429%2007-52%20PM/CTC-XXVI_03_ocred.pdf.

US-DOD. (2016). *Department of Defense press briefing by Secretary Carter and Gen. Dunford in the Pentagon briefing room from February 29, 2016.* Washington, USA. Retrieved from https://www.defence.gov/News/Transcripts/Transcript-View/Article/682341/department-of-defence-press-briefing-by-secretary-carter-and-gen-dunford-in-the/.

US-ICS-CERT. (2016). Alert (IR-ALERT-H-16-056-01) *Cyber-attack against Ukrainian critical infrastructure.* The U.S. Industrial Control System Computer Emergency Response Team. Retrieved from https://ics-cert.us-cert.gov/alerts/IR-ALERT-H-16-056-01.

UNIDIR. (2013). *The cyber index—International security trends and realities*, Geneva, Switzerland. Retrieved from www.unidir.org/files/publications/pdfs/cyber-index-2013-en-463.pdf.

Wassenaar. (1996). *The Wassenaar Arrangement on export controls for conventional arms and dual-use goods and technologies, Initial Elements.* Retrieved from https://www.wassenaar.org/docs/IE96.html.

Wassenaar. (2017). *The Wassenaar Arrangement on export controls for conventional arms and dual-use goods and technologies—List of dual-use goods and technologies and munitions list.* Wassenaar Arrangement Secretariat.

Wehberg, H. (1959). Pacta sunt servanda. *The American Journal of International Law, 53*(4), 529–551.

Dipl.-Inf. Thomas Reinhold is a peace and security researcher and an expert for the challenges of the militarization of the cyberspace. As a graduated computer scientist, he works on technical measures for trust and security building for this domain like verification, arms control and non-proliferation. He is a Non-Resident Fellow at the Institute for Peace Research and Security Policy (IFSH) and a Ph.D. candidate at the research group Science and Technology for Peace and Security (PEASEC) at TU Darmstadt. He is also a member of the Transatlantic Cyber Forum and the Research Advisory Group of the Global Commission on the Stability of Cyberspace.

Chapter 21
Trust in the Digital Age—The Case of the Chinese Social Credit System

Peter Leibkuechler

In the near future, it is likely that cameras at restaurants, subways and airports can automatically identify your credit status. People will be able to go out without a mobile phone, cash or even ID card. They can go anywhere with only their "face" and the big data of creditworthiness behind it. Creditworthiness is becoming everyone's "passport" in society; the trustworthy will be welcomed everywhere, while the untrustworthy will be rebuffed at any step.

[Ant Financial 2016 Sustainability Report (p. 19) headlined "Moving towards a better society for the future" (未来好社会)]

Introduction

Trust among society is crucial not only for successful economic development but also for a healthy society that ensures a comfortable and enjoyable life for its members. The People's Republic of China (in the following: PRC) has identified trust and creditworthiness as one of the key factors on its long way to becoming the world leading power in an ever more digitized world. Surveys conducted by the Institute of Sociology under the Chinese Academy of Social Sciences have shown a steady decrease in trust among the Chinese society over the past years, a perception already reflected in the media as can be seen in the report by He Dan in February 2013 on China Daily and in academic writings (Wang 2013; Rao et al. 2013). According to the Annual Report on Social Mentality of China (2016), only 38.4% of the respondents agreed to the statement that "most people can be trusted".

Comparing these findings to a recent survey requested by the European Commission and published as the "Special Eurobarometer 471" in April 2018 on "Fairness, Inequality and intergenerated mobility" one can indeed see considerable differences.

P. Leibkuechler (✉)
Sino-German Institute for Legal Studies, Nanjing University, Nanjing, China
e-mail: dcir.nanjing@hotmail.com

© Springer Nature Switzerland AG 2020
D. Feldner (ed.), *Redesigning Organizations*,
https://doi.org/10.1007/978-3-030-27957-8_21

The stronger European economies, determined by gross domestic product per capita, see an average of approximately 65% of citizens agreeing to the statement that "Generally speaking, most people in (our country) can be trusted".

> The numbers for the respective countries are: Finland (85%), Denmark (82%), Sweden (76%), Netherlands (69%), Austria (67%), Germany (58%), Belgium (52%), UK (50%). Of the bigger economies within Europe, only Italy (47%) and Spain (46%) show a result similar to the Chinese survey and France suffering from a low trust of only 31%.

Even though these surveys are not conducted by the same entities and were using different methods, the differences are at least remarkable and indeed hint at a lack of trust within Chinese society.

The Chinese government has decided to tackle this issue with an immensely wide-ranging project generally referred to as the "Social Credit System" (in the following: SCS). The SCS aims at monitoring, assessing and ultimately steering the behaviour of all natural and legal persons in the PRC by making full use of centralized big-data to an extent unprecedented in human history. This project has seen its first preliminary foundations laid already in the late 1990s with pilot projects starting in the early 2000s (Meissner 2017). But it considerably gained momentum and speed during the last years. On June 14th 2014 the State Council has released a notice on its "Plan for the Establishment of a Social Credit System (2014–2020)" (in the following: SCS-Plan) aiming at the establishment of the system as of 2020. In the following the author will try to assess how close the project will bring the Chinese society to the scenario described in the introductory quote.

The Implementation of the SCS

Aims

According to para. 1 of the preamble of the SCS-Plan its "…inherent requirements are establishing the idea of sincerity culture, and carrying forward sincerity and traditional virtues, it uses encouragement to keep trust and constraints against breaking trust as incentive mechanism, and its objective is raising the honest mentality and credit levels of the entire society".

The SCS, No. 1, sect. 2, para. 3, shall provide "…an effective method to strengthen social sincerity, stimulate mutual trust in society, and reducing social contradictions, and is an urgent requirement for strengthening and innovating social governance, and building a Socialist harmonious society".

It needs to be stressed that the SCS by no means restricts itself to aiming at more honest behaviour of natural persons. The SCS-Plan clearly spells out other key areas in which it sees an increased trust as being fundamental. These concern among others:

– the construction of *government affairs* sincerity by all political and administrative actors (No. 2, Sect. 1),
– the deepened establishment of sincerity in *commercial affairs* by companies including the areas of production, construction, logistics, finance, taxation, pricing, government procurement, bidding and tendering, traffic and transportation, e-commerce, statistics, the intermediary services sector, exhibitions and advertising (No. 2, sect. 2),
– credit construction in the areas of health care, hygiene and birth control, social security, labour and employment, education and scientific research, culture, sports and tourism, intellectual property rights, environmental protection and energy saving, internet applications and services (No. 2, sect. 3),
– the establishing of *judicial credibility* encompassing courts, prosecuters, public security and law enforcement, as well as judicial administration and judicial enforcement (No. 2, sect. 4).

Ohlberg, Ahmed, and Lang in their 2017 study "Central Planning, Local Experiments—The complex implementation of China's Social Credit System rightly recognized that with a look at the enormous scope of the areas addressed, the aims of the SCS can be summarized as being set up to become the "cure-all solution to a multitude of disparate societal and economic problems".

This article, however, will focus its analysis on what may be referred to as social governance, i.e. the assessment and steering of the "creditworthiness as the new passport" of natural persons as mentioned in the introductory quote above. The SCS-Plan (No. 2, sect. 1, para. 1) refers to this at different occasions stressing that a fundamentally sincere interpersonal behaviour within society is a precondition in order to create harmonious and amicable relationships. The SCS-Plan (No. 2, sect. 3, para. 10) wishes to fulfil this aim by addressing the establishment and perfection of natural persons' credit records "in economic and social life". Some clues on what may be meant by the reference to "social life" here can be drawn from No. 3 SCS-Plan that aims at the construction of sincerity education and a sincerity culture. It stresses the need for moral cultivation, civil virtue, social morals, professional ethics and household virtue culminating in the call for "seeing sincerity and trust-keeping **as glorious** [emphasis added by author], and seeing the loss of integrity to temptation and gains **as shameful** [emphasis added by author] across the entire society".

Obviously, the concrete meaning of many of these colourful terms remains unclear and vague. What are the implications for natural persons, what behaviour is "expected", what incentives will be given, what kind of punishments are to be seen and how will implementation look like, especially how is data collection organized? In any case, from what is said so far it can clearly be seen that—different from systems more familiar to foreign observers—the SCS is by no means limited to a purely financial assessment of creditworthiness. It goes way beyond and this is what constitutes its unique character.

Status Quo and Actors

Looking at the current state of the system, public and private sector both need to be taken into account. From its outset—as can been seen in the SCS-Plan—the system relies on state actors of all levels and areas to establish the necessary technical infrastructure and streamline processes of data gathering, exchange, evaluation and the respective punishment and reward system which often relies on a black list system (No. 5, sec. 1, para. 2). While full implementation is foreseen for the year 2020 only, a total of more than 40 municipalities and districts have been chosen as pilots to experiment with the assessment of natural and legal persons' creditworthiness.

These include Beijing Haidian District (北京市海淀区), Hohot and Wuhai of Inner Mongolia (内蒙古自治区呼和浩特市, 乌海市), Dalian, Anshan and Liaoyang in Liaoning Province (辽宁省大连市, 鞍山市, 辽阳市), Suifenhe in Heilongjiang (黑龙江省绥芬河市), Shanghai New Pudong District and Jiading District (上海市浦东新区, 嘉定区), Suzhou, Suqian and Nanjing in Jiangsu Province (江苏省苏州市, 宿迁市, 南京市), Taizhou, Wenzhou, Hangzhou and Yiwu in Zhejiang Province (浙江省台州市, 温州市, 杭州市, 义乌市), Anqing and Huabei in Anhui Province (安徽省安庆市, 淮北市), Fuzhou, Xiamen and Putian in Fujian Province (福建省福州市, 厦门市, 莆田市), Weifang, Weihai, Dezhou and Rongcheng in Shandong Province (山东省潍坊市, 威海市, 德州市, 荣成市), Zhengzhou and Nanyang in Henan Province (河南省郑州市, 南阳市), Wuhan, Xianning, Yichang and Huangshi in Hubei Province (湖北省武汉市, 咸宁市, 宜昌市, 黄石市), Guangzhou, Shenzhen, Zhuhai, Shantou and Huizhou in Guangdong Province (广东省广州市, 深圳市, 珠海市, 汕头市, 惠州市) as well as Luzhou in Sichuan Province (四川省泸州市).

All these cities and districts have to different degree enacted local regulation on the SCS implementation itself or included references to it into legislation in different fields; some examples for several of these cities are given by Ohlberg, Ahmed, and Lang in their study. These differences are most probably intended by the central government serving best the purpose of trial and error. Experience gathered here will be the basis for the actual nationwide system to be implemented in 2020. In order to facilitate national implementation between the many different bureaucracies each natural and legal person is to be identified by a unified 18-digit identification number (the basis for this can be found in No. 5, sec. 2, para. 5). For legal persons this new number will replace the different numbers formerly used for financial and tax issues, business licensing, social security services and other relevant regimes. For natural persons the number will be identical with their already existing identity card number as Creemers confirms in his article "China's Social Credit System". The fairly detailed implementation rules published by one of the smaller pilot cities concerned, Rongcheng in Shandong Province, will be analyzed in more detail below (see below, Section "The Example of Rongcheng City, Shandong").

While the state actors play a significant role, the introductory quote of this article is not an official government announcement but has been taken from a financial report published by Ant Financial, a private company running Alipay, one of the biggest Chinese mobile and online payment providers. Alipay runs a non-mandatory credit scoring system called Sesame Credit. Users are given a score according to five categories: personal information, payment ability, credit history, social networks and behaviours. However, the exact details remain vague and the algorithms unrevealed. As Cheang Ming observes in a 2017 CNBC article "FICO with Chinese characteristics: Nice rewards, but punishing penalties", high score (maximum are 950 points) will grant the user certain rewards such as deposit free rentals of hotel rooms, e-bikes or umbrellas, discounted car rental or even facilitated visa application for certain countries. Negative influence of a low score is not clearly announced, but at least exists in the denial of the aforementioned benefits.

Whereas Sesame Credit at this point of time arguably affects the lives of more citizens than the state-run system, it remains uncertain whether Sesame Credit and other similar systems run by private companies will in a later stage be integrated into the centralized state system. Up to today for Creemers there is no evidence for Sesame Credit actually sharing their data with governmental entities without prior consent of its users and there exists at least no automatism that it will do so in the future.

The Example of Rongcheng City, Shandong

Rongcheng City of Shandong Province was one of the first pilot cities elected. The local government has been comparatively open with publishing its implementation rules for the social credit system and may serve as an example of what could be expected by the system to be implemented in 2020. The city government has been particularly active in regulating and two legal documents are of particular interest as they make it possible to outline in detail the system Rongcheng set up during the last years: Firstly the "Rongcheng City Rules for the Evaluation of Credit for Natural Persons and Societal Legal Persons" (Original Chinese title "荣成市自然人和社会法人信用信息评价规定", Reference No. RCDR-2016-0010003, in the following referred to as "Rongcheng Evaluation Rules"). Secondly the "Rongcheng City Method for the Administration of Personal Credit Reward and Punishment" (Original Chinese title "荣成市个人信用奖惩管理办法", Reference No. RCDR-2016-0010004, in the following referred to as "Rongcheng RP-Method").

First of all, the RP-Method clarifies that the system is applied to all citizens of Rongcheng City above the age of 18 who have lived there for more than a period of one year. Foreigners are explicitly included (Art. 3 Rongcheng RP-Method). Citizens are distinguished into four different categories of declining credit worthiness: A (good credit), B, C and D (bad credit).

The system sets its own theme as "encouragement as a priority, penalization as an auxiliary" and "warning first, penalization second" ("激励为主, 惩戒为辅" and "先警示, 后惩戒", Art. 6 Rongcheng RP-Method). At the same time it leaves no doubt about its comprehensive approach when postulating that for persons that "keep their word everywhere, anything will go smoothly" whereas those "who broke trust at one occasion will face obstacles everywhere" ("…守信者"处处守信, 事事方便", 失信者 "一处失信, 处处受制"…).

Persons categorized as A citizens are being included into a positive "red list" and do obtain rewards such as priority treatment when it comes to schooling and welfare (Art. 11 Rongcheng RP-Method). Even more trustworthy citizens, rated AA or AAA, enjoy monetary contribution when making their payments to public services (e.g. reduction of water supply costs), social security insurance (the government pays up to 200 RMB, Art. 12 Nr. 1 and 2 Rongcheng RP-Method) and need to pay less for stationary hospital treatment (the bill can be reduced by 5%, cf. Art. 12 Nr. 3 Rongcheng RP-Method). Persons up to the age of 64 get financial help with public transportation tickets, persons 65 or older are rewarded with certain free of charge health insurance offers (Art. 12 Nr. 4 Rongcheng RP-Method). Being rated AA or AAA also allows one to apply for interest free government loans of up to 100.000 RMB (app. 13.500 €) with a retention period of two years (Art. 12 Nr. 6 Rongcheng RP-Method).

Citizens rated B will be the addressee of creditworthiness campaigns and encouraged to improve their creditworthiness for instance by discretionally being granted the benefits foreseen for A citizens (Art. 13 Rongcheng RP-Method).

Being rated a C citizen is already followed by clearly perceptible consequences: Social welfare and support as well as government preferential treatment may be reduced or completely withdrawn and one's status may be publicly displayed and one may be included into a "yellow list" for 2 consecutive years. Eligibility for public procurement may be abolished. Even more severely, C category citizens are not eligible to apply for a position as a public servant in the city's administration (Art. 14 and 20 Rongcheng RP-Method).

In addition to these punishments, D category citizens face harsher consequences: state funded assistance or subsidies may be withdrawn completely, and D-citizens are being included into publicly accessible "black lists" for a period of five years. Business licenses may be withdrawn and the eligibility for taking up loans limited (these restrictions can be found in Art. 15 and 20 Rongcheng RP-Method).

Concerning social welfare withdrawal Art. 30 RP-Method opens the possibility for temporary suspension of the punishment in the case of severe circumstances in the yellow- or black-listed C- or D-category citizen or their family members.

As for how a citizen is categorized into one of the mentioned categories it is clearly stated in the Evaluation Rules. The basis for this is information gathered within the city's own administration, from all levels of public or business branch credit information sources as well as other organizations and associations not further specified (Art. 2 Rongcheng Evaluation Rules). For a start, each citizen is granted 1000 points of creditworthiness. According to the citizen's behaviour he or she can win or lose points. The categories (Art. 3 Rongcheng Evaluation Rules) are related to the points as follows: AAA (1050 points or above), AA (1049-1030), A (1029-960), B (959-850), C (849-600), D (599 and below).

The annex to the Rongcheng Evaluation Rules foresees detailed descriptions of well over 200 relevant behaviors as well as the respective points to gain or lose. The maximum number of points to be gained or lost by one single behavior is 100 points. The behaviours followed by a deduction of 100 points mainly include criminal offenses such as serious tax fraud, the falsification of documents or seals or the use of force or threat against administrative or judicial enforcement as well as false statements or the delivery of false evidence to judicial organs.

Theft of publicly supplied resources such as water, gas, electricity results in a deduction of 50 points, whereas delayed payment for these services will be followed by a reduction of 20 points. Traffic violations are also included into the system: Depending on the respective sentence obtained by the traffic authorities, the amount of points reduced changes. 5 points are gone for minor violations (sentence of up to 500 Yuan by the traffic authorities), whereas 60 points will be deducted from your credit score in case that your driving license got suspended. Including traffic rule violation into the system has occasionally been criticized, as Mu Xuchong (木须虫) says 2018 in Legal Daily: "Credit loss punishment in public traffic needs to be clearly defined" (交通失信惩戒须把握好界定). Posting inappropriate content using internet communication services may as well result in a deduction of 50 points. Obviously, the term "inappropriate" gives an enormous room for different interpretations by the authorities in this sphere. Academic fraud such as plagiarism and fraud in examinations leads to a minus of 20 points. Interestingly, behaviour considered less severe but still included into the system encompass several activities that cause annoyance to others due to the noise they produce: renovation activities in one's house or apartment, dancing on public squares and playing music with loudspeakers, playing with gyroscope spinners (5 points deduction each). Others include pedestrians not using zebra lines or drivers not respecting them, parking your car wildly in a closed living community, littering or planting your own plants in a public spot (minus 10 points each). Endangering the social stability by getting drunk, steering up fights or quarrels, insulting and defamation will give you a minus of 20 points in your social credit score. Not caring for the elderly, maltreating them or abandon family members is followed by a deduction of remarkable 50 points. This most probably refers to the duty of children to care for their parents that can be found in Art. 14 et seq. Law of the People's Republic of China on the Guarantee of the Rights and Interests of the Elderly (中华人民共和国老年人权益保障法).

Apart from the behaviour that negatively influences one's score, the annex to the Evaluation Rules also states clearly what acts may help you to raise your score. The highest score will be gained by becoming an organ donor (+100) or making a bone marrow donation (+50 points). Making financial donations will also earn you points depending on the amount given, starting from 5 points (donation above 1000 RMB) to a maximum of 50 points (donations above 500.000 RMB). Other encouraged behaviours include: returning money that has been found on the street (+5 points), doing volunteer work at least ten times a year (+5 points), supporting old and weak non-consanguineous persons for a long period of time (+20 points), taking part in emergency rescue activities following major incidents endangering national security and the lives and property of the masses (+20 points). There is also encouragement for reporting to the relevant authorities: Reports about big scale illegal sales or production of counterfeit products will be rewarded with 30 points, reporting on potential food safety issues that might have a relatively big effect on the society will gain a citizen 5 points.

The annex contains many more specific descriptions, but the examples taken here may sufficiently illustrate how far the system in Rongcheng reaches into daily, non-economic behaviour of citizens including it into their personal score affecting it to the negative or positive. Considering the benefits for highly rated citizens seen above the system functions like a game, rewarding the behaviour that the law maker deems preferable.

Societies Perception

According to the author's personal perception as well as according to extended recent research the SCS is not receiving as much attention among the Chinese society as one would expect (see also Ohlberg, Ahmed, and Lang). News coverage mainly stresses the advantages of a strengthened system of trust and critical views are rare, even in social media. The potential annoyances and restrictions on personal conduct seem not to be regarded as a problem or at least as one that is outweighed greatly by the advantages of the system making sure that many more people will act in a trustworthy way.

Concluding Considerations

It is surely not exaggerated to claim that the Chinese government is engaging in the biggest experiment to centralize personal information and steer individual behaviour of its citizens by giving incentives and immediate punishments through the SCS. An experiment unprecedented in world history. However, it needs to be stressed that the example of Rongcheng City given above displays only one of many different approaches that the chosen pilot cities have taken. It cannot be deducted that this is how the system is going to look like in the end. Most presumably it will be a mix of all the pilots enacted.

It is therefore at this point of time too early to conclude that we are witnessing the rise of an "Orwellian state" as expected by some commentators. However, it can certainly be expected that the final system will indeed—much more than credit rating in other countries—also encompass "normal social behaviour", even though its final shape is still unclear.

Questions and doubts, also especially remain on the broader concept of the system itself:

Firstly, it seems doubtful that actual trust is strengthened by the system. Trust is an interpersonal concept mainly relying on personal experience between the individuals involved and only gained over time. A good score in the SCS is merely the formal shell of trust lacking all the material implications normally present when feeling trust towards someone. It is doubtful that a society that is lacking trust, will develop real trust by a scoring system.

Secondly, a high score might not necessarily be a reliable basis for trust expectation. Taking the example of Rongcheng, donating organs or bone marrow will raise a person's score. While this behaviour is probably showing that person's affection to their fellow citizens, there is not necessarily any connection to that person behaving trustworthy in other areas of economic or social life. In addition, people aware of the scoring scheme might become donors purely for the sake of gaining points which leads the implication that a donor is a trustworthy person ad absurdum.

Thirdly, financial donation raising a person's score as foreseen in the Rongcheng system, make it possible to literally buy a high score. Tax fraud reducing your score by 100 points and financial donation of above 500.000 RMB gaining you 100 points might make it a pure matter of maths whether rule abiding behaviour is beneficial or not; speaking of trustworthiness under these preconditions seems pointless.

Finally, it still remains to be seen how society reacts if knowledge about score punishment and assessment for "normal" behaviour in life raises. Up to today, Chinese citizens seem not to be too worried about this kind of social governance. This might change, however, when the actual system is installed in 2020.

References

Studies

Ant Financial 2016 Sustainability Report, "Moving towards a better society for the future" (未来好社会). For further information see www.antfin.com. Last visited May 5, 2018.

Meissner, M. (2017). China's Social Credit System—A big-data enabled approach to market regulation with broad implications for doing business in China. Mercator Institute for China Studies (MERICS) China Monitor of May 24 2017, available under https://www.merics.org/en/microsite/china-monitor/chinas-social-credit-system. Last visited May 5, 2018.

Ohlberg, M., Ahmed, S., & Lang, B. (2017). Central planning, local experiments—The complex implementation of China's Social Credit System. Mercator Institute for China Studies (MERICS) China Monitor of December 12th 2017. Available under https://www.merics.org/en/microsite/china-monitor/central-planning-local-experiments. Last visited May 5, 2018.

Special Eurobarometer 471, Fieldwork December 2017 Publication April 2018 on "Fairness, Inequality and intergenerated mobility" available under http://ec.europa.eu/commfrontoffice/publicopinion/index.cfm/Survey/getSurveyDetail/instruments/SPECIAL/surveyKy/2166. Last visited May 5, 2018.

Wang, J. (王俊秀) (Ed.). (2016). Blue Book of Social Mentality (社会心态蓝皮书), Annual Report on Social Mentality of China (2016) (中国社会心态研究报告 [2016]).

Articles

Creemers, R. (2018). *China's Social Credit System: An evolution practice of control*, p. 12 et seq. Available at https://papers.ssrn.com/sol3/papers.cfm?abstract_id=3175792. Last access May 24, 2018.

Dan, H. (2013, February). Trust among Chinese 'drops to record low'. *China Daily*. Available under www.chinadaily.com.cn/china/2013-02/18/content_16230755.htm. Last visited May 5, 2018.

Ming, C. (2017, March 16). FICO with Chinese characteristics: Nice rewards, but punishing penalties. *CNBC*. Available at http://www.cnbc.com/2017/03/16/china-social-credit-system-ant-financials-sesame-credit-and-others-give-scores-that-go-beyond-fico.html. Last access May 2018.

Mu, X. (木须虫). (2018). Credit loss punishment in public traffic needs to clearly defined (交通失信惩戒须把握好界定). *Legal Daily* (法制日报) of 11 April, 2018, p. 7.

Rao, Y., Zhou, J., Tian, Z., & Yang, Y. (饶印莎, 周江, 田兆斌, & 杨宜音). (2013). Investigation Report on the on the state of societal trust among city dwellers (城市居民社会信任状况调查报告). Democracy and Science (民主与科学) 2013, issue 3, pp. 47 et seq (47, 51, 52).

Wang, J. (王俊秀). (2013). Pay attention to the society's mind – Promote society's approval—Aggregate society's common understanding" (关注社会情绪-促进社会认同-凝聚社会共识), Democracy and Science(民主与科学) 2013, issue 1, pp. 64 et. seq.

Legislation

A German translation can be found in Zeitschrift für Chinesisches Recht (ZChinR), vol. 25, issue 1, pp. 45 et seq.

An English translation by Rogier Creemers can be accessed under https://chinacopyrightandmedia.wordpress.com/2014/06/14/planning-outline-for-the-construction-of-a-social-credit-system-2014-2020/. Last visited May 5, 2018.

Law of the People's Republic of China on the Guarantee of the Rights and Interests of the Elderly (中华人民共和国老年人权益保障法) as of 24 April, 2015, http://www.pkulaw.cn/fulltext_form.aspx?Db = chl&Gid = 03800bb68eb6807fbdfb&keyword =%e4%b8%ad%e5%8d%8e%e4%ba%ba%e6%b0%91%e5%85%b1%e5%92%8c%e5%9b%bd%e8%80%81%e5%b9%b4%e4%ba%ba%e6%9d%83%e7%9b%8a%e4%bf%9d%e9%9a%9c%e6%b3%95&EncodingName=&Search_Mode=accurate&Search_IsTitle=0. Last visited May 24, 2018.

Reduction of Water Supply Costs: http://www.creditchina.gov.cn/lianhejiangcheng/difangtuijinqingkuang/201804/t20180419_113595.html. Last visited 25 May, 2018.

Rongcheng City Method for the Administration of Personal Credit Reward and Punishment, Original Chinese title荣成市个人信用奖惩管理办法 (Reference No. RCDR-2016-0010004) issued in 2016. Available at http://www.rccredit.gov.cn/rccreditweb/web/cont_22b4a3a6373049929ee94575229c14fa.html. Last access May 24, 2018.

Rongcheng City Rules for the Evaluation of Credit for Natural Persons and Societal Legal Persons, Original Chinese title 荣成市自然人和社会法人信用信息评价规定 (Reference No. RCDR-2016-0010003) issued in 2016, available at http://www.rcwhg.com/art/2016/1/8/art_66_97347.html. Last access May 24, 2018.

State Council document Guofa 2014/21 (国发 [2014] 21号), 国务院关于印发社会信用体系建设规划纲要 (2014—2020年) 的通知, available in Chinese under www.gov.cn/zhengce/content/2014-06/27/content_8913.htm. Last visited May 5, 2018.

The Sesame Credit, www.xin.xin. Last access May 24, 2018.

Sources on the list of pilot cities:

www.hbcredit.gov.cn/xyjs/wjzl/20152016/201801/t20180104_25499.shtml. Last access May 28, 2018.

http://www.creditchina.gov.cn/hangyexinyong_824/zonghedongtai/zhengfubumen/201801/ t20180110_106069.html. Last accessed May 28, 2018.

Dr. Peter Leibkuechler is currently Vice-Director of the Sino-German Institute for Legal Studies of Göttingen University and Nanjing University at Nanjing University, China. He is Editor-in Chief of the German Journal of Chinese Law (Zeitschrift für Chinesisches Recht), the only German law journal focusing exclusively on Chinese law. Besides the Social Credit System, his recent research interests touch upon the Cyber Security Law, the General Part of Civil Law and new developments in Private International Law. He studied in Germany and holds a Ph.D. in Law (Hamburg University), an LL.M. (China-EU School of Law, Beijing), and he is a Wirtschaftsjurist (University of Bayreuth, Germany).

Chapter 22
Trust in the Functioning of Technology and Criminal Liability Based on the Example of Driving Automation

Nadine Zurkinden

Introduction

Arguably, trust in the functioning of a new technology is an essential prerequisite for its legalization. If a new technology is considered being too dangerous, people will first push for its prohibition. Using driving automation as an example, this chapter looks at the relationship between trust in the functioning of a new technology and criminal liability in case a novelty malfunctions and causes harm.

Comparing the past when the automobile first appeared, and our era's driving automation supports the argument that trust is an essential prerequisite for legalizing new technology (II). Once a new technology was officially accredited, the question arises whether this fact constitutes a permission to use that technology also as a defence against incurring criminal liability should an accident happen (III) or, on the contrary, trust in its functioning may rather become a criminal liability trap (IV).

Furthermore, dilemma situations may lead to different trust issues given that the decision on how to solve such a situation is not taken by the driver anymore but rather ahead of time which may put the car occupant at risk (V).

N. Zurkinden (✉)
Substantive and Procedural Criminal Law, University of Zurich,
Faculty of Law, Zurich, Switzerland
e-mail: nadine.zurkinden@rwi.uzh.ch

© Springer Nature Switzerland AG 2020
D. Feldner (ed.), *Redesigning Organizations*,
https://doi.org/10.1007/978-3-030-27957-8_22

Trust as an Essential Prerequisite for Legalizing New Technology

A Glance at the Past—The Rise of the Automobile

Technological progress always bears new risks—sometimes technology may even endanger lives (Von Bar 1871). When, for example, automobiles began to replace horse-drawn carriages at the beginning of the 20th century, petrol engines were not only considered being noisy and fumy but also very hazardous as they enabled drivers to speed up and endanger pedestrians, horses, horse-drawn carriages and cattle. Interestingly, some contemporaries (such as Emperor Wilhelm II) did not worry since they were convinced that the automobile was just a passing fad.

Others, however, worried enough to prohibit automobiles: for example, the Swiss canton Grisons prohibited all automobile traffic in 1900. The administration argued that automobiles would endanger postal transport in particular, including passenger transport. Four years later, lorries and public transport busses received licences to circulate on the canton's roads with a speed limit of mere 12 km per hour (Schwarzenbach 2016). It took ten popular votes until the canton Grisons finally abolished this prohibition 25 years later, in 1925 (Gisler-Jauch 2015). The change of the public's mind can arguably be ascribed to economic reasons given that the canton risked isolating itself (Neue Zürcher Zeitung 1925).

This short historical review illustrates two things. One: it may take people quite some time to entrust their safety to a new, disruptive technology such as the automobile. Two: trust in the benefits of new technology such as allowing to move faster between point A and point B as well as the economic progress implied let society accept a certain risk profile, which the legislator eventually legalizes: even though people die in road traffic accidents, cars are not prohibited anymore.

A Glance at the Present—The Rise of Driving Automation

Today, we are facing driving automation that is subsequently accredited.

Driving Automation—From Zero Automation to Driverless Cars

There are different levels of driving automation for on-road vehicles, for particular systems within such vehicles, and for the operation of such vehicles. SAE International (formerly known as Society of Automotive Engineers) provided a taxonomy widely recognised (SAE 2016). The Swiss Federal Roads Office (i.e. the federal authority responsible for road infrastructure and private road transport) too refers to this taxonomy in its guidelines on how to receive special permits for test runs using cars with high degrees of driving automation on Swiss public roads (Bundesamt für Strassen (ASTRA) 2017).

The SAE taxonomy covers driving automation ranging from level zero (no automation) to level five (full automation). For the levels zero, one (assisted) and two (partial automation), monitoring of the driving environment fully remains with the human driver. From a technical and legal point of view, the driver thus remains liable for the driving performance and accordingly always serves as fall back when the automation fails. For the levels three (conditional automation), four (high automation) and five (full automation), the dynamic driving task is entirely performed by an automated driving system during a given driving mode or trip. Thus, the automated driving system monitors the driving environment. For level three however, human drivers still serve as fall back, so they must appropriately respond to intervention requests of the system. Only levels four and higher will provide technology allowing drivers to sit back and relax or use the time spent behind the wheel in a productive way. From a legal point of view, however, the human in the car is not necessarily "off the hook"; they are the one entrusted with driving and thus still liable.

Driving Automation and the (International) Law

In order to understand when the human in the car is "off the hook"—in other words in how much driving automation the legislator trusts today, one must look at the law, but not merely the domestic law. Uniform traffic rules in order to facilitate international road traffic and increasing road safety were first adopted in the Vienna Convention on Road Traffic (see its preamble), concluded on 8 November 1968 (in the following referred to as the Vienna Convention). 75 parties—mostly European states with a few exceptions such as the UK—signed and ratified this convention (for a detailed list of states see https://treaties.un.org/Pages/ViewDetailsIII.aspx?src=TREATY&mtdsg_no=XI-B-19&chapter=11&Temp=mtdsg3&lang=en. Accessed April 2018).

Human driver or autopilot system?

Article 8(1) of the Vienna Convention broadly states that every moving vehicle or combination of vehicles must have a driver. But what exactly is a "driver"? Can only a human be a driver? Or could a self-driving system (aka driverless car) also be considered being a "driver"? The U.S. arguably follows the latter approach: In February 2016, the National Highway Traffic Safety Administration (NHTSA) informed Google in a letter that they will interpret "driver" in the context of Google's described motor vehicle design as referring to the self-driving system and not to any of the vehicle occupants (Shepardson and Lienert 2016). Digital natives probably trust in computer systems and accept machines as drivers too, whereas older people seem more reluctant to do so. They mostly agree that only humans qualify as drivers (please note that this impression is the result of only a handful of spontaneous opinion polls during lectures and presentations). This approach is also taken by the Vienna Convention, which defines a driver as any *person* who drives a motor vehicle (Article 1 lit. v Vienna Convention). From a legal point of view, only two kinds of personhoods may be exposed to criminal liability: natural and legal persons. There are, however, discussions in academia and politics whether an "electronic personhood" should be created in order to ensure rights and responsibilities for robots (European Parliament 2017; Gless 2017).

Another solution for allowing computer systems as drivers would be to simply amend the definition of *driver* in the Vienna Convention. This was proposed by Sweden and Belgium. The suggested provision read: "*driver means any person who drives or a vehicle system which has the full control over the vehicle from departure until arrival*" (United Nations Economic Commission for Europe (UNECE) 2015a).

However, the Vienna Convention has not been amended respectively. As long as the notion of a *driver* is not redefined by way of amendment and as long as the electronic personhood is not established, the combination of the notions *person* and *driver* in the Vienna Convention entrusts only humans with driving.

The clash—driver duties versus technical developments

Looking more closely at the road traffic rules, it becomes clear that road traffic laws are currently designed for human drivers who must at all times be alert and in control of the vehicles, they drive (Articles 8(5) and Article 13(1) of the Vienna Convention). Yet, as soon as the driver hands control over to the vehicle, they are no longer in control at all times. Such discord must be reconciled.

A first amendment to the Vienna Convention paves the way for more driving automation: The new paragraph 5bis was added to Article 8 and entered into force on 23 March 2016 (United Nations Economic Commission for Europe 2014, 2016; Lutz 2014; Lohmann 2015):

> Vehicle systems which influence the way vehicles are driven shall be deemed to be in conformity with paragraph 5 of this Article and with paragraph 1 of Article 13, when they are in conformity with the conditions of construction, fitting and utilization according to international legal instruments concerning wheeled vehicles, equipment and parts which can be fitted and/or be used on wheeled vehicles.

> Vehicle systems which influence the way vehicles are driven and are not in conformity with the aforementioned conditions of construction, fitting and utilization, shall be deemed to be in conformity with paragraph 5 of this Article and with paragraph 1 of Article 13, when such systems can be overridden or switched off by the driver.

According to Article 8(5bis), driving automation technologies are thus deemed to be in conformity with Article 8(5) and Article 13(1) and allowed on Europe's roads if they can either be overridden or switched off by the driver or if they are in conformity with certain international legal instruments, particularly with regard to the so-called ECE-Regulations (Lohmann 2015). States that are party to the so-called 1958 Agreement (Agreement concerning the Adoption of Uniform Technical Prescriptions for Wheeled Vehicles, Equipment and Parts which can be fitted and/or be used on Wheeled Vehicles and the Conditions for Reciprocal Recognition of Approvals Granted on the Basis of these Prescriptions, of 20 March 1958) must implement the latter. All European states with the exception of Ireland are parties to this agreement (http://www.unece.org/trans/maps/un-transport-agreements-and-conventions-18.html. Accessed April 2018).

The amendment, however, does not clarify if those states have more trust in human drivers or in computer cars, as they do not define the particular duties of human drivers in relation to automation technologies (Zurkinden 2017). It is, for example, unclear in which cases the human driver must override the technology and in which they must switch it off (United Nations Economic Commission for Europe (UNECE) 2015b). Legal scholars take different approaches. According to some, the theoretical possibility to switch off the systems suffices (Lohmann 2015 with further references). Other scholars and Swiss authorities take a more restrictive approach stating that the driver remains responsible at all times (Bundesamt für Strassen (ASTRA) 2015) and that Article 8(5bis) does not allow for driverless cars (Von Bodungen and Hoffmann 2015 with further references).

The driver must be able at all times to immediately assume control over the vehicle (United Nations Economic Commission for Europe (UNECE) 2015b). The fact that driver duties in general have not been amended yet supports the understanding that we still trust humans more than technology. For example, Article 8(6) of the Vienna Convention has not been amended, and Article 8(5[bis]) provides no exception for it. According to Article 8(6), a driver of a vehicle must minimize any activity other than driving such as using their cell phone while the vehicle is in motion (Lohmann 2015). Thus, the driver still has observation duties even if Articles 8(5) and 13(1) no longer need to be complied with by the driver (Zurkinden 2017).

Lutz, however, is of the opinion that the newly added Article 8 paragraph 5[bis] takes precedence over other requirements aimed at the driver by the Vienna Convention (Lutz 2016). Furthermore, some scholars argue that driver duties no longer apply for level three vehicles (observation of the driving environment by the system, yet the human car occupant must intervene at the request of the system) as soon as such vehicles are admitted to traffic because level three vehicles and higher have no driver anymore (Riedo and Maeder 2016). This argument misses that as long as Article 8(1) of the Vienna Convention is not amended, every vehicle must still have a human driver. Accordingly, driver duties would still apply.

This divergence in opinions illustrates how important it is to solve the clash between human driver duties and technical developments within legal rules. Therefore, in order to allow more driving automation or even driverless cars on Europe's roads and to clarify road user duties, the Vienna Convention requires further amendments. The Global Forum for Road Traffic Safety (WP.1, formerly known as Working Party on Road Traffic Safety) is working on further amendments to Article 8 of the Vienna Convention (http://www.unece.org/trans/themes/transits/selfdriving/next-steps.html. Accessed April 2018). It stresses the importance of cooperation between the different forums working on regulations regarding technology on the one hand and road users on the other hand. While WP.1 deals with the latter, the World Forum for Harmonization of Vehicle Regulations (WP.29) issues the so-called ECE-

Regulations. WP.1 argues that technology (field of WP.29 activity) and road users (field of WP.1 activity) cannot be regulated in isolation from each other when it comes to automated driving (United Nations Economic Commission for Europe (UNECE) 2015b).

Even though driverless cars are currently not allowed by the Vienna Convention, national road traffic laws might provide for exemptions, i.e. the possibility to grant special permits for test runs with cars possessing high degrees of driving automation such as driverless vehicles on public roads. Such special permits have been issued in Switzerland, for example for test projects implying driverless passenger busses (Bundesamt für Strassen (ASTRA) 2018). When such special permits are granted, a qualified person (*Begleitperson*) always remains the person of trust responsible for the driving performance of the vehicle. They may escape criminal liability though if the risks taken during the test project amount to so-called permissible risks (Zurkinden 2016a, b).

Legally Permitted Trust as a Defence Against Criminal Liability?

Road Traffic—A Permissible Risk

Today, we cannot imagine life without motorized traffic even though—according to statistics—a road accident fatality occurs every 41 h on Switzerland's roads (Bohnenblust and Pool 2017). Thus, society accepts manifold risks when it comes to road traffic. This risk acceptance coins the scholarly debate. In Swiss (and German) criminal law, road traffic is one of the rarely disputed examples of a so-called permissible risk (e.g. Stratenwerth 2011). More critical, however, is Schubarth arguing that the concept of permissible risk allows for a license to massive deaths in road traffic (Schubarth 2011). Yet an abstract definition of what a permissible risk constitutes is still missing (Zurkinden 2016a). What amounts to a permissible risk is, first of all, defined by the legislator who balances all interests involved (e.g. the benefits of motorized traffic vs. its risks) before issuing legal provisions (e.g. in the Road Traffic Act). The legal provisions define the rules for dealing with these risks in general (e.g. by establishing minimum technical requirements) as well as with risks arising from specific situations (e.g. precedence rules and speed limits) (Zurkinden 2017). In general, the amount of permissible risks is the amount resulting when risks are reduced to a level where further risk reduction would require disproportionate efforts (Stratenwerth 2011) or result in the uselessness of the technology (for example, there would not be much benefit to a car that only moves at walking speed).

In sum, for "traditional" road traffic, society and the legislator have accepted that certain road traffic risks are considered permissible risks. No criminal liability attaches to taking a risk by driving a car and complying with all rules, even if an accident should happen.

Driving Automation—Another Permissible Risk?

Driving automation is a new disruptive technology challenging today's society, just as the rise of the automobile did more than a hundred years ago. Some may argue that research in the area of driving automation leading up to driverless cars has already been performed since the 1920s (Sinclair and Schafer 2017 with reference to Lafrance 2016). However, only today almost every major car manufacturer and other companies such as Google and Uber that have not been engaged in the automobile industry before the rise of driving automation are committed to develop their own driverless cars. Driving automation comes with the promise that it will enhance mobility especially for elderly and disabled persons and children (Wohlers 2016), while at the same time reducing (fatal) crashes because machines need no sleep and are never stressed, drunk or distracted. Furthermore, driverless cars would enhance convenience and allow for more productivity or quality time (e.g. for family and friends) by freeing up time spent behind the wheel. At the same time, this will benefit economic efficiency (Lohmann 2016b; Thierer and Hagemann 2015).

Driving automation will reduce crashes attributable to human error. However, this does not necessarily mean that there will be less or no (fatal) crashes anymore. It might well be that errors caused by machines will simply replace human errors (Sander and Hollering 2017 with further references). In the past, driving automation systems such as autopilots already made fatal mistakes (Vlasic and Boudette 2016; Curtis 2016; Levin and Wong 2018).

The question thus arises which risk profile the legislator is willing to accept when it comes to driving automation, now and in the future, bearing in mind that zero tolerance regarding fatal accidents probably would bring innovation to a complete halt.

Trust in Technology as a Criminal Liability Trap

It is possible that, in the future, the human in the car will trust the autopilot system. The duty to remain in control of the vehicle at all times may be abolished and the human in the car would not incur criminal liability if the system fails and an accident occurs.

However, it is difficult to make predictions and laws will not necessarily follow technology. Existing legal provisions even apply if technology evolves. Trust in technology may then become a criminal or civil liability trap. This is especially true for autopilot systems in cars. Car occupants start trusting the autopilot and relax instead of keeping a close eye on the driving environment and staying in control of the car (Cassart 2017: "la responsabilité du conducteur est aggravée par le système qui devrait pourtant lui faciliter la vie").

As long as road traffic laws oblige human drivers to be in control of the vehicle at all times, a driver cannot defend himself against incurring criminal liability by claiming he trusted the autopilot (Hofstetter 2017). Nevertheless, trust in the functioning of technology was taken into account to mitigate the sentence for a driver in a Swiss case (the judgment was published in extracts and discussed in Zurkinden 2018).

Manufacturing companies though may trust that drivers will comply with road traffic laws (*Vertrauensgrundsatz*) and that they will therefore not incur criminal liability for failures of their autopilot systems when the driver fails to comply with his duty to stay in control of the vehicle.

However, if a car manufacturer markets a driving assistance system as an autopilot, alluding thus to the competence of autonomous driving, they might violate their instruction duties (Lohmann and Müller-Chen 2017; Lohmann 2016a). In its close resume on the investigation of the fatal Tesla accident, the NHTSA stated: "*Although perhaps not as specific as it could be, Tesla has provided information about system limitations in the owner's manuals, user interface and associated warnings/alerts, as well as a driver monitoring system that is intended to aid the driver in remaining engaged in the driving task at all times. Drivers should read all instructions and warnings provided in owner's manuals for ADAS technologies and be aware of system limitations*". However, the NHTSA specified in a footnote of the report: "*While drivers have a responsibility to read the owner's manual and comply with all manufacturer instructions and warnings, the reality is that drivers do not always do so.* **Manufacturers therefore have a responsibility to design with the inattentive driver in mind.** *See Enforcement Guidance Bulletin 2016-02: Safety-Related Defects and Automated Safety Technologies, 81 Fed. Reg. 65705*" (NHTSA 2017, accentuation by the author). Thus, if car manufacturers violate their instruction duties, they may incur civil but not necessarily criminal liability.

To prevent customers from falling into a liability trap, some car manufacturers try to foster trust into their driving automation technology by promising to accept full liability whenever their cars are involved in accidents while circulated in autonomous mode (Atiyeh 2015). Although such promises might be legally possible regarding civil liability cases, they have no validity whatsoever with regard to criminal cases where the state determines who incurs criminal liability. Furthermore, and as a rule, there is no compensation of guilt in criminal law (as opposed to other areas of law). This means that even if the car manufacturer incurs criminal liability, the car owner, driver, or occupant can incur criminal liability at the same time anyway.

When the legislator trusts the human more than the driving automation and driver duties consequently remain while technology evolves, trust in technology can indeed become a criminal liability trap.

Dilemma Questions—On a Side Note

Humans do not trust machines taking a definite decision on them. This becomes obvious when one looks at the dilemma debate that covers a disproportionate part of the general discussion if humans should entrust machines with decisions over life and death.

A famous dilemma situation is the so-called trolley problem: A trolley is travelling down a track at high speed towards a number of railroad workers on the track. The switchman standing at the tracks operating the turnout can save the lives of the railroad workers by switching the track. This, however, would kill another railroad worker working on the other track. Kohler discussed a (car-related) version of the problem in Germany as early as in 1915 (Kohler 1915). Engisch (1930) and Welzel (1951) also discussed this problem, and in the U.S., Foot (1967) introduced the same problem which can also be applied to cars: When a crash is inevitable and either decision will lead to the death of a person, should the car driver, for example, save a mother crossing the road with her baby by manoeuvring the car and killing four pedestrians? As long as human actors who are in a dilemma situation must take a decision, this is a problem of a justificatory collision of duties or duress, which will mitigate or exclude punishment (Wohlers and Hörnle 2018 with further references). Furthermore, the drivers are never obliged to sacrifice their own life (Stratenwerth 2011).

However, driverless cars will lead to more difficulties to determine whether criminal liability attaches to someone (and if yes, to whom?) after a dilemma situation occurred. Given that the decision on how the car should react is taken in advance, namely, when the car is programmed, it is not the driver anymore who acts out of a stress situation and whom we can easily forgive. It now is the person programming the car (or their boss, or one day maybe the legislator) who takes this decision way before the dilemma situation occurs (see also Engländer 2016; Weber 2016; Hilgendorf 2017; Wohlers and Hörnle 2018 regarding dilemma situations and driverless cars).

This adds a new approach for solving dilemma situations: the car could swerve in order to avoid killing pedestrians and hit a wall instead, thus killing its occupant (Millar 2014, for further examples of dilemma situations with driverless cars see http://moralmachine.mit.edu/. Accessed April 2018). In such an event, a new trust issue arises: imagine that you are one of the car occupants and you know that the car could sacrifice you when getting into a dilemma situation. Would you still trust and use it? A survey conducted regarding dilemma situations found that people would not buy a car that could sacrifice them when in a dilemma situation (Bonnefon et al. 2016). Therefore, if the legislator would ever decide that a car should be able to sacrifice its passengers when in certain dilemma situations, people would probably only trust and use such cars when they are certain that the vehicle will recognise and avoid any dilemma situations thus making it highly unlikely to be sacrificed.

Conclusion

Disruptive technologies such as the invention of the automobile more than a hundred years ago or today's driving automation challenge the society as much as the legislator. Trust in risky new technologies (and their economic benefits) is a crucial element for accrediting such technologies. If a technology and its risk profile are "legalized", a certain risk is permissible and the user of the technology complying with the rules cannot incur criminal liability. However, since technology evolves faster than the law, one must be aware that trust in technology might also be a criminal liability trap, for instance when people begin to trust the autopilot system and therefore fail to comply with their duty to control the car. The fact that humans do not want to entrust life or death decisions to machines is illustrated by the debate about driverless cars in dilemma situations. This might impede trust in driving automation when cars are set to sacrifice their occupants in a situation where there is no other way out.

References

Atiyeh, C. (2015, October 8). Volvo will take responsibility if its self-driving cars crash. *Car and Driver.* https://www.caranddriver.com/news/volvo-will-take-responsibility-if-its-self-driving-cars-crash . Accessed April 2018.

Bohnenblust, D., & Pool, M. (2017). In Bundesamt für Statistik (2017) Verkehrsunfälle in der Schweiz 2016. https://www.bfs.admin.ch/bfs/de/home/statistiken/mobilitaet-verkehr/unfaelle-umweltauswirkungen.assetdetail.3103126.html. Accessed April 2018.

Bonnefon, J.-F., Shariff, A., & Rahwan, I. (2016, June 24). The social dilemma of autonomous vehicles. *Science, 352*(6293), 1573–1576.

Bundesamt für Strassen ASTRA. (2015, July 1). Bundesrat genehmigt Anpassung des Wiener Übereinkommens über den Strassenverkehr. Medienmitteilung. https://www.astra.admin.ch/astra/de/home/dokumentation/medienmitteilungen/anzeige-meldungen.msg-id-57943.html. Accessed April, 2018.

Bundesamt für Strassen ASTRA. (2017). Automatisiertes Fahren. Merkblatt zur Durchführung von Pilotversuchen in der Schweiz. https://www.astra.admin.ch/astra/de/home/themen/intelligente-mobilitaet/pilotversuche.html. Accessed April, 2018.

Bundesamt für Strassen ASTRA. (2018, March 27). Bund erteilt Bewilligung für weiteren selbstfahrenden Bus. Medienmitteilung. https://www.admin.ch/gov/de/start/dokumentation/medienmitteilungen.msg-id-70243.html. Accessed April 2018.

Cassart, A. (2017). Aéronefs sans pilote, voitures sans conducteur: la destination plus importante que le voyage. In H. Jacquemin & A. De Streel (Eds.), *L'intelligence artificielle et le droit.* Bruxelles (pp. 319–339).

Curtis, J. (2016, September 15). Shocking dashcam footage shows Tesla 'Autopilot' crash which killed Chinese driver when futuristic electric car smashed into parked lorry. *Daily Mail.* http://www.dailymail.co.uk/news/article-3790176/Shocking-dashcam-footage-shows-Tesla-Autopilot-crash-killed-Chinese-driver-futuristic-electric-car-smashed-parked-lorry.html#ixzz5B7ro9fAc. Accessed April, 2018.

Engisch, K. (1930). Untersuchungen über Vorsatz und Fahrlässigkeit im Strafrecht. Berlin.

Engländer, A. (2016). Das selbstfahrende Kraftfahrzeug und die Bewältigung dilemmatischer Situationen, *Zeitschrift für Internationale Strafrechtsdogmatik (ZIS)*, issue 9/2016, pp. 608–618.

European Parliament. (2017). Press release. Robots: Legal Affairs Committee calls for EU-wide rules. http://www.europarl.europa.eu/news/en/press-room/20170110IPR57613/robots-legal-affairs-committee-calls-for-eu-wide-rules. Accessed April, 2018.

Foot, P. (1967). The problem of abortion and the doctrine of the double effect in virtues and vices. *Oxford Review* (5), 5–15.

Gisler-Jauch, R. (2015). Automobil. In *Historisches Lexikon der Schweiz.* http://www.hls-dhs-dss.ch/textes/d/D13901.php. Accessed April, 2018.

Gless, S. (2017). Von der Verantwortung einer E-Person. *Goltdammer's Archiv für Strafrecht*, pp. 324–329. http://edoc.unibas.ch/55627/. Accessed April, 2018.

Hilgendorf, E. (2017). Autonomes Fahren im Dilemma. Überlegungen zur moralischen und rechtlichen Behandlung von selbsttätigen Kollisionsvermeidesystemen. In E. Hilgendorf (Ed.), *Autonome Systeme und neue Mobilität* (pp. 143–175). Baden-Baden.

Hofstetter, J. (2017, November 30). Hightech schützt vor Strafe nicht. *Berner Zeitung*. https://www.bernerzeitung.ch/region/emmental/teslafahrer-kaempft-nach-unfall-um-mildere-strafe/story/21683947. Accessed April, 2018.

Kohler, J. (1915). Das Notrecht. *Archiv für Rechts- und Wirtschaftsphilosophie*, (4), 411–449.

Lafrance, A. (2016, June 29). Your Grandmother's Driverless Car. *The Atlantic*. http://www.theatlantic.com/technology/archive/2016/06/beep-beep/489029/. Accessed April, 2018.

Levin, S., & Wong, J. C. (2018, March 19). Self-driving Uber kills Arizona woman in first fatal crash involving pedestrian. *The Guardian*. https://www.theguardian.com/technology/2018/mar/19/uber-self-driving-car-kills-woman-arizona-tempe. Accessed April, 2018.

Lohmann, M. F. (2015). Erste Barriere für selbstfahrende Fahrzeuge überwunden – Entwicklungen im Zulassungsrecht. *sui-generis* 2015. http://sui-generis.ch/article/view/sg.17. Accessed April, 2018.

Lohmann, M. F. (2016a). Automatisierte Fahrzeuge im Lichte des Schweizer Zulassungs- und Haftungsrechts. Baden-Baden.

Lohmann, M. F. (2016b). Liability issues concerning self-driving vehicles. *European Journal of Risk Regulation*, (2), 335–340. Special Issue on the Man and the Machine. https://www.robotics.tu-berlin.de/fileadmin/fg170/Publikationen_pdf/2016_Lohmann-EJRR.pdf. Accessed April, 2018.

Lohmann, M. F., & Müller-Chen, M. (2017). Selbstlernende Fahrzeuge – eine Haftungsanalyse. *Schweizerische Zeitschrift für Wirtschafts- und Finanzmarktrecht (SZW)*, issue 1/2017, pp. 48–58.

Lutz, L. (2014). Die bevorstehende Änderung des Wiener Übereinkommens über den Straßenverkehr: Eine Hürde auf dem Weg zu (teil-)autonomen Fahrzeugen ist genommen! *DAR-Deutsches Autorecht* 2014, 446–450.

Lutz, L. S. (2016, March). Automated Vehicles in the EU: A Look at Regulations and Amendments. *Gen Re Casualty Matters International*, pp. 1–7. http://de.genre.com/knowledge/publications/cmint16-1-en.html. Accessed April, 2018.

Millar, J. (2014). An ethical dilemma: When robot cars must kill, who should pick the victim? *Robohub*, http://robohub.org/an-ethical-dilemma-when-robot-cars-must-kill-who-should-pick-the-victim/. Accessed April 2018.

National Highway Traffic Safety Administration (NHTSA). (2017). Office of Defects Investigation Close Resume of Investigation PE 16-007. https://static.nhtsa.gov/odi/inv/2016/INCLA-PE16007-7876.pdf. Accessed April, 2018.

Neue Zürcher Zeitung. (1925). Das Auto im Bündnerland, 23. Juni 1925, Morgenausgabe. https://static.nzz.ch/files/2/8/4/Graub%c3%bcnden+Automobilverbot+23_1.18765284.6_1.18765284.1925+zusammen_1.18765284.pdf. Accessed April 2018.

Riedo, Ch., & Maeder, St. (2016). Die Benutzung automatisierter Motorfahrzeuge aus strafrechtlicher Sicht, in: Probst Th, Werra F, *Strassenverkehrsrechtstagung* 21 – 22 Juni 2016. Bern, pp. 85–120.

SAE. (2016). Taxonomy and definitions for terms related to driving automation systems for on-road motor vehicles J3016. http://standards.sae.org/j3016_201609/. Accessed April, 2018.

Sander, G. M., & Hollering, J. (2017). Strafrechtliche Verantwortlichkeit im Zusammenhang mit automatisiertem Fahren. *Neue Zeitschrift für Strafrecht (NStZ)*, issue 4/2017, pp. 193–205.

Schubarth, M. (2011). Gedanken zur Risikogesellschaft und zum Recht auf Leben im Strassenverkehr. *Strassenverkehr* 2/2011.

Schwarzenbach, R. (2016, July 4). Der Kampf ums Automobil. *Neue Zürcher Zeitung*. https://www.nzz.ch/schweiz/schweizer-geschichte/sonderfall-graubuenden-der-kampf-ums-automobil-ld.103634. Accessed April, 2018.

Shepardson, D., & Lienert, P. (2016, February 10). Exclusive: In boost to self-driving cars, U.S. tells Google computers can qualify as drivers. *Reuters*. https://www.reuters.com/article/us-alphabet-autos-selfdriving-exclusive/exclusive-in-boost-to-self-driving-cars-u-s-tells-google-computers-can-qualify-as-drivers-idUSKCN0VJ00H. Accessed April, 2018.

Sinclair, J., & Schafer, B. (2017, November 23). Autonomous vehicles: The path to liability is still unclear. *Jusletter IT*. https://jusletter-it.weblaw.ch/issues/2017/23-November-2017/autonomous-vehicles-_687b1213fc.html__ONCE. Accessed April, 2018.

Stratenwerth, G. (2011). Schweizerisches Strafrecht. Allgemeiner Teil I, 4. Auflage, Bern.

Thierer, A., & Hagemann, R. (2015). Removing roadblocks to intelligent vehicles and driverless cars. *Wake Forest Journal of Law & Policy*, 2015, 339–391.

United Nations Economic Commission for Europe (UNECE). (2014). Inland Transport Committee, Working Party on Road Traffic Safety, Sixty-eighth session, Geneva, 24–26 March 2014. Report of the sixty-eighth session of the Working Party on Road Traffic Safety. https://www.unece.org/fileadmin/DAM/trans/doc/2014/wp1/ECE-TRANS-WP1-145e.pdf. Accessed April, 2018.

United Nations Economic Commission for Europe (UNECE). (2015a). Inland Transport Committee, Working Party on Road Traffic Safety, Seventieth session, Geneva, 23–26 March 2015, Autonomous Driving, Submitted by the Governments of Belgium and Sweden. http://www.unece.org/fileadmin/DAM/trans/doc/2015/wp1/ECE-TRANS-WP1-INT-2e.pdf. Accessed April, 2018.

United Nations Economic Commission for Europe (UNECE). (2015b). Inland Transport Committee, Working Party on Road Traffic Safety, Seventy-first session, Geneva, 5–7 October 2015, Automated driving. http://www.unece.org/fileadmin/DAM/trans/doc/2015/wp1/ECE-TRANS-WP1-2015-8e.pdf. Accessed April, 2018.

United Nations Economic Commission for Europe (UNECE). (2016). Press release. UNECE paves the way for automated driving by updating UN international convention. https://www.unece.org/?id=42459. Accessed April, 2018.

Vlasic, B., & Boudette, N. E. (2016, June 30). Self-Driving Tesla was involved in fatal crash, U.S. Says. *The New York Times*. https://nyti.ms/29dJjPp. Accessed April, 2018.

Von Bar, C. L. (1871). Die Lehre vom Causalzusammenhange im Rechte, besonders im Strafrechte. Leipzig.

Von Bodungen, B., & Hoffmann, M. (2015). Belgien und Schweden schlagen vor: Das Fahrsystem soll Fahrer werden!, *Neue Zeitschrift für Verkehrsrecht (NZV)* 2015, 521–526.

Weber, P. (2016). Dilemmasituationen beim autonomen Fahren. *NZV* 2016, pp. 249–254.

Welzel, H. (1951). Zum Notstandsproblem. *Zeitschrift für die gesamte Strafrechtswissenschaft (ZStW)*, issue 1/1951, pp. 47–56.

Wohlers, W. (2016). Individualverkehr im 21. Jahrhundert: das Strafrecht vor neuen Herausforderungen. *Basler Juristische Mitteilungen (BJM)*, issue 3/2016, pp. 113–137.

Wohlers, W., & Hörnle, T. (2018). The Trolley Problem Reloaded. Wie sind autonome Fahrzeuge für Leben-gegen-Leben-Dilemmata zu programmieren?, *Goltdammer's Archiv für Strafrecht*, issue 1/2018, pp. 12–34.

Zurkinden, N. (2016a). Strafrecht und selbstfahrende Autos – ein Beitrag zum erlaubten Risiko. *recht*, issue 3/2016, pp. 144–156.

Zurkinden, N. (2016b, November 24). Crash beim Testbetrieb selbstfahrender Fahrzeuge. Unrecht oder strafrechtlich erlaubtes Risiko?, *Jusletter IT*. https://jusletter-it.weblaw.ch/issues/2016/24-November-2016/crash-beim-testbetri_91fd98526d.html. Accessed April 2018.

Zurkinden, N. (2017). AI and driverless cars: From international law to test drives in Switzerland to criminal liability risks. In: H. Jacquemin & A. De Streel (Eds.), *L'intelligence artificielle et le droit*. Bruxelles (pp. 341–355).

Zurkinden, N. (2018, December 3). Vertrauen in Fahrzeugautomatisierung als strafmindernder Umstand? Anmerkungen zur Urteilsbegründung des Regionalgerichts Emmental-Oberaargau vom 30. Mai 2018, PEN 17 16 DIP, *Jusletter*. https://jusletter.weblaw.ch/services/login.html?targetPage=http://jusletter.weblaw.ch/en/juslissues/2018/960/vertrauen-in-fahrzeu_15fcbd5df7.html__ONCE&handle=http://jusletter.weblaw.ch/en/juslissues/2018/960/vertrauen-in-fahrzeu_15fcbd5df7.html__ONCE. Accessed December 2018.

Dr. Nadine Zurkinden works at the Faculty of Law of the University of Zurich (Switzerland) since June 2018. Her research currently focuses on innovative technology and criminal liability risks. She earned a Master of Law degree from the University of Bern (Switzerland) in 2007 and a doctorate from the University of Zurich in 2013. For her doctoral dissertation "Joint Investigation Teams" she received the Professor Walther Hug Prize (dissertation award) in 2014. From 2006 to 2009, she was an assistant at the chair of Prof. Schwarzenegger at the University of Zurich. From April 2009 to March 2011, a scholarship for prospective researchers from the Swiss National Science Foundation enabled her to conduct research for her doctoral dissertation at the Max Planck Institute for Foreign and International Criminal Law in Freiburg i. Br. (Germany, as a visiting researcher) and at the Catholic University of Leuven (Belgium, as an International Scholar). From 2011 to 2015, Nadine Zurkinden was initially a researcher and later a senior researcher at the Max Planck Institute in Freiburg. From April 2015 to May 2018, she was a PostDoc assistant at the University of Basel (Professorship Gless).

Chapter 23
Integral Corporate Cyber Security—Challenges and Chances for Showing the Way Towards Effective Cyber Governance

Hans-Wilhelm Duenn and Lukas W. Schaefer

Companies Facing the Cyber Threat Situation

Between 2015 and 2017, more than every second company has been attacked by cyber criminals in Germany, which caused a national economic damage of €55 billion (Berg and Maaßen 2017). The cyber threat situation is real and equally affects politics, business and society. Companies of any size constitute a popular target for cyber criminals, as they hold valuable specific knowledge on products, processing operations and marketing strategies, among others. Due to the digital transformation of in-house workflows and communication, manufacturing and customer interaction, the digital attack surface grows. The process of digitization continues: a survey by PricewaterhouseCoopers forecasts that the average level of digitization will increase from 33 to 72% by 2020, and companies will budget around five percent of their yearly turnover for this purpose (2016).

This essay identifies the challenges of cyberspace and outlines approaches toward a corresponding integral corporate cyber security. It will also be revealed that companies are not entities merely responding to evolving challenges; instead, they are able to actively shape developments in cyberspace and contribute to the redesigning of institutions toward effective cyber governance, which is currently hindered by the decentralized, unregulated and private sector dominated nature of cyberspace, eroding traditional forms of governance (Jayawardane et al. 2015).

H.-W. Duenn
German Cyber Security Council e.V., Berlin, Germany
e-mail: duenn@cybersicherheitsrat.de

L. W. Schaefer (✉)
Research and Public Relations, German Cyber Security Council e.V., Berlin, Germany
e-mail: schaefer@cybersicherheitsrat.de

© Springer Nature Switzerland AG 2020
D. Feldner (ed.), *Redesigning Organizations*,
https://doi.org/10.1007/978-3-030-27957-8_23

Challenges for Companies in Cyber Security

According to the *Routledge Handbook for New Security Studies* (2010),

> [i]t is argued (truthfully) from the side of government that it is technically and economically impossible to design and protect the infrastructure to withstand any and all disruptions, intrusions or attacks, i.e., that absolute security in the field of cyber-security is not possible (Cavelty 2010: 161).

This assertion claims validity not only for the political, but also for the economic and societal spheres. The starting position for establishing a stable cyber security architecture is challenging for various reasons.

Cyber Security as a Field of Self-help

One of the reasons is the fact that most entities that engage in cyberspace are dependent on self-help when it comes to security. States claim their legal systems to be valid in the physical and digital space, resulting in conventional crimes like narcotrafficking, arms trade or murder conducted or organized online being equally punishable as pure cybercrimes. However, the cyber realm is unregulated and prevents states from an effective governance. It is accordingly difficult for states to fulfil their protection mandate and defend their economy and society against cyber threats. Furthermore, the surface web, deep web and, especially, dark web provide a vast variety of cyberweapons, as well as tactics for successful and anonymous cyberattacks. Thus, possibilities for effectively governing cyberspace—for instance, by issuing binding legislation and enforcing concomitant prosecution—are limited, particularly since the global character of cyberspace demands international cooperation, which is currently hindered by the lack of trust. There is no internationally recognized code of conduct for cyberspace, and together with the difficulties of attribution, non-state or state(-sponsored) perpetrators can plausibly deny committing cyberattacks (Townsend 2017). There have been attempts to establish common norms, for example, by the Unites Nations Group of Governmental Experts on Developments in the Field of Information and Telecommunications in the Context of International Security (UN GGE). However, "[t]he problem with norms is that they must first be agreed by everyone, and then obeyed by everyone before they can be called 'norms'" (ibid.).

Another reason for the relative absence of state power in cyberspace is the original idea of the Internet. Although it arose from a U.S. Department of Defense project, it was eventually opened to the public, and responsibilities were withdrawn from the government. This process was encouraged "[…] by the assumption of the dot-com era, when Internet pioneers proclaimed repetitively that the Internet and the World Wide Web were new phenomena; old rules did not apply because it is untrammelled, borderless and without need for traditional government" (Lewis 2010: 56). Contrary to the belief in an open web solely promoting freedom and generating wealth, the last decade revealed that cyberspace has partially turned into an arena where both state and non-state actors seek to exploit the provided IT infrastructures: according to the Europol Serious and Organized Crime Threat Assessment, cyber criminality has superseded international drug trafficking as the most profitable channel for organized crime, with an annual turnover of around US$3 trillion in 2013 (Europol 2013); authoritarian regimes use the Internet for surveillance or massive censorship; and states of all kinds conduct cyber espionage and warfare—inter alia via state-sponsored cyber criminals—causing geopolitical tensions.

Given this increasingly Hobbesian state of cyberspace, states meanwhile look for possibilities for stronger regulation and interference. The application of conventional security paradigms—in this case, meaning the strengthening of security agencies and their cyber capacities and mandates—however, is likely to undermine global cyber security. In May 2017, the global ransomware epidemic WannaCry demonstrated the devastating effects of highly developed cyberweapons and exploits falling into wrong hands. Moreover, recent attempts to establish state sovereignty online mostly resulted in the fragmentation of cyberspace, also referred to as the "Balkanization of the Internet"; extreme examples include the North Korean Internet consisting of less than 50 websites and the "Great Firewall of China".

The necessity for self-help, particularly for companies, is likewise based on the fact that the majority of IT expertise lies within the economy and not the state. States are even dependent on these companies as they provide the necessary infrastructure and software for governmental networks. Additionally, as the networks of private actors like large-scale companies, small and medium-sized enterprises (SMEs) or operators of critical infrastructures are the main targets of cyber criminals, they possess and steadily develop know-how on the current cyber threat situations and effective defenses. Countermeasures of intelligence agencies hence rely on the input of the private sector (Bendiek 2016: 17).

The Asymmetric Setting of Cyber Threats

Another big challenge for cyber security is the asymmetric setting of cyber threats, which stems from the advantages cyber criminals have over the attacked entities. Cyber criminals can commit extortion, espionage or sabotage campaigns cost-effectively from anywhere in the world, and there are many ways of anonymizing attacks—for instance, via hijacked IT infrastructures. Above all, while potential targets must provide comprehensive security and secure all possible gateways, cyber criminals only need to identify one weak point to successfully infiltrate the targeted network. Thus, the proceeding digitization of corporate processes plays into cyber criminals' hands, as attack surfaces are likely to expand due to evolving digital surfaces, the integration of smart items or the online access to sensitive corporate databases and networks by negligently acting employees.

Another asymmetric dimension evolves when there is a mismatch regarding the offensive and defensive cyber security capacities. As mentioned in the previous section, cyber threats spring from various actors. There are single perpetrators, criminal organizations, professionally organized hacker groups, as well as military and intelligence entities carrying out attacks because of ideological conviction, self-enrichment or governmental order (Dickow and Bashir 2016). Furthermore, the emergence of cybercrime being offered as a service results in even fewer IT-skilled criminals entering the cybercrime industry thanks to "client-friendly" hacking tools and lucrative promises (Amann 2017: 38). Following this, it is probable that large companies, often equipped with well-trained Computer Emergency Response Teams (CERTs), can easily defend their networks against untargeted cyberattacks—for instance, in terms of unsophisticated phishing campaigns. However, it is also possible that well-organized, even state-sponsored hacker groups attack less-prepared entities like SMEs, which often lack respective personnel and awareness. SMEs typically possess highly specified know-how for certain products inevitable for industrial manufacturing chains, turning some of them into hidden champions and making them interesting targets for cyber criminals. Besides, it has become a common pattern of cyber criminals to take a detour via less-protected SMEs before attacking large companies (supply chain attacks).

Altogether, it is improbable that the asymmetric setting of cyber threats—one of the main reasons why complete cyber security cannot be achieved—will be overcome. It appears unlikely that states can establish their monopoly on violence in cyberspace, providing that all actors would finally subordinate to an international binding legal framework. Simultaneously, the point of no return in terms of "resetting" cyberspace was passed by the pace of digital transformation a long time ago. For companies, this implies that corporate cyber security strategies should focus on appropriate IT security precautions and "best possible managing" residual risks that are caused by non-technical factors.

Approaches Toward Integral Corporate Cyber Security

The management of technical and non-technical safety factors can only be accomplished by an integral cyber security concept. This requirement correlates with the proceeding digital transformation of all company departments, no matter if they are responsible for legal affairs, human resources, marketing or facility management. In so doing, departments and employees with less understanding and awareness about cyber security come into contact with cyberspace, bringing along not only faster communication and higher productivity, but also a higher possibility of carelessly opened unknown mail attachments and data being exposed and accessed by successful phishing mails. Consequently, it is vital to take the human factor into account and to anchor cyber security within the company's business culture. In addition to technical measures—such as consistent access management, strong spam filters and regular data backups on storage places separated from the Internet—trainings and briefings on basic IT security knowledge and cyber hygiene are necessary.

Another element of an integral corporate cyber security is openness for cooperation with state authorities and even competitors. At first glance, this requirement appears counterproductive, as it calls for sharing sensitive information on IT security incidences, undermining in-house information security. However, an integral cyber security strategy claims sustainability, which can only be ascertained by cyber security concepts that exceed a company's premises. That is because cyber security requires comprehensive networks, while large gaps between entities regarding their cyber security levels are counterproductive. Cyber criminals have proven that they are able to identify and then successfully attack the least secure entities of "cyber ecosystems"—for example, industrial sites, governmental agency networks or research clusters. Cooperation is thereby indispensable, not only within the economic level, but also between business and politics.

> Within national cybersecurity, the importance of the private sector [...] is obvious. The private sector is responsible for virtually all of the software and hardware that is exploited for cyberattacks, maintains most of the network infrastructure over which these attacks are conducted, and often owns the critical infrastructure against which these attacks are directed (Klimburg 2011: 29).

Notwithstanding, cooperation has to provide benefits for all participating actors. States should thus seek for cross-sectoral cooperation. These could be a transparent processing of information on IT security incidents provided by companies, the sharing of findings resulting from respective analyses of intelligence services or the provision of support in case of IT emergencies by Computer Emergency Response Teams (CERTs). For now, most platforms aimed at fostering cross-sectoral cooperation—for instance, Germany's IT Security Law for the protection of critical infrastructures and the US Cybersecurity Information Sharing Act, which were both passed in 2015—are based on compulsion, and it has been criticized that some provisions can even cause competitive disadvantages for affected companies. Instead, there are many examples for mutual advantageous cooperation.

Firstly, there is a massive and still growing IT security skills shortage, resulting in a fight for talents between the economy and the state. Simultaneously, respective solution approaches—like the incorporation of IT literacy and programming into scholar and academic curricula—need another decade before results will become visible. Hence, the connection of IT security experts, common emergency exercises and trustful exchange are vital. With regards to the dependency between state and business in cyber security, an imbalance of IT security manpower is disadvantageous for both.

Secondly, companies can strive for strategic partnerships with research institutions based on the model of financial funding in return for exclusive access to and patents on research outcomes, if politicians hesitate to invest appropriate volumes into cyber security.

Thirdly, with the emergence of forward-looking technologies, such as artificial intelligence (AI) or quantum computing, a debate on common values and ethical responsibilities must take place. The international community holds a wide range of different corporate cultures, ideologies and systems of government, indicating that they will take advantage of these technologies in different ways to assert their interests. To prevent the abuse of technology for criminal reasons or violations against civil liberties, a transnational dialogue is imperative.

As a final example for mutual advantageous cooperation, there is a need for a common development of information and communications technology (ICT) certification schemes. "ICT cybersecurity certification becomes particularly relevant in the view of the increased use of technologies [...] such as connected and automated cars, electronic health or industrial automation control systems [...]" (European Commission 2017) and has the potential to counteract geopolitically caused mistrust in ICT, such as the ban by U.S. and U.K. authorities of Kaspersky Lab IT products and solutions made in Russia in 2017. In addition, ICT featuring certified information security and data privacy provides competitive advantages for companies: customer interaction via apps, wearables or online shopping has become a prerequisite for business success online and is best built on secure and trusted communication channels.

On the same note, integral corporate cyber security strategies mean responsible handling of new technologies, especially with the advancement of the Industrial Internet of Things (IIot). Considering the lack of ICT certification, together with widespread politically intended "digitization pressure" on national industries, the number of cyberattack surfaces is likely to increase. It will thus turn into an "Industrial Internet of Threats" when IT security concerns are subordinated to the goals of productivity and state-of-the-art business culture. Generally, the intra-corporate discourse on security versus productivity is a discourse over conflicting goals, but must be overcome for the sake of cyber security. IT security is often considered a tedious over-expenditure and inhibitor of in-house projects. This narrative must be transformed into the conception that IT security is a progress enabler and a competitive advantage. IT security officers should be integrated into all projects from start to finish according to the production principle "Secure by Design" in order to prevent the cancellation of long-developed initiatives due to security concerns. Furthermore, a partner-like relationship between the IT security department and decision makers is necessary to enable constructive dialogues on cost estimates and impact assessments (Tschersich 2017).

Integral Corporate Cyber Security as a Blueprint for Redesigning Institutions Toward Effective Cyber Governance

According to the "Outcome Document of the High-Level Meeting of the General Assembly on the Overall Review of the Implementation of the Outcomes of the World Summit on the Information Society", the United Nations General Assembly recognizes "[...] the leading role for Governments in cybersecurity matters relating to national security" (United Nations General Assembly 2015: 11). However, as outlined before, states struggle with establishing effective governance in cyberspace, which hinders them in fulfilling their assigned leading role. At the same time, "[…] the important roles and contributions of stakeholders, in their respective roles and responsibilities […]" (ibid.: 11) are also recognized, indicating the joint responsibility of all entities to contribute to a stable cyber security architecture. The contribution of the private sector to effective cyber governance, however, requires the adoption of a multistakeholder approach, which appears advantageous for the international community considering the described unregulated and private sector-dominated nature of cyberspace. This approach, as suggested by The Hague Institute for Global Justice, envisages the assignment of responsibility to different actors according to their expertise. The private sector should thus handle the technical and operational aspects of a working and reliable cyberspace, while states are to negotiate on legally binding treaties.

> Multistakeholderism in cyberspace can increase representativeness and effectiveness in the governance […] by levelling the playing field, preventing the capture of cyberspace by any one type of stakeholder, and allowing different types of stakeholders authority over aspects of governance that they are best equipped to handle (Jayawardane et al. 2015: 7).

Thus, an integral organization of corporate cyber security has the potential to serve as a blueprint for a successful redesign of (political) institutions toward effectiveness in cyber governance. That is because companies, as drafted in the sections above, possess a hierarchically organized IT security department, allowing for consistent chains of action even in case of an emergency. Moreover, their corporate culture envisages the active participation in pan-state tasks and debate on the future design of cyber security. In many states, to the contrary, cyber responsibilities are distributed over several authorities with different mandates. Although they are able to pursue cyber forensics and prosecution, they lack channels for merging respective insights. A duplication of efforts or even the loss of pieces of evidence is probable. Also, the fragmentation of competences is likely to cause a slowing of digitization itself and greater bureaucratic expenditure. On the German governmental level, for instance, there is a total digitization staff of 482, consisting of 244 teams in 76 departments in 14 ministries (Christmann 2018). Certainly, digital matters and cyber security must be organized holistically and, hence, require the incorporation of all policy areas represented by ministries. Nonetheless, a supervising and coordinating office is absolutely essential. Consequently, it is obvious that companies cannot contribute only to their own corporate cyber security via an integral approach, but also to a reliable pan-state cyber security architecture. Therefore, they should not only focus on in-house security and best manage technical and non-technical risks as outlined in this essay, but also be aware of their indispensable role in countries' cyber ecosystems and be open for cooperation and pan-state learning processes.

References

Amann, P. (2017). Den Feind im Blick: Cyber-Kriminalität als Businessmodell. In: Palo Alto Networks (Germany) GmbH (ed) Wegweiser in die Digitale Zukunft – Praxisrelevantes Wissen zur Cyber-Sicherheit für Führungskräfte. Planet c GmbH, Düsseldorf.

Bendiek, A. (2016). Sorgfaltsverantwortung im Cyberraum – Leitlinien für eine deutsche Cyber-Außen- und Sicherheitspolitik. Stiftung Wissenschaft und Politik. https://www.swp-berlin.org/publikation/sorgfaltsverantwortung-im-cyberraum/. Accessed May 4, 2018.

Berg, A, & Maaßen, H.-G. (2017). Wirtschaftsschutz in der digitalen Welt. bitkom. https://www.bitkom.org/Presse/Anhaenge-an-PIs/2017/07-Juli/Bitkom-Charts-Wirtschaftsschutz-in-der-digitalen-Welt-21-07-2017.pdf. Accessed May 7, 2018.

Cavelty, M. (2010). Cyber-security. In: J. Peter Burgess (Ed.), *The Routledge handbook of new security studies*. New York: Routledge.

Christmann, A. (2018). Antwort BMVI Digitalisierung in Bundesministerien. http://annachristmann.org/wp-content/uploads/2018/03/180223-Antwort-BMVI-Digitalisierung-in-Bundesministerien.pdf. Accessed May 8, 2018.

Dickow, M., & Bashir, N. (2016). Sicherheit im Cyberspace. Bundeszentrale für Politische Bildung. http://www.bpb.de/apuz/235533/sicherheit-im-cyberspace?p=all. Accessed May 4, 2018.

European Commission. (2017). Proposal for a regulation of the European Parliament and of The Council on ENISA, the "EU Cybersecurity Agency", and repealing Regulation (EU) 526/2013, and on Information and Communication Technology cybersecurity certification ("'Cybersecurity Act'"). http://eur-lex.europa.eu/legal-content/EN/TXT/?uri=COM:2017:0477:FIN. Accessed May 7, 2018.

Europol. (2013). *EU serious and organised crime threat assessment (SOCTA 2013)*. https://www.europol.europa.eu/activities-services/main-reports/eu-serious-and-organised-crime-threat-assessment-socta-2013. Accessed May 4, 2018.

Jayawardane, S., Larik, J., & Jackson, E. (2015). Cyber governance: Challenges, solutions, and lessons for effective global governance. The Hague Institute for Global Justice. http://www.thehagueinstituteforglobaljustice.org/wp-content/uploads/2015/12/PB17-Cyber-Governance.pdf. Accessed May 9, 2018.

Klimburg, A. (2011). Lessons from the Comprehensive Approach for Whole of Nation Cybersecurity. per Concordiam, issue 2/2011, pp 29–33.

Lewis, J. (2010). Sovereignty and the Role of Government in Cyberspace. *Brown Journal of World Affairs, (2),* 55–65.

PricewaterhouseCoopers. (2016). *Industry 4.0: Building the digital enterprise*. https://www.pwc.com/gx/en/industries/industries-4.0/landing-page/industry-4.0-building-your-digital-enterprise-april-2016.pdf. Accessed May 8, 2018.

Townsend, K. (2017). The increasing effect of geopolitics on cybersecurity. *Security Week*. https://www.securityweek.com/increasing-effect-geopolitics-cybersecurity. Accessed May 7, 2018.

Tschersich, T. (2017). IT-Sicherheit: Vom Bremsklotz zum Umsatzbringer. In: Palo Alto Networks (Germany) GmbH (ed) Wegweiser in die Digitale Zukunft – Praxisrelevantes Wissen zur Cyber-Sicherheit für Führungskräfte. Planet c GmbH, Düsseldorf.

United Nations General Assembly. (2015). Outcome document of the high-level meeting of the General Assembly on the overall review of the implementation of the outcomes of the World Summit on Information Society. unctad.org/en/PublicationsLibrary/ares70d125_en.pdf. Accessed May 7, 2018.

Hans-Wilhelm Duenn holds a diploma in administrative sciences from the University in Potsdam. From 2007 to 2009, he worked as a Personal Consultant to the Minister's Office of Economic Affairs and the Deputy Minister-President of the Federal State of Brandenburg, who was also Chairman of the German Federal Network Agency. From 2009 to 2010, he was a Member of the Supervisory Board Energie and Wasser Potsdam GmbH. From 2010 to 2012, he was Managing Director of Security and Safety in Brandenburg e.V. In addition, from 2011 to 2014, he was also part of the Supervisory Board of VIP Verkehrsbetrieb Potsdam GmbH. Furthermore, Hans-Wilhelm Dünn is a member of the Supervisory Board at the Klinikum Ernst von Bergmann GmbH and the Lausitz Klink Forst GmbH, as well as part of the Advisory Council of Luftschiffhafen Potsdam GmbH. As Co-founder and President—previously Vice-president and General Secretary—of the Cyber-Security Council Germany (CSCG), he consults operators of critical infrastructure about cybersecurity. He imparts his expertise at high ranking conferences and events and is often a contact person for TV channels or a guest author for several magazines and journals.

Lukas W. Schaefer has a Bachelor in Political Science and completed a Master program in International Peace Studies at Trinity College Dublin. He then worked as Editor for National and European Politics, Energy and Pharmacy in the area of media analysis, before he joined the Cyber-Security Council Germany as Head of Research and Public Relations in March 2017. His main responsibility is the internal analysis of developments in cyber space and the external communication of respective political, economic and social implications.

Chapter 24
Cyber Security… …by Design or by Counterplay?—Enabling and Accelerating Digital Transformation Through Managing Information Security Technology, Risk and Compliance at the Right Place

Thomas Hemker

The world is facing a huge change in businesses through digital transformation projects. In the information security realms, change is inevitable anyway due to the everchanging threat landscape. According to the *2018 Internet Security Threat Report* by Symantec, in 2017, there was a 600 percent increase in IoT (Internet of Things) attacks compared to the previous year. With the world becoming more and more "smart" through endless connectivity and everything being computerized, the attack surface gets bigger and bigger. While in the past it was mainly the owners and users of vulnerable technology who became the victims of attackers, it is now possible for attackers to hijack millions of connected vulnerable (consumer) IoT devices, in order to attack nation states, large organizations and critical infrastructure all over the world. The Mirai Botnet in 2016 is obviously the most famous proof for this so far.

Research by Charlie Miller and Chris Valasek on the potential for remotely hacking connected cars provides another example of how the increased connectivity and digitization of the world we live in could have an impact on our safety in the physical world (Greenberg 2016). This research led to a recall of 1.4 million vehicles. For present and future DX projects, the implication is that the extent to which security must be an essential component of the supporting technologies and processes is a question that has to be answered, so that transformation will not put the security and safety of people and society at risk. From an information security standpoint, there needs to be clear guidance on how confidentiality, integrity and availability can enable and accelerate safety and digital transformation in general.

T. Hemker (✉)
Security Strategy, EMEA CTO Office
Symantec, Hamburg, Germany
e-mail: THe@guterhafen.de

© Springer Nature Switzerland AG 2020
D. Feldner (ed.), *Redesigning Organizations*,
https://doi.org/10.1007/978-3-030-27957-8_24

When digital transformation (DX) is the subject of a discussion among cyber security people with a certain level of business acumen today, these are often the key topics:

- The radical rethinking of how an organization pursues new revenue streams, reduced operating costs or new business models using technology
- The drivers tend to be disruption from market newcomers or innovation from rivals seizing the opportunity to win new customers and steal market share
- It can be commercially focused externally to drive revenue, or operationally focused internally to reduce cost.

  but also

- It has budgets that are different from those of traditional incremental security-improvement projects
- It is driven by the board or an executive (CxO), rather than within a security organization.

If we try to amalgamate security and digital transformation, we face the situation that the concept of digital transformation might be quite well developed, but the role security plays in it is less well discussed. As mentioned before, the application of security is fundamental for a safe and connected world. Therefore, the security functions within organizations need to be transformed and empowered. Security by design, the right decisions regarding the supply chain (e.g., vulnerable IoT devices) and managing the risks of the new and transformed business processes will not be the only objectives for Chief Information Security Officers (CISOs) and their teams.

From Security Organizations to Secure Organizations

A lot of organizations fall into four stages of maturity related to how security integrates with and supports digital transformation (Knott 2018 and Picture 24.1).

The lowest level of maturity is usually "Siloed". The DX programs are run from the board in isolation without any interaction with the security staff. There is little to no input and influence by information security to shape these programs and address possible associated risks early. The security teams are uncomfortable finding themselves involved in projects run by business leaders, and some teams lack the skills to articulate security risk to business leaders. At the same time, traditional security projects are concerned with incremental improvements, so teams are often unable to appreciate the goals of transformation.

The CISO is also at risk of being overwhelmed by DX. The regulatory pressures to make the business compliant are high, and the DX projects are often an addition to their daily activities without an increase of their budgets. This sometimes leads to instances where CISOs become subservient to transformation and are seen as operational and not transformational.

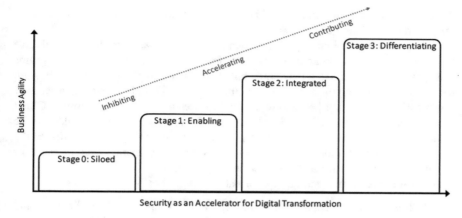

Picture 24.1 Maturity stages. *Own Source* Knott (2018)

Security is—in most of these organizations—just an afterthought and is often seen as an inhibitor. The lack of adequate security is often an accepted risk. Otherwise, CISOs have to address the risk after key decisions were made, either because of the significant business benefits that can be gained, or because there are founded or unfounded concerns that by engaging the security teams, their initiatives might be blocked.

The security mindset is not aligned with the transformation process or vice versa, because security used to be a defensive and reactive activity in the past. This is obviously seen as the opposite of "forward thinking" about the innovative maximization of an opportunity. Needless to say, these different mindsets and expectations often lead to conflicts in such environments.

This is typically seen when a CISO reports to the Chief Information Officer (CIO) or other IT functions within an organization. Often, the Chief Financial Officer (CFO), along with the CIO, probably signed a large deal for cloud services without fully consulting the security team—which subsequently did not accept the risk associated with untested and unapproved security controls and, therefore, delayed the migration until approved controls were deployed. A transformation to a digital supply chain is in process; however, after months of planning, the security teams review the designs and find inadequate information protection, which requires third party assessments that cause a near-complete redesign and a significant delay in the project. These are just a few examples of what can happen in a siloed organization.

The "Enabling" level of maturity is where security is seen as an integral part of the transformation process and as a key supporting function. This is how security becomes an enabler for the transformation. To move from siloed to enabling, the CISO position needs an elevation from an operational role to be a part of the business discussion. In an ideal situation, the CISO reports directly to the board or the equivalent of a Chief Digital or Digital Transformation Officer. Alternatively, a transformation working group could be established, so that the CISO has a view of the programs in their earliest stages.

Being proactive and anticipating demand is something security leaders will have to improve on. Providing an integration platform—so that transformation programs can operate quickly and independently, but also reduce complexity and benefit from, for example, threat intelligence and enrichment—could be a blueprint for a security organization. Deploying key controls ahead of demand—like, for example, Cloud Access Security Broker (CASB) or Data Loss Prevention (DLP)—are likely to be reused across multiple transformation programs. If security leadership drives the innovation along with the executive board, they can recommend areas for transformation to ensure involvement and be secure by design.

Typical concerns preventing transformations might be:

- For Cloud Adoption: Data security and regulatory compliance, loss of control and visibility, identity and access concerns, insufficient knowledge and experience
- For the Digital Workplace: Unproven/inadequate security, employee freedom versus corporate protection, mobile data security and user experience, identity management complexity
- For Operational Technology (OT), IoT and Edge Computing: Increased attack surface (botnets, backdoors, vulnerabilities), unauthorized surveillance/invasion of privacy, demands of availability, realtime and air gapped systems, non-standard architectures and operating systems
- For Big Data and Analytics: Data privacy (personal data), data security (intellectual property), vulnerability to fake data, inadequate technical and analytical knowledge.

So what does an "Enabling" organization look like in security? A look at the above list of four typical areas for digital transformation and some examples of how security and security technology can play a key role there might clarify things here.

The Cloud Adoption is today often demanded by a "cloud first" strategy. There are some key technologies that, if deployed early, can accelerate its adoption. Setting up a CASB solution ahead of demand would allow new SaaS solutions to be secure from day one, without adding additional delays. Protecting cloud workloads ahead of projects that use infrastructure or platform as a service can be achieved by integrating security into DevOps and training them to use the native security controls provided by the cloud platform and compliant suppliers.

Security teams can be seen to be proactive by implementing email security and DLP ahead of a move to Office 365 to ensure that risk is not increased by the move.

For OT/IoT and edge computing, manufacturing companies have challenges to secure their legacy—but now connected—equipment, as well as the Industry 4.0 generation of devices. Protection and hardening for embedded devices, detection of anomalies and malware/attacks and mature management of assets—together with the adoption of IT security standards, especially in areas where OT is connected to internal IT systems (e.g., ERP) and cloud platforms—is critical in this area.

Mobility is also a key area of transformation, allowing organizations to become more agile and productive, which in turn enables better business outcomes. Due to the rise of mobile business apps, security requirements for devices have advanced beyond configuration management. Smartphones should be treated as endpoints, just as you would your laptop, and therefore have the same controls. Security teams should recognize this and make recommendations to leadership. The sharing of data is fundamental to digitalization. Information protection solutions allow data to be tracked and encrypted as they move between authorized parties. Ensuring that the controls are deployed will minimize the risk of a data breach.

Big data analytics allows organizations to make meaningful, strategic adjustments that minimize costs and maximize results. Very often, customer and employee data—as well as intellectual property (e.g., for predictive maintenance)—is involved here. Protecting the vast amount of data should be seen as a priority. Most organizations are planning to transform in these areas, so if security functions anticipate this demand, they could accelerate the programs. Digital transformation is about change, agility, speed, connectivity, real-time economy and customer expectations. If the security teams and their leadership want to be seen positively, then they must also meet the same goals. Proactively deploying an integration platform allows transformation programs to run independently of each other while still benefiting from and contributing to the overall security posture of the organization.

In the Enabling level, security should begin to add value to the organization by supporting and encouraging digital transformation.

The "Integrated" level is focused more intrinsically on the security function itself. It needs to be transformed as well. There are some operational issues to take into account. Like the market for business software in the mid-1990s, the security technology market is also demanding change. The application vendors that specialized in one area—PeopleSoft owned HR, Baan manufacturing apps, JD Edwards finance—were replaced with enterprise-class ERP (Enterprise Resource Planning) solutions from Oracle and SAP. The objective was to centralize all business data into a common repository that could anchor the business and be updated and used for various departmental functions and business processes in real time. Although the ERP journey was a bit painful for some companies, the transition resulted in a steady increase in business productivity, enhanced efficiency and better decision making. On the supply side of the equation, the ERP evolution led to industry consolidation as large software vendors acquired smaller ones. By the early 2000s, just a few enterprise-class business application software vendors remained, while other specialists became ecosystem partners for large vendors, adding niche value in specific areas. According to the *Gartner Market Share Analysis: Security Software Worldwide* 2017 report, the security market has reached a level of maturity where consolidation is now a high priority for organizations. If operational issues within security are not resolved, then it will be impossible to mature into a transformation-differentiating function.

To solve those problems, security departments need to adopt a holistic integrated approach. There are some objectives for security integration that business leaders should be aware of:

- Deploying best-of-breed threat protection technology to stay ahead of threats and to automate protection and security updates. Protection against targeted attacks across all control points like endpoint, applications, network and cloud.
- Reducing the risk of a data breach. Prevention of accidental data loss and protection of sensitive and regulated data.
- Securing the modern workplace. Reduce the risk of cloud and mobile usage, secure and accelerate critical and web applications. Control the data in the cloud and optimize security performance. Enable DevOps to build security directly into service deployment workflows.
- Ensuring Regulatory Compliance and avoiding fines. Rapid identification and assessment of policy violations with real-time visibility into infrastructure changes. Demonstrate compliance and efficiency of security controls.
- Managing Risk. Assess risk for security operations and prioritize remediation. Orchestrations and integration with third parties. Visibility across control points and into targeted attack campaigns to improve productivity of threat investigators and incident response teams.
- Integration of security telemetry, threat intelligence, data management and business intelligence. Deploy middleware and message bus services to support heterogeneous cyber security tools. Select the right architecture, standards and APIs.

In the desired "Differentiating" level, security is seen as a competitive advantage in the marketplace. Building and leveraging innovative technology and a level of trust in an organization has always been a differentiator. This move to security as being a positive influence on the brand will also shift the focus of a compromise and successful attack away from threats and toward the actual business outcome within the security teams. In this level, security is not only seen within the company, but also extended to the supply chain and the customers. Those in the supply chain will probably consider opening their vulnerability and threat data to partners so that they can be seen as better and, therefore, be preferred over competitors (see Picture 24.2).

Technology Management Transformation

To support the digital transformation process for security, it is critical to understand some aspects of implementing cyber security technology in an organization today. Here are some (not all) aspects described to guide non-cyber security people in the coordination of experts in the respective areas. The classical governance-oriented approach to IT and information security—driven by COBIT (Control Objectives for Information and Related Technologies) and formalized in 2008 through ISO/IEC 38500, the international standard regarding governance of IT for organizations—defines responsibilities for the board if the business is strategically dependent on IT. In the context of digital transformation, IT and IT security is therefore not solely a technical task but clearly a strategic one for businesses. A case from the University of Antwerp illustrates how organizations where the board can participate in the digital

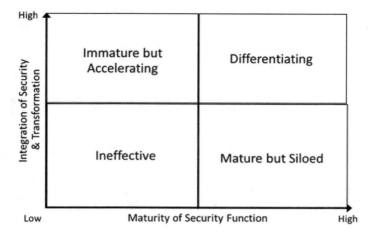

Picture 24.2 Matrix of security versus transformation. *Own Source* Hemker (2018)

debate could look (De Haes and Van Grembergen 2015). New structures created two committees with board engagement steering information technology spending and development and digital transformation.

Again, like in any enterprise transformational program, people, processes and technology are the key components for success. Two different examples should demonstrate how important the governance and technology integration approach is today and will be in the future.

Information Governance, as an example of this (classic) approach, manages the creation, storage, usage, archival, deletion and the valuation of data within an organization. It requires collaboration between security, IT, compliance, legal and the business owners the data belongs to. This collaboration requires at least a board-level mandate.

On the technology side, it is quite easy to map tons of technical controls to the different objectives and functions of such an information governance program (see Picture 24.3), but the challenge here is that the technology ideally has to support the protection of intellectual property, as well as the requirements of the EU General Data Protection Regulation (GDPR).

Leveraging an ERP type of integrated cyber defense platform would be easier for board members to understand how to apply security to meet regulatory requirements on the one hand, and to secure new risk scenarios like big data, artificial intelligence, distributed ledger technologies, cloud and edge computing, leading to "dark data" and disorganized complexity.

An example that's different from Information Governance is the detection of a cyber security event as it is mandated by the EU-NIS Directive—the Directive on security of network and information systems—protecting critical infrastructure. The following use-cases might be reasonable here:

- Real-time detection of security incidents on premise and cloud

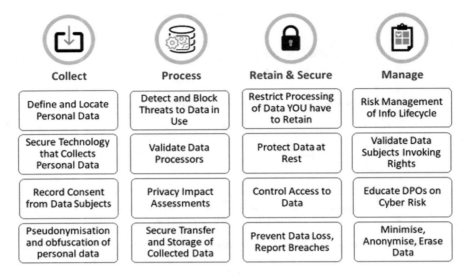

Picture 24.3 Information governance for GDPR. *Own Source* Hemker (2018)

- Reduction of incident severity by reduced time-to-detection ensures that attackers have less time to act and response teams can act faster
- Faster incident detection can reduce regulatory scrutiny and improve public perception
- Understanding malware and attack characteristics, as well as a holistic view of the security landscape, to proactively identify unique targeted attacks
- Contextual understanding of the attack
- Detailed detection process involving coordination between multiple stakeholders and security tools to ensure that events are detected and escalated appropriately
- Ongoing training and process gap analysis with continuous refinement
- Identification of potential events through the use of network and security data collation and correlation across multiple vectors
- Full incident information necessary to notify regulators and affected stakeholders.

Obviously, that's a lot of use cases, and some of them probably need technology investments, while others require staffing and organizational changes. So, without standards, best practices and control frameworks like ISO/IEC 27000:2018—titled "Information technology—Security techniques—Information security management systems—Overview and vocabulary"—and the National Institute of Standards and Technology (NIST) Cybersecurity Framework, the different objectives of security controls and their deployment would be difficult to manage.

A selection and combination of these frameworks could look like this:

– Prepare: Managing cyber risk
– Protect: Protecting against cyberattacks and data breaches
– Detect: Detecting cyber security events

– Respond: Minimizing the impact of a security incident.

While this is a good approach for managing cyber security in traditional businesses, this can also be a good foundation for digitally transformed business processes. On the cyber security governance level, the board defines how the implementation of a control framework—to manage cyber security as an integrated platform—could support business transformation processes.

Preparation for the Unknown

At the Information Security Forum (ISF) World Congress 2017 in Cannes, a group of industry experts discussed how to secure new disruptive technologies and if it would need a new approach to cyber risk assessments. The summary of the discussion contributed to the research for *Threat Horizon 2020: Foundations Start to Shake*, a report by the ISF. Things like artificial intelligence, blockchain, drones, machine learning, 3D printing, virtual and augmented reality (VR and AR), natural language processing, natural language generation, chatbots, robotic process automation, beacons, the internet of things, cognitive computing among others were named as disruptive technologies. The panelists saw the balance of confidentiality, integrity (CI&A) and availability shifting significantly: integrity is becoming more important across all new technologies. As an example, with the move of VR and AR to more mission critical applications, trust-your-screen attacks could result in fatalities.

Internet of Things (IoT) devices are less likely than corporate systems to have a lot of confidential information on them, but trusting who's controlling the device is essential. And of course, the importance of availability will continue to increase in proportion to our dependence on technology. During the discussion, the so-called CI&A triad—confidentiality, integrity and availability—a staple of cyber security and information risk management, was challenged as too limited, while it was agreed that speed, ease-of-use, activity and scale should also be considered. Security and digital transformation need to amalgamate.

Furthermore, these new technologies are not only increasing security risks and the threat surface but could also be useful for security. Machine Learning (ML) is a good example of how things can have at least two sides. On the one side, today, ML is running on more than 100 million endpoints, already detecting unknown malware and targeted attack campaigns (Hemker 2018). In Security Operation Centers (SOCs), trained virtual analysts analyze trillions of rows of telemetry data worldwide to find stealth attacks and to counter the cyber security skills gap as it is outlined in the *2018 Global Cybersecurity Workforce Study* by ISC2. On the other side is the use of ML by adversaries to create and automate attacks and modify malware. So-called adversarial ML is also targeting AI -based security—and, for example, autonomous cars—causing big security and safety risks (Eykholt et al. 2018). Therefore, the resilience of new technology and business processes needs to be established, constantly evaluated and improved. Transparency, standards, best practices and regulation come into place, again.

Conclusion

The connected world—where everything is based on computer technology—needs more security, not less. Otherwise, the safety of people and the security of society is at risk. There will be certification, as well as new regulation, in this area. In organizations, the board and executive level is the right place for cyber security management. The state of maturity of the security function and the alignment to the business defines the success of digital transformation projects. Differentiating businesses through security is desired, and CISOs and their teams should start to enable digital transformation. The deployment of security technology as an integrated platform will empower businesses to act proactively and support new processes. The constantly changing and growing threat landscape, the associated risk and regulatory requirements mandate not only "state of the art", but also the "...by design" implementation of technical security controls. For new disruptive technologies, built-in security might often be the only option anyway. It is fundamental for a successful digital transformation.

References

arXiv:1707.08945v5 [cs.CR] 10 Apr 2018; https://arxiv.org/pdf/1707.08945.pdf.
COBIT. https://www.isaca.org/cobit/pages/default.aspx.
De Haes, S., & Van Grembergen, W. (2015). *Enterprise governance of information technology: Achieving alignment and value, featuring COBIT5*. Switzerland: Springer International Publishing.
EU NIS Directive. https://ec.europa.eu/digital-single-market/en/network-and-information-security-nis-directive.
Gartner Market Share Analysis: Security Software Worldwide. (2017). https://www.gartner.com/doc/3889275?ref=mrktg-srch.
Hemker, T. (2018). *Datenschutz Datensich, 42*, 629. https://doi.org/10.1007/s11623-018-1014-1.
Internet Security Threat Report ISTR23 (p. 80) https://www.symantec.com/security-center/threat-report?inid=globalnav_scflyout_istr.
(ISC)2. *2018 Global Cybersecurity Workforce Study*; https://www.isc2.org/News-and-Events/Press-Room/Posts/2018/10/17/ISC2-Report-Finds-Cybersecurity-Workforce-Gap-Has-Increased-to-More-Than-2-9-Million-Globally.
ISF Information Security Forum, *Threat Horizon 2020*. https://www.securityforum.org/research/threat-horizon-2s-start-to-shake/.
ISO/IEC 38500:2008 Corporate Governance of Information Technology, 2008—new version 2015. https://www.iso.org/standard/62816.html.

ISO/IEC 27000:2018 *Information technology—Security techniques—Information security manage-ment systems.* https://www.iso.org/standard/73906.html.
Knott, P. (2018). *Security strategist at symantec UK at Gartner Security Summit UK.*
Mirai Botnet Explained. https://krebsonsecurity.com/tag/mirai-botnet/.
NIST Cybersecurity Framework. https://www.nist.gov/cyberframework.
WIRED Magazine. https://www.wired.com/2016/08/jeep-hackers-return-high-speed-steering-acceleration-hacks/.

Thomas Hemker CISSP, CISM, CISA, Is Director Security Strategy at Symantec with over 23 years of experience in information security. When joining the company in 2010, he also became part of the Symantec CTO Office and is responsible for exchanging ideas with security profes-sionals in the corporate world, the public sector, and actors of the security industry in general. He advises customers regarding their security strategy, cyber resilience, current threat scenarios and technological development. Prior to his position at Symantec, Thomas Hemker was in charge of the Central-European presales department at PGP (PGP Deutschland AG). He also gathered expe-rience as a systems engineer, consultant and product manager at Network Associates and a secu-rity distributor, with a focus on encryption, PKI and key management, as well as network security and analysis (Firewall, IDS/IPS). Thomas is a distinguished speaker at security conferences and holds several security industry certifications, for instance the Certified Information Systems Secu-rity Professional (CISSP). He represents Symantec at ISF, TeleTrust e.V. Bitkom, ENISA and ITU. As a security professional he is a member of ISACA and (ISC)2, where he co-founded the (ISC) 2 Chapter Germany and served as member of the board.

The Future of Education and Work

Chapter 25
Redesigning Traditional Education

Martina Francesca Ferracane

Nothing is more powerful than an idea whose time has come.
—Victor Hugo

The Start of a New Era

We are experiencing a historical transformation that opens up incredible opportunities to make the world a more equal, inclusive and fulfilling place. However, this moment is filled with uncertainty and a feeling of vulnerability among those who see themselves as victims of this change. To avoid a future of growing inequality, mass unemployment and, in turn, political instability, there is a lot that governments can do. In the short term, reskilling and support for those whose jobs are being automated will be inevitable. However, if we want to build strong foundations to prepare our society for the future, we need to think about the long-term and start with rethinking our education and make sure it prepares our students for the new digital era.

While the ways in which people learn, work, travel and communicate have been shaken up, schools have not yet changed in a substantial way over the last decades, or even the last century. It is finally time to adopt educational policies that can prepare today's children for a reality that is evolving at an exponential speed. In this way, the next generation can be assured to rise *with* the machines and not against them. Failure to do so can result in a deepening of inequalities and an inevitable clash between a group of privileged, who can make use of digital solutions to express their

M. F. Ferracane (✉)
Oral3D, Milan, Italy
e-mail: martina.ferracane@gmail.com

© Springer Nature Switzerland AG 2020
D. Feldner (ed.), *Redesigning Organizations*,
https://doi.org/10.1007/978-3-030-27957-8_25

329

full potential, and a group of people left behind, who see their lives lacking prospects because of technology.

Some Jobs Are Disappearing, Most Jobs Are Transforming

It is not a secret anymore that technology and automation are destroying jobs. The World Economic Forum (WEF (2016b) estimates that current trends could lead to a net employment impact of more than 5.1 million jobs lost to disruptive labor market changes over the period 2015–2020, especially in administrative and routine white-collar office functions. An additional study from 2013 estimates that roughly 47% of U.S. jobs could be lost over the next decade or two because they involve work that can easily be automated. Another recent study by the OECD estimates that around 9% of U.S. jobs are at high risk (Rotman 2017).

What is even more worrying is that low-paying jobs are those particularly vulnerable to automation in the future. An estimate published in the *MIT Technology Review* shows that 83% of low-paid jobs are likely at risk of automation, compared with only 4% of high-paid jobs.

However, job destruction is only one part of the story. Technology and automation are also creating opportunities for new jobs, and the reality is that most of the jobs today are actually transforming and requiring new skills. In particular, we are seeing a sharp decrease in the share of tasks requiring routine cognitive skills, as well as both routine and nonroutine manual skills. On the other hand, the share of tasks requiring nonroutine analytical and interpersonal skills keeps rising. For those with the right knowledge, skills and character to adjust to a world with fewer jobs (in particular, fewer routine jobs), this can be liberating and exciting. But for those who are insufficiently prepared, it can mean the scourge of vulnerable and insecure work, and life without prospects (Elliot 2017).

A popular estimate finds that 65% of children entering primary school today will ultimately end up working in jobs that do not exist yet (WEF 2016a). Another recent estimate by McKinsey suggests that between 400 and 800 million individuals could be displaced by automation by 2030, and up to 375 million workers (that is 14% of the global workforce) might need to switch occupational categories (McKinsey 2017b). On the other hand, the same study suggests that, with the right policies job growth could more than offset jobs lost to automation. In addition, about half of the activities people are paid to do globally could be automated using currently demonstrated technologies (McKinsey 2017a).

We can only guess which it will be the actual share of jobs lost and jobs transformed. However, the extent of this transformation is clear from the fact that already today, in many industries, the most in-demand occupations did not exist just a few years ago. Adjusting won't be an easy task. In addition, employment growth is expected to derive disproportionately from smaller, high-skilled jobs that might be unable to absorb the jobs lost in other parts of the economy (WEF 2016a). This is a threatening prospect in a reality in which we are already witnessing rising inequality, often associated with automation and technology (Rotman 2014).

To avoid a future of mass unemployment, governments need to act. These actions should start with education. Education at all levels should be supported, from primary schools to re-skilling of those people who are not well equipped to work in the digital era. In this article, I focus on how we can rethink education in schools so that our students see technology as a means of creation and empowerment, rather than something to consume and use only passively.

Failure to equip our students with the skills necessary to thrive and express themselves in the digital age can result in a world where digitalization benefits a few at the expense of the masses. The results of a "wait-and-see" approach by policymakers at the forefront of this transformational change would be a missed opportunity for growth and creation of wealth and could lead to "a failure to embrace and invest in technology's abundant possibilities" (Rotman 2017). The way the transition is handled will be crucial to defining the outcome of the digital transformation of society.

The Skills of the 21st Century

The ability to prepare students for change and to provide them with the right skillset is critical to mitigate undesirable outcomes. In a reality changing at an exponential speed, trying to equip students with a specific set of theoretical and specialized knowledge is bound to fail. In fact, currently, nearly 50% of subject knowledge acquired during the first year of a four-year technical degree is already outdated by the time students graduate (WEF 2016a).

This means that teaching knowledge—as traditional education does—isn't enough for the future of work. It will not be enough to teach our students how to code or to use certain digital services passively. What really matters is preparing children to be creative, to adapt to a changing environment, to learn how to learn and to take initiative and be entrepreneurial. In this sense, technology and digital skills should not be seen by students as an end, but rather as a tool to learn, to create, to solve problems and to express themselves. Students should prepare themselves for jobs that do not exist, technologies that have not yet been invented and problems that are not yet recognized as such (Dumont and Istance 2010).

The workers of the future need to be able to solve unstructured problems, work with new information and carry out nonroutine tasks (Levy and Murnane 2013). It is expected that a wide range of occupations will require a higher degree of cognitive abilities, such as creativity, logical reasoning and problem sensitivity as part of their core skills (WEF 2015), and the value of skills needed for non-automatable tasks, such as social skills, will also increase (Autor 2015; Deming 2015). But this goes beyond pure work. This is about learning to live within new environments shaped by technology. It is about what citizens in the digital era would need in order to actively participate in society, reach their potential and be able to thrive in a digital society. We are shifting from industrial- to knowledge-based economies and societies whereby

Foundational Literacies	Competencies	Character Qualities
Literacy	Creativity	Initiative
Numeracy and Mathematics	Collaboration	Curiosity
ICT literacy	Communication	Persistence
Financial literacy	Critical thinking and problem-solving	Adaptability
Scientific literacy		Leadership
Cultural and civic literacy		Social and cultural awareness

Fig. 25.1 21st Century skills (Own table based on WEF, 2015)

knowledge becomes central and needs to be continuously generated through learning (Dumont and Istance 2010). In this sense, the ability to use technology today should be "both a requirement and a right" for citizens (EC 2018).

So, what are the skills that should be taught to students today? In a landmark report in 2015, the World Economic Forum (WEF) identified the 16 most critical 21st century skills (Fig. 25.1). These are grouped in three pillars. The first consists of foundational literacies. These represent how students apply core skills to everyday tasks and serve as the base upon which students need to build more advanced and equally important competencies and character qualities. This category includes skills of literacy and numeracy, which remain extremely relevant in a digital society as a prerequisite to acquiring other digital skills and for lifelong learning (OECD 2016b).

Before acquiring ICT skills, in fact, good literacy and numeric skills remain critical as a foundation for any student to learn new skills and breach the achievement gap. Foundational skills are also scientific literacy, ICT literacy, financial literacy and cultural and civic literacy.

A basic ICT literacy, however, might not be enough in a digital world. Education today needs to teach students how to be active users and digital creators rather than passive consumers of technology. For students to turn into active users of technology, feel more empowered and use technology as a means of creation, they should develop additional skills. First of all, they should be able to critically analyze information and therefore develop the ability to search for information and data and to evaluate and judge it; to communicate and collaborate in various forms through digital means; to create, edit and improve digital content; to stay safe in the digital sphere; and to solve problems through digital means and innovate through technologies (EC 2018).

Already in 1999, the National Research Council of the United States recognized the difference between technological literacy (a general set of skills and intellectual dispositions for all citizens) and technical competence (in-depth knowledge that professional engineers and scientists need to know to perform their work) (National

Research 1999, cited in Blikstein and Krannich 2013). In order to teach technical competence, education needs to include "intellectual capabilities to empower people to manipulate the medium to their advantage and to handle unintended and unexpected problems when they arise" (National Research 1999).

Therefore, the second pillar of the WEF concept covers the competencies. These describe how students approach complex challenges and critically evaluate and convey knowledge, as well as work well with a team. For that, critical thinking is key. It is the ability to identify, analyze and evaluate situations, ideas and information in order to formulate responses to problems. A precondition for critical thinking is creativity. Creativity itself is the ability to imagine and devise innovative new ways to address problems, to answer questions or express meaning through the application and to synthesize or to repurpose knowledge. Communication and collaboration involve working in coordination with others to convey information or tackle problems.

Finally, the third pillar to prepare the workers and citizens of the 21st century includes personality characteristics. These describe how students approach their changing environment. Amid rapidly changing markets, personality characteristics such as persistence and adaptability ensure greater resilience and success in the face of obstacles. Curiosity and initiative serve as starting points for discovering new concepts and ideas. Leadership and social and cultural awareness involve constructive interactions with others in socially, ethically and culturally appropriate ways.

The second and third pillars in the WEF 21st century skills concept more generally cover social and emotional learning (SEL). In another study, the WEF (2016a) finds that students that receive SEL education tend to have higher achievement scores and SEL can potentially lead to long-term benefits, such as higher rates of employment and educational attainment. The OECD also identifies foundational skills, digital literacies, as well as social and emotional skills as being crucial to enable effective use of digital technologies and to adapt rapidly to new and unexpected occupations and skills needs (OECD 2016b; Elliott 2017). It recognizes the importance of three sets of skills: technical and professional skills (including ICT specialist skills to program, develop applications and manage networks); ICT generic skills to use such technologies for professional purposes; and ICT complementary skills (leadership, communication and teamwork skills; problem solving communication).

The Digital Divide

These skills are not the skills of the future, but the skills of today. It is already the case that workers with higher digital skills earn on average 27% more than those with basic computer skills (OECD 2016b) and 42% of people in OECD countries with no digital competences are unemployed (OECD and European Union 2015).

Also, there are large gaps today between developed and developing economies and within countries in the same income group (WEF 2015). At the level of national skills development systems, very few countries have developed strategies to foster the acquisition of 21st century skills in formal education (OECD 2016a).

And young people with an economic or social disadvantage tend to have weaker digital competences (EC 2018). This implies that too many students are not developing the necessary skills not only to find a job, but to thrive and feel empowered in a digital society. We are witnessing a growing digital divide—that is a "growing gap between the underprivileged members of society, especially the poor and rural portion of the population who do not have access to computers or the internet, and the wealthy and middle-class living in urban and suburban areas who have access" (from the Plugged In-program at Stanford University).

Particularly worrying, too, is the already significant gap in the share of females in an area where most of the jobs will be created in the future. In OECD countries, the share of male workers who are ICT specialists in 2014 was 5.5%, compared to 1.4% of female workers, and on average across OECD countries, less than 5% of girls contemplate pursuing a career in engineering and computing (OECD 2016a). In the U.S., only 11% of engineers are women (Sterling 2013), while in the EU, only 16% of employed ICT specialists are female (EC 2018). Girls start losing interest in math and science already at the age of six (Sterling 2013) and therefore special attention should be put on how technology can contribute to nurturing the interest of girls in these subjects already in primary school.

How Digital Tools Can Support Education

While identifying the skills of the future is itself an important challenge, their implementation in the education curriculum is likely to be even harder. Very little has changed in the curriculum of most schools over the last century, and we are still educating our students for the past rather than the future (Robinson 2006). Lots of the competences taught in schools today are those that are the easiest to digitize, automate and outsource.

There are different ways in which technology can support the learning of 21st century skills and support a modernization of education. One of them is the so-called "flipped classroom" model. Herein technology can support delivery of knowledge so that, for example, the standard curriculum can be taught through videos, and then the teachers can spend their entire time offering personalized support for kids. A typical example of the benefits of this model is the application of Khan Academy in academia. Khan Academy is a nonprofit educational organization that produces short lessons in the form of YouTube videos (Khan 2011). According to this model, direct instruction is partially replaced by video lectures that can be played anytime and anywhere. The learner can thus decide when to stop and rewind the lecture and focus on what he/she finds harder to understand. Then, the teachers can spend their entire time on complementing this standard basic education in ways that respond to each student's interests and attitude.

Digitalization can also support traditional education through Massive Open Online Courses (MOOCs) and open platforms for learning. These courses and platforms can be especially valuable in this time of change, but they are still underutilized.

MOOCs can support the delivery of standard educational content, but can also enable certain students to deepen their understanding of specific topics if they wish to. Online courses and platforms can also support personalized and adaptive content and curricula by allowing differentiated learning with one-on-one computer learning tailored to individual student needs (EC 2018).

Digital technologies can also help students gain access to digital education when they are in remote areas or are not able to be physically present in school. In general, technology can help lower the costs and raise the quality of education. It permits personalization of learning, engagement with the disengaged, complementing what happens in the classroom, extension of education outside the classroom and the delivery of access to learning to students who otherwise might not have enough educational opportunities (WEF 2016b).

These technologies should be further explored to find new, innovative ways to support education. Other interesting applications to support learning use virtual reality tools and online games (OECD 2016b). For example, during a class on geography, students could use virtual reality (VR) solutions to actually visit (virtually) the countries they are studying and see how things might have looked in the past when they study history. Virtual environments can also allow access to experiences like flying to the moon or exploring a molecule from the inside (EC 2018).

These tools can be very powerful to enhance learning, but they are not enough to prepare our students for the digital era. Technology can be a tool to engage in creative, productive, life-long learning, but most importantly should also become a tool of creation and of change, rather than simply being consumed passively.

Students need to learn how to make use of technology to shape their surroundings, to think creatively about how to solve problems and to be creators. They should develop a vision of technology as a tool to make things and to design a better society. By doing so, we can expect a more equal and democratic society in which the majority of citizens are able to contribute actively and can feel part of the digital revolution.

Making: The Tool To Redesign Traditional Education

Now, with the tools available at a makerspace, anyone can change the world. Hatch (2014)

What if, instead of being simply passive users of digital solutions and digital contents, students were taught how to use technology to build something? For example, to build their own Do-It-Yourself (DIY) virtual reality glasses to use during class. This would be an important twist in the perception that students have of technology. Technology would not be perceived as something the students passively consume, but as something they can actively use and shape to their preference to eventually create something.

Let's go a step further. Imagine if, for example, during a science class about bacteria, students learned to build their own microscope to look at bacteria. Creating a DIY microscope can be very simple when students have access to certain tools of digital fabrication (Ferracane 2016). This simple creation would enable students to see technology as a tool to create something useful and to investigate their surroundings. Also, the creation of tools for scientific experimentation can have a strong democratizing effect on science. In fact, digital microscopes can be expensive for schools to buy. But making them is only about 10% of the cost of buying them and, on top of this, the process of making it is per se an incredible learning opportunity for the kids.

The idea of learning-by-doing as a medium to favor better learning outcomes is not new. Experiential education, constructionism and critical pedagogy are theoretical and pedagogical pillars that are over three decades old (Blikstein and Krannich 2013). Papert's theory of constructionism places embodied, production-based experiences at the core of how people learn (Harel and Papert 1991). Constructionism is about "learning by constructing knowledge through the act of making something shareable" (Martinez and Stager 2013) and emphasizes the role that students can take in constructing their own learning through direct physical engagement with phenomena and problems in the world (Bevan 2017). In the digital era, however, this concept acquires new important applications in connection with digital fabrication and the "Maker Movement". It can be described as a grassroots culture dedicated to hands-on making and technological innovation (Peppler et al. 2016).

The Maker Movement

Digital fabrication is the use of certain digital tools—such as 3D printers, CNC machines and electronics—to create almost anything. The first time this equipment was packaged in a standardized low-cost lab was in 2002 at MIT (Blikstein and Krannich 2013). This is when the concept of the Fabrication Laboratory or "FabLab" was born. The founder of the FabLab concept, Prof. Neil Gershenfeld, describes this digitalization of fabrication as the process "where you don't just digitize design, but the materials and the process" (Solon 2013), so digital fabrication is about "bringing programmability to the real world".

Since then, the concept of FabLabs spread all over the world, not only in universities (more rarely in schools), but also as private or public spaces designed for entrepreneurs or more widely for anyone interested in creating something. The spread of digital fabrication labs has been triggered by a drop in the cost of digital fabrication tools. In fact, in the early 2000s, prototyping equipment, such as laser cutters and 3D printers, dramatically dropped in price. In 2009, the expiration of patents on FDM 3D printing led to a further drop in the price of desktop 3D printers (Bensoussan 2016). And open source software further popularized these technologies.

This radically transformed the nature of product engineering as it became much cheaper and quicker to test new ideas and introduce tweaks in the prototype until the optimal result was reached. Beyond applications in prototyping, digital fabrication labs, or "makerspaces", have become a reference point for Do-It-Yourself (DIY) enthusiasts that were looking for the right tools and environment to experiment with their creativity. In these labs, anyone can make use of digital fabrication tools to create anything they consider useful, valuable or simply fun. Projects done in FabLabs range from jewelry to furniture and all the way up to entire houses. In FabLabs, the maker can also create something that does not exist today, but that they feel would be useful to make (Ferracane 2017).

Tinkering is an important component of the making process. It is about solving problems related to the development and realization of innovative ideas in a creative, iterative and open-ended manner (Petrich et al. 2016). An important part of the makers culture is that ideas and projects are freely shared among makers, and this makes it possible to access a vast amount of know-how online, including the projects mentioned earlier on DIY virtual reality glasses or on how to build a DIY lab for scientific experiments. The Waag's Biohack Academy is a wonderful example of open and accessible information on how to create a biolab, including DIY microscope, spectrometers and many other interesting tools.

Most of the projects FabLabs has been used for, however, have remained in the realm of prototyping and higher education. It was only in 2008 that the space saw the first conceptualization of a project with an explicit focus on primary and secondary education—with Stanford University's FabLab@School project. Since then, especially in a selected number of countries, there has been a growing interest among educators in primary and secondary schools regarding how to incorporate "making" into the classroom.

Despite the novelty of this approach, making has already been shown to support the development of an array of learning dispositions, including resourcefulness, creativity, teamwork and forms of adaptive expertise (Bevan 2017; citing Martin and Dixon 2016; Peppler 2016; Ryan et al. 2016). In fact, the idea of "playful experimentation" with tools and materials is a powerful one in the context of learning (Regalla 2016; Resnik 2017). When children make things with their hands, they are engaged in active learning while having fun (Ibid.).

The recent studies on bringing a culture of making into schools through workshops and the adaptation of the academic curriculum suggest that students develop proficiencies, as well as interest, in design and engineering practices (Berland 2016; Kafai et al. 2014). And students also develop identities as creative thinkers and problem solvers (Martin and Dixon 2016). Making is also frequently interdisciplinary in nature. An interdisciplinary learning is another important element in the education of 21st century citizens and workers, who will be faced with increasingly complex and interdisciplinary challenges. This approach can leverage a reform of today's highly siloed education that reflects an older vision of the world, where workers were more often than not doing the same routine tasks for their entire career. But the reality is now different, and technological trends are creating new cross-functional roles, for which employees will need to adapt to new roles with the help of technical, social and analytical skills (WEF 2016b).

Through the process of digital design and fabrication, students experience "novel levels of team collaboration" (Blikstein 2013), and learners frequently directly request help from or offer it to one another, inspire or are inspired by others new ideas or strategies for troubleshooting, and physically build on or connect their own work to the existing body of work of a fellow tinkerer (Bevan 2017; citing Gutwill et al. 2015). More generally, a recent review of the literature on making in education by Bevan (2017) shows that there is a growing body of evidence on the many ways in which making can motivate and support learners' activity, position STEM practices (that is science, technology, engineering and mathematics) as a powerful tool to engage in interest-driven activity, and leverage cultural resources with the goal of deepening engagement and learning.

Making in education can therefore be a tool to prepare our students with critical foundational competencies, skills and personality characteristics needed in the 21st century.

However, the application of making in schools is still very recent, and today the vast majority of the makers activities are primary located in private and affluent schools, museums and higher education (Bevan 2017; Blikstein and Worsley 2016). This needs to change if we want to avoid that the benefits of digitalization remain concentrated in a few countries and in the hands of a few privileged students. We should make sure that all students have access to digital fabrication as part of their academic curriculum so that the next generation can be equipped with 21st century skills and students are prepared to express themselves to their full potential in the digital era.

Making can be associated not only with teaching of STEM-based activities, but with virtually any subject taught in schools today. It can promote learning in an innovative and entertaining way: "by setting aside time for play, free exploration, iteration, reflection, and sharing, (…) youth will have more opportunities to develop self-awareness, collaboration skills, and decision-making abilities" (Regalla 2016). By learning *how* to learn and make use of the process of tinkering, students also learn the power of learning from mistakes, and they refine their skills through experience and persistence. By cultivating a nimble perspective toward problem-solving, developing curiosity and becoming comfortable with "not knowing", students develop a maker's mindset that will be extremely valuable in an ever-changing job market (Ibid.).

Digital fabrication should also be a crucial extracurricular activity for students to experiment with entrepreneurial ideas after school. Not only do FabLabs provide the creativity mindset and environment to innovate, but they also considerably reduce the time and cost of prototyping. Therefore, these labs lower the barriers for students to prototype an innovative idea and reduce the financial exposure needed to test the idea in the market. The students can make use of the tools in FabLabs to rapidly build a prototype within a few hours and with costs below a few hundred euros—depending on the complexity and size of the product. In the absence of a FabLab, the alternative would be to rely on specialized machine shops, with each iteration costing up to thousands of dollars and requiring weeks or even months for production.

More generally, access to creative laboratories outside the classroom would provide additional ways for students to express their creativity and let their imaginations flourish in a less time-constrained environment.

The Way Forward

The challenges ahead are not trivial. Digitalization can lead to a society that is more equal, more just and where everybody can be supported to live up to their potential and be able to express themselves. Digitalization can also lead to a society where there are, on the one hand, a majority of students who have access to the benefits of digitalization and, on the other hand, a big portion of students (most probably from smaller cities and with an already lower average income) who keep receiving the same education their grandparents but need to cope with a completely different reality. It is inevitable that these students will find it harder to find a job and feel fulfilled in the digital era, in turn developing a feeling of being left behind. Governments should act boldly and timely to prevent this outcome.

All students need to receive access to 21st century skills to avoid an exacerbation of inequality and deepening of the education gaps. Governments need to bring digital fabrication into the curricula of *all* schools, starting not from the most open to change, but rather from the periphery and those environments that are the least likely to change without government intervention. As stated by some of the pioneers in bringing FabLabs into schools: "we have the opportunity to give to millions of children a new entry point into the world of knowledge and science, and give them a much richer palette of expressive media for their ideas to become true, creating much more sophisticated 'objects to think with'" (Blikstein and Worsley 2016).

Finding a way to make this technology accessible across the board must be a goal for all countries. More pilot projects should be welcome in this regard to investigate how making improves education, how it raises interest (especially that of girls) in STEM fields, how it can shape students' choices regarding higher education, how it can integrate students that do not conform with the current learning setting, and how making can be integrated into the national curricula. An interesting analysis by Samuelson and Brahms (2016), for example, finds that the success of bringing digital fabrication into the classroom relies heavily on the school's leadership, allocation of space and integration with the existing curriculum. In doing so, governments can rely of the network of FabLabs, makerspaces and similar projects already active in their countries. These initiatives are a powerful source of lessons learnt and can contribute to design policies that respond to the idiosyncratic nature of the national educational curricula.

It will also be extremely important to take into account the socio-economic contexts of each school and to start by facilitating universal student access to the Internet (EC 2018; Dotter et al. 2016). Ensuring that all students have access to computers and the Internet is "no longer a luxury, it is a 21st century necessity" (Dotter et al. 2016). In addition, all schools should be supported financially to create a digital fabrication lab in their premises. Digital fabrication tools can be very affordable and virtually any school today could have a basic creative lab with an investment below 1000 Euros per school. A social project run in Sicily, for example, has brought creative digital education to over 850 students in the academic year 2018–2019 with overall costs below 10,000 Euros. The project is called Teens4Kids and it is run by the FabLab Western Sicily (Ferracane 2018).

Also, the resources available online today related to the Maker Movement (DIY projects, workshops and pilot projects, for example) are mostly in English, and efforts should be welcomed to make sure information, workshops and best practices are accessible to all teachers and all students in their own language. To give an example, when searching on Coursera.org (one of the most used online platforms for MOOCs), over 90% of courses are provided in the English language (own calculation). This can be done by creating national platforms on which teachers and students can rely to start experimenting with digital fabrication in their schools.

In addition, most of the material available online today is not created by educators, but rather by DIY enthusiasts. The result is a wealth of information that often is not shared in the most intuitive format, instead using rather technical terms and taking a basic understanding of certain topics for granted. This can obviously scare away students that do not have the basic skills to approach the material and, therefore, inhibit the engagement of newcomers. This is why it will be crucial that governments create new national curricula that are designed to respond to national priorities and systems—so that all children in the country have the chance to start from scratch together.

More generally, each country needs to identify those policies that create a solid basis for their country to thrive on digital education. The WEF (2015) identifies four key country-level educational areas that strongly diverge among countries and can make a difference for promoting digital skills education: enabling policies (standards that govern K-12 education), human capital (teacher quality, training, expertise), financial resources (importance of education in public budgets), and technological infrastructure (access to new digital tools and content via the Internet). With the right complementary policies, the inclusion of digital fabrication into today's curricula for all schools can be revolutionary. Introducing digital fabrication into today's curriculum as a medium to support education in all subjects can be a powerful tool to teach our students the skills of the future and to make sure the next generation leverages digitalization as a tool to shape a better, more equal and collaborative society.

References

Autor, D. (2015). Why are there still so many jobs? The history and future of workplace automation. *Journal of Economic Perspectives, 29*(3), 7–30.

Bensoussan, H. (2016). The history of 3D printing: 3D printing technologies from the 80s to today. *Blog post at Sculpteo,* December 2016, https://www.sculpteo.com/blog/2016/12/14/the-history-of-3d-printing-3d-printing-technologies-from-the-80s-to-today/.

Berland, M. (2016). Making, tinkering and computational literacy. In *Makeology: Makerspaces as learning environments* (Vol. 2, pp. 196–205). New York, NY: Routledge.

Bevan, B. (2017). The promise and the promises of making in science education. *Studies in Science Education, 53*(1), 75–103.

Blikstein, P. (2013). Digital fabrication and 'Making' in education: The democratization of invention. In J. Walter-Hermann & C. Buching (Eds.), *FabLabs: Of machines, makers and inventors* (pp. 1–21). Bielefeld: Transcript Publishers.

Blikstein, P., & Krannich, D. (2013). *The makers' movement and FabLabs in education: Experiences, technologies and research*. IDC Jun 24–27 2013.

Blikstein, P., & Worsley, M. (2016). Children are not hackers: Building a culture of powerful ideas, deep learning, and equity in the maker movement. In *Makeology: Makerspaces as learning environments* (Vol. 1, pp. 64–79). New York, NY: Routledge.

Deming. (2015). *The growing importance of social skills in the labour market*. NBER Working paper no. 21473.

Dotter, G., Hedges, A., & Parker, H. (2016, January 14). *The digital divide in the age of connected classroom: How technology helps bridge the achievement gap*.

Dumont, H., & Istance, D. (2010). Analysing and designing learning environments for the 21st century. *The nature of learning: Using research to inspire practice* (pp. 19–34). Paris: OECD Publishing.

Elliott, S. (2017). Computers and the future of skill demand. *Educational Research and Innovation*. Paris: OECD Publishing. http://dx.doi.org/10.1787/9789264284395-en.

European Commission. (2018). *Commission staff working document accompanying the document communication from the commission to the European Parliament, the Council, the European Economic and Social Committee and the Committee of the Regions on the Digital Education Action Plan*, Brussels, January 17, 2018.

Ferracane, M. F. (2016). *Portare la creatività tra i banchi scuola—il microscopio fai da te!*, Blog article at the FabLab Western Sicily website, March 1, 2016. http://www.fablabws.org/index.php/2016/03/01/portare-la-creativita-banchi-scuola-microscopio-fai/.

Ferracane, M. F. (2017). *Talk on the title "1school-1FabLab initiative"*, EPP Congress, March 2017. https://www.youtube.com/watch?v=CD3qCWk5OG0.

Gutwill, J. P., Hido, N., & Sindorf, L. (2015). Research to practice: Observing learning in tinkering activities. *Curator: The Museum Journal, 58,* 151–168.

Hatch, M. (2014). *The maker movement manifesto*. New York: McGraw Hill.

Harel, I. E., & Papert, S. E. (1991). *Constructionism*. Norwood, NJ: Ablex.

Kafai, Y. B., Fields, D. A., & Searle, K. A. (2014). Electronic textiles as disrupting designs: Supporting challenging maker activities in schools. *Harvard Educational Review, 84,* 532–556.

Levy, F., & Murnane, R. (2013). Dancing with robots: Human skills for computerized work. *Third Way*. http://content.thirdway.org/publications/714/Dancing-With-Robots.pdf.

Martin, L., & Dixon, C. (2016). Making as a pathway to engineering and design. In *Makeology: Makerspaces as learning environments* (Vol. 2, pp. 183–195). New York, NY: Routledge.

Martinez, S. L., & Stager, G. (2013). *Invent to learn: Making, tinkering, and engineering in the classroom*. Torrance, CA: Constructing Modern Knowledge Press.

McKinsey Global Institute. (2017a). *A future that works: Automation, employment and productivity*, January 2017.

McKinsey Global Institute. (2017b). *Jobs lost, jobs gained: Workforce transitions in times of automation*, December 2017.

National Research Council. (1999). *Being fluent with information technology*. The National Academies Press.

OECD and European Union. (2015). *The missing entrepreneurs 2015: Policies for self-employment and entrepreneurship*. Paris: OECD Publishing.

OECD. (2016a). *Skills for a digital world*. 2016 Ministerial meeting on the digital economy— Background report. OECD Digital Economy Policy Papers, No. 250, 2016.

OECD. (2016b). Skills for a digital world. *Policy Brief on the Future of Work*, December 2016.

Peppler, K. (2016). ReMaking arts education through physical computing. In *Makeology: Makerspaces as learning environments* (Vol. 2, pp. 141–157). New York, NY: Routledge.

Peppler, K., Rosenfeld Halverson, E., & Kafai, Y. B. (2016). Introduction to this volume. In *Makeology: Makerspaces as learning environments* (Vol. 1, pp. 1–11). New York, NY: Routledge.

Regalla, L. (2016). Developing a maker mindset. In *Makeology: Makerspaces as learning environments* (Vol. 1, pp. 257–272). New York, NY: Routledge.

Petrich, M., Bevan, B., & Wilkinson, K. (2016). Tinkering with MOOCs and social media. In *Makeology: Makerspaces as learning environments* (Vol. 1, pp. 175–189). New York, NY: Routledge.

Resnik, M. (2017). *Lifelong kindergarten cultivating creativity through projects, passion, peers, and play*. MIT Press, August 2017, ISBN: 9780262037297.

Ryan, J. O, Clapp, E. P., Ross, H., & Tishman, S. (2016). Making, thinking, and understanding: A dispositional approach to maker-centered learning. In *Makeology: Makerspaces as learning environments* (Vol. 2, pp. 29–44). New York, NY: Routledge.

Rotman, D. (2014). Technology and inequality. *MIT Technology Review*, October 21, 2014.

Rotman, D. (2017). The relentless pace of automation. *MIT Technology Review*, February 13, 2017.

Samuelson Warprid, P., & Brahms, L. (2016). Taking making to school: A model for integrating making into the classrooms. In *Makeology: Makerspaces as learning environments* (Vol. 1, pp. 97–106). New York, NY: Routledge.

Solon, O. (2013). Digital fabrication is so much more than 3D printing. *Wired*, March 13, 2013. http://www.wired.co.uk/article/digital-fabrication.

World Economic Forum. (2015). New vision for education: Unlocking the potential of technology. *Industry Agenda*.

World Economic Forum. (2016a). The future of jobs: Employment, skills and workforce strategy for the fourth industrial revolution. *Global Challenge Insight Report*, January 2016.

World Economic Forum. (2016b). New vision for education: Fostering social and emotional learning through technology. *Industry Agenda*, March 2016.

TED Talks

Ferracane, M. F. (2018). TEDx Talk titled "A Recipe for a Generation of Creators". https://www.tedxlakecomo.com/2018/.

Khan, S. (2011). TED Talk titled "Let's Use Video to Reinvent Education". https://www.ted.com/talks/salman_khan_let_s_use_video_to_reinvent_education.

Robinson, Sir Ken. (2006). TED Talk titled *Do Schools Kill Creativity?* https://www.ted.com/talks/ken_robinson_says_schools_kill_creativity.

Sterling, D. (2013). TED Talk titled *"Inspiring the next generation of female engineers"*. https://ed.ted.com/on/EpAqFHAm.

Web References

Plugged-In program at Stanford University. https://cs.stanford.edu/people/eroberts/cs181/projects/digital-divide/start.html.

The Waag's Biohack Academy. http://biohackacademy.github.io/.

Martina Francesca Ferracane is passionate about policy-making and technological innovation. She is currently working as a consultant on digital trade policy issues and she is a Ph.D. candidate at Hamburg University. During her Ph.D. she has been affiliated with the European University Institute, Columbia University and the California International Law Center. She is particularly interested in digital entrepreneurship and was recently selected in Forbes 30 Under 30 list for Science and Healthcare for her work with Oral3D, a start-up she co-founded in the area of 3D printing and dentistry. Martina also founded and manages FabLab Western Sicily, a non-profit organization which brings digital fabrication to Sicilian kids. For her work, she was listed in 2018 among the 15 most influential Italian women on digital issues. She is also a Research Associate at the think-tank European Centre for International Political Economy (ECIPE) in Brussels, where she writes about digital trade issues and data flows. Previously, she worked at the European Commission and at the United Nations. To know more about her work, visit the website www.martinaferracane.com.

Chapter 26
Mind the Gap!—Speed Matters in Education: Relating Technology to Human Capacities

Fré Ilgen

Fré Ilgen—"Just You and I", bronze, 2018, H48 × L43 × W19.5 cm, No. 1.5 private collection Berlin

F. Ilgen (✉)
Berlin, Germany
e-mail: freilgen@t-online.de

© Springer Nature Switzerland AG 2020
D. Feldner (ed.), *Redesigning Organizations*,
https://doi.org/10.1007/978-3-030-27957-8_26

While it is clear that technology is growing in importance in contemporary society, it is *un*-clear how realistic the current debate on digitalization is. The debate seems to simply split the target groups into hypers and victims. Understandably, the immediate corporate interests of high-tech companies and the energy sector, as well as international competition, have a strong influence on the drive and speed of this debate. Speed is fast becoming the new paradigm for policies by decision-makers and for possibilities offered by new technologies. Ironically, the speed of human processing (involving also handling of materials and tools—technological devices are also kinds of tools) and human creative behavior, are too often neglected or entirely ignored. This essay is an attempt to urge anyone involved in the digitalization process to keep in mind that, whatever technology can offer, the end user continues to be human. My focus is on the methods and motivations for new directives to enhance the use of technology in education to include a proper understanding of natural human learning capacities, including the role of creative behavior. For this objective, we need to improve our general understanding of the biological characteristics fundamental to how humans perceive, act and react. These characteristics are common in all persons, and are, in addition, the features that make each single person a unique individual.

Artists who create their works themselves, offering their work to other persons for visual experience, play a key role in understanding these features. This is to be distinguished from artists who may design their work on a computer and have third persons or machines produce the work, or who only offer some concept that will be executed by another person, an industry or machine. Real creative artists have much experience in creating artifacts that, in and of themselves, by their visual and tactile presence, naturally appeal to a viewer—an appeal that stimulates brain processes involved in focusing attention comparable to how music affects attention. It is crucial to acknowledge that the visual experience of works of art trains a viewer's general ability to focus attention. The value of the experience of the artwork is not limited to the artwork itself, and it is important to distinguish between what we are told art would be about and the biological impact of experiencing art.

This may seem irrelevant to the subject of digitalization but is actually key to all technologies that involve the human end user. Human perception, experiences and subsequent behavior are reflective processes. Many technological devices demand quick jumps in attention, hardly allowing reflection, thus ignoring the complexities of our mind/body by focusing only on speed. The natural need for reflection is obviously also a main reason why creativity or creative behavior plays such a key role in society.

Managing Digital Technologies—In Sync with Different Purposes and with Human Capacities

To open the debate on technology, one may first distinguish between technologies evidently useful for doing things better, faster and on a wider scale—while improving data storage and data communication—and technologies that interact directly with humans.

While, in the first case, speed obviously matters for economic and geopolitical reasons, in the latter, speed in implementation of technology depends on the (re-)active involvement of a human being. This is important in the debate about technological upgrading and the way we as humans respond in the age where advanced technologies are increasingly important in everyday life. In light of the common understanding of AI (artificial intelligence), the way digital products interact with humans should be critically reviewed from the perspective of human abilities. However, one must question if what is presented as AI has anything in common with human intelligence or natural human behavior. Unquestionably, very smart technological advancements evolve all the time, yet there continues to be a large gap between what is actually possible and what would be theoretically possible. This is comparable to the issue of quantum computers: though in theory, such fast computers are possible, many complex problems still have prevented the realization of this technology, even though the word *quantum* is frequently (and incorrectly) applied to market technological applications.

Because creative behavior is a natural and basic aspect of human life, creativity is an essential part of any person's ability to properly function in social or work situations. In situations where the same action is repeatedly demanded, or merely a limited amount of variations of the same action are called for, such actions can obviously be replaced or even done better by a computer, a robot or other machine. In most professions—and in society in general—however, individual intelligence and therefore creative behavior cannot entirely be replaced by technology. This is one explanation for why professionals in the high-tech industries continue to fly around the globe to meet other individuals in person instead of interfacing by technological means. An interesting observation is, that in some discourses "machine learning" and "machine intelligence" is preferred over "Artificial Intelligence". "Machine learning" clearly indicates "what a machine can learn" which is obviously not the equivalent of human learning, while it might be more sophisticated than human learning regarding a specific task (like fast communication).

Keeping Humans in the Loop—Natural Interaction Between Humans and Machines

Human brain processes are tremendously fast, but human (re)action involves looping processes within the brain, a necessary repetition of perception, involving more than just one particular part of the brain and involving the rest of the body, as well (the brain being a part of the body). The use of one's arms and hands, for instance, is not a handicap in any desired (re)action but rather enhances those brain processes naturally involved in focusing attention. In education, it is important to make new technologies available to young people, to extend their abilities to contribute to contemporary and future society and to allow them to add value to many professions in higher education, industry and commerce. New visions for the education of coming generations in an enhanced digitalized society can only be useful when these perspectives take into consideration the natural parameters of the student's capacity for perception, experience and learning. These natural parameters have not, in fact, changed much over time.

A decline in the capacity to focus attention in any substantial way has been observed in the members of more recent generations, especially those who have grown up with a large amount of screen time, obviously including smartphones and tablets. This clearly must alert policy-makers involved in setting new parameters for a digital education following technological developments. Digital devices serve as valuable props for the development of young people's personalities in the hybrid age and are essential to extending capabilities in order to contribute to contemporary and future society and to prepare for a variety of professions. However, these devices cannot replace those human features that are basic for the individual functioning in society.

Policymakers who wish to reform person-machine interfaces in the future should therefore reflect on findings from various fields of science to consider human behavior but especially to learn from neuroscience. As opposed to the current tech debate among policymakers, a more widely reflected approach is required for reforming digital education.

How to Adjust Education

Policymakers direct schools to make young people work with digital devices through-out their education, replacing activities such as writing by hand. In the meantime, cognitive neuro-researchers have discovered that the use of smartphones and tablets in schools has caused new generations to grow up with a lack of substantial and longer-focus attention. In the media and on the internet, one can find an abundance of information about the impact on learning and social behavior of uncontrolled or excessive exposure to and focus on smartphones, tablets and computers (including on disrupted development of speech, sleep and anxiety disorders). There is, obviously, also information disregarding or even denying dangers to society, an impulse moti-vated by the industries that benefit from a growing market. Although policymakers certainly have to support the interests of such industries, the proven problems should motivate them to set priorities and limits. Predictably, at some universities, this lack of attention has been observed in undergraduates. This inability to focus for longer periods of time will prove disastrous in many businesses, research and teaching posi-tions, as well as in many other professional occupations. This development should alarm anyone with a serious interest in professional education. These neuroscientific findings and general observations do not mean one should not include digital devices in education, but rather it should make the determination of the right balance the pith of the matter.

Studies demonstrate that optimal parameters for educating young people should include training in mind-hand control. Handwriting, for instance, has the underestimated advantage of a reduced speed that works naturally well for the brain to focus and process attention. The culture of handwriting by itself is not the topic here, but how the speed and character of mind-hand control matches learning capacity. The repetition of focusing on the same action causes the involved brain circuits to process the perceived information to such a degree that focus may become strong enough to be impregnated in long-term memory; in other words, a physical action like writing by hand allows real learning—an individual's acquisition of knowledge and abilities. This observation has everything to do with biology. When thinking about digitalization in education, one should find the right combination of continuing training in handwriting and other focused physical training—such as drawing or the manual creation of 3D objects or dance—along with the use of digital devices. Debates on education and learning processes frequently include references to creativity.

Creativity—One of the Least Understood Features of Natural Human Behavior

But what is human creativity? Avoiding mere philosophical debate on defining human intelligence, one may nonetheless define creativity and the focus of attention as key notions in a human-centered debate about digital transformation. Creativity is the process of finding new solutions—which are mostly variations of known, established solutions—and is not limited to, nor the equivalent of, fine arts but is a core feature of human behavior.

Unfortunately, the ongoing tech debate focuses mainly on the advantages of technological devices, not on technologies as perceived and experienced from the perspective of the end user. The designs of wireless remote controls in consumer electronics of the 1990s—with too many buttons—are a classic example, teaching that ignoring the behavior and preferences of the consumer makes corporations lose their market shares to corporations with an advanced interest in human semantics. To become successful, a technology must offer space for stimulating human creativity and natural focus of attention, instead of obstructing both. Acknowledging natural, human behavior not only helps people, but it also supports increases in sales for tech companies.

In the tech-education debate, some experts focus on *homo ludens* (the playful human)—a model that underscores the importance of elements of play in a culture and society. Although humans are fascinated by gaming, it demands less focus than creativity, which requires substantial and deep attention significantly different from attention in gaming. The brain's incredible speed in absorbing information is helpful in gaming, but this activity only summons a focus of attention that does not linger enough to contribute to actual learning. For learning and creativity, information must be processed in a sequence of repeated observations that loop within the brain. This process demands a different level of concentration than a game allows.

A holistic education must enhance creativity in a person, not by putting pressure on the student to reach a pre-defined result—as is the case in gaming—but by focusing on fragments of the subject, gradually building toward the generally described objective or task. Creativity requires reduced speed. Many artists know that creativity is a process of finding and reflection, not of searching with a fixed goal. Most artists are aware that they think with their hands (*nelle mani*), not only with their minds. "Nelle mani" is a term originally coined in the Italian Renaissance, referring to the way that experienced artists allow their hands to lead what they do in painting or sculpture (Ilgen 2014). Additionally, creativity is known to be stimulated by pragmatic parameters, for instance, set by the limited range of physical properties of the human body, or by the conditions of certain circumstances in the real world. Of course, experience, knowledge and training additionally play crucial roles for artists, like for any person.

In the later 1980s, when personal computers were introduced at universities for architecture—obliging students to only work with CAD programs—the professors soon discovered that the students had not developed any sense for space in their designs. Understanding the dilemma, the staff decided to hire an art professor to teach the students free-hand sketching from life. This is an exercise that stimulates the brain to focus on spatiality. In drawing from life, eyes-hand-mind control, and the slow speed involved, allow the brain to process various features of three-dimensionality; staring at a computer image does not encourage the development of these skills.

Focus of Attention—The Key Factor for Education

Focus of attention is a relatively young field in cognitive neuroscience (Posner 2012). Here, too, we may distinguish between different but overlapping and dialectical (mutually influencing) ways of focusing of attention: one may be described as biological by nature, the other as psychological. The biological factors involved in the focus of attention deserve to be taken seriously, because the impact of biological processes on our behavior is, although not the only influence, still much larger than most people and certainly most policymakers are aware of. Biological processes within our bodies—caused by the complex ways our mind/body responds to and interacts with our immediate environment—come into play in everyday behavior (otherwise a person, for instance, could never walk up or down stairs) and, obviously, in handling any technological device. As early as the 1990s, high-tech developers liked to talk with and study creative persons such as artists, designers and architects, to understand their creative processes in order to improve human interface technology (Candy and Edmonds 2002 [2019]).

The psychological focus of attention is influenced by one's personal history, and this includes a person's family background, culture, society and education. It must be noted that the majority of research on the focus of attention is done in controlled, even confined laboratory settings, excluding many features of natural behavior and perception. Still, the obtained knowledge is useful for understanding focus of attention and might help to improve technological applications. A large part of the research about attention is naturally about vision and the involvement and influence of eye movements on the shifts of attention. Attention is also stirred by other senses, such as sound, smell, or feeling heat, cold, drafts or touch. Events in our field of vision may trigger eye motions but do not necessarily influence our attention on a one-to-one level. This phenomenon is the cause of many car accidents. A driver may have his eyes open, and his eyes may move in response to all that is happening on the road, but when he is too absorbed in thought, or is distracted in any substantial way (like by using a cell phone), he may not consciously notice dangerous situations.

It is, of course, possible that a mature artist—or scientist, or economist, or CEO—may think that when he or she brings a higher amount of attention to bear on those features important to his or her profession, then attention for other things diminishes. Some capacities are delegated from the brain to other parts of the body. Trained, repetitive physical actions involved in attention are stored as motor reflexes. Examples include walking stairs, the professional use of tools or quasi-tools in sports, like in golf, tennis or hockey. Technology can replace or enhance certain components of human behavior, but it cannot replace the fundamentals of being human. A neat example, as mentioned earlier, is that even leading specialists in advanced technologies need to fly around the globe for their profession, regardless of cutting-edge communication technologies: they prefer personal encounters to video conferencing.

Lessons Learned from the Art World—Applied to Digitalization of Education

Though obviously more complex, here I would like to select and emphasize two kinds of focus of attention: the focus of attention in learning, training, assembling knowledge and creative abilities, and the focus of attention on a fixed mission. In other words, the first is the focus of attention of target groups, the second is the focus of decision-makers.

While this subject is relevant for the discussion on digitization and digitalization, the following remarks may demonstrate the gap between policymakers' good intentions and the views of their target-groups. This gap turns on differences in focus of attention. Most persons who decide to visit a museum have a pre-set focus to enjoy a nice day looking at many things to distract their main focus away from their everyday concerns (job, family, finances, politics). The curators of that museum have an entirely different focus, mainly aimed at creating exhibitions with subjects intended for lecturing the public and aiming for their own peer recognition (with regard to their possible next jobs). The art world repeatedly demonstrates that such differences in basic focus collide (Ilgen 2017). Art professionals who wish to be of contemporary relevance are convinced that their exhibitions should include new technologies as typical main interests of our time. Though it is correct that people use technologies at work and at home, the implementation of such kinds of technology in museum exhibitions is a misjudgement, since it does not correlate with the reasons why people visit a museum or gallery. It is no surprise that the attendance is globally declining for museums of contemporary art, while attendance is increasing for museums showing art from history until Modernism.

At the 2015 Venice Biennale, for instance, the majority of exhibitions showed digitalized videos displayed on large screens in darkened spaces. Many visitors quickly went in and immediately out, often with some irritation. After all, visitors traveling far to Venice with the purpose of enjoying the sun, the city and the Bienniale are hardly inclined to be captured for long in such darkened spaces. While from an art professional's point of view there was nothing wrong with this choice of artworks, it clearly collided with their target audiences' focus of attention. Equally, at the opposite side of the spectrum, when one decides to go to a cinema, one focuses on the enjoyment of sitting in a darkened space, immersed in the large screen projection and sound. Now imagine how that viewer would feel if it was not a movie that was displayed but a painting, or a person on the stage reading aloud from a book? For the Biennial's own statistics, every visitor that went through the entrance door was counted—a normative way of surveying attendance and delivering proof of a successful exhibition.

Video art and other digital art have their reasons for being, but clearly do not work in all circumstances, and one understandably sees a steady decline of such (digital) technologies in exhibitions. All professionals in the art world are challenged to understand their target audiences, to improve their understanding of the naturalness of the focus of attention of exhibition visitors and to learn from neurobiology why the traditional art media (painting, drawing and sculpture) continue to be successful in terms of natural visual appeal. With justification, one should pose questions about purpose, time and place that have much to do with focus and attention, and therefore, are fundamental for setting parameters for each new technology.

Basic Human Capacities—In Art as Well as in Education

In general, many feel that coming generations will be different, better informed and smarter, certainly when it comes to handling new technologies. While some of this will be true, this exaggeration must be set against the acknowledged fact that the human brain biologically does not offer reasons to assume that the next generations of people will be very different. The brain has vast but limited possibilities. In very simple terms, the brain's capacities can be imagined as a table loaded with many objects. If one object is enlarged, taking up more space, some other objects will fall off the table. A person may accomplish specialization—collecting, building and maintaining vast knowledge on some specialized subject—but this comes with a price. Extreme specialization on, for instance, highly abstract thinking may lead to deviant social behavior.

The basic features and capacities of the human organism have not fundamentally changed since early homo sapiens—meaning, for instance, that artists have a very large choice of artistic expressions but still must remain within constraints set by the limits of their bodies. Having two arms and two hands offers us a large but still limited range of movements. It is fascinating that exactly these limitations set the parameters for the cohesion and analogies between creative expressions in art, which, again by these limitations, appeal naturally by analogy to any other person. This provides an important reason why traditional painting and sculpture through natural appeal spark and stimulate focus of attention. Also, the matter of natural limitations is an important reason why creative expressions in which the human body is not involved may attract some attention but do not appeal to the same degree. This observation is not limited to fine art but also applies to music.

Art is strictly linked to our natural perceptual preferences—not merely those of the mind, but also those that involve the whole body—and therefore remains close to the natural, human behavior and needs. It is fascinating that people throughout human history have needed creative expressions as a virtual intermediary for interacting with the world. Though it is beyond the scope of this essay to explore this observation in depth, for stimulating more thought on creativity (and technology) it is useful to remark here that artistic expression revolves around a limited range of variations. Sameness can be found on the level of visual appeal, tactility and compositional preferences; differences are expressed in verbal explanations following local (cultural) and individual preferences. Such sameness is by approximation, otherwise we could not distinguish between a work by Michelangelo, a Bernini or a Rodin. Artworks are figurative or abstract, may be described as beautiful or ugly, but these features are all less important than their visual appeal and thus, like music, offer the brain a stimulus to temporarily find some biological relief important for the brain to stay healthy. This is not a matter of culture, nor of criticism on what technology can offer, but a matter of how human beings function. This is a crucial lesson to be learned in any thinking about education.

When one realizes that art professionals do not consider the natural side of experiencing art, it becomes clear why the drive for endless disruptions of supposed traditions in art has led art professionals astray, away from the basic interests of their target audiences. It is revealing that in all debates about technology and art in all the many discussions—from the innovation of tools, the subjects of artworks, the process of making art as a technological process, up to the psychological understanding of how people verbalize their art experiences—there is little reference to the biological processes within our mind/body evoked by the artwork. It is possible that the real purpose of experiencing art may be different from what it seems to be at first glance, involving just the senses. The natural chain reaction in our organism is, perhaps, what art actually is for. Comparably, we may discuss culinary culture by analyzing the structure and ways of the preparation and presentation of food, or the enjoyment of the glamour of a star restaurant, but we would miss the real natural need to eat (energy) and to socialize with others (self-reflection). The restaurant that does not take into account the limits and abilities of the human palate is bound to fail.

The processes of naturally experiencing art (for a large part a biological process) explain why, for instance, new technologies—including digital and AI-manufactured art—can never appeal beyond the degree of any screen-saver or decoration. Artworks that are not manually made do not have the capacity for holding anyone's attention long enough to affect long term memory, while classical media such as paintings and sculptures do. This is one of the reasons for the growing gap between the art that professionals continue to promote in their institutions and the art that people like to include in their own environment: their homes.

The capacity of holding one's attention refers to the way our eyes, without any conscious decision, can be attracted by something we look at—for instance, evoked by the visual features of an artwork. When the artwork offers the right visual stimuli, the eyes may roam over the work repeatedly and the focus of the brain follows suit, focusing our attention. This biological process does not deny the influence of art historical knowledge or fame of an artwork or the artist, but plays a more crucial role across the board than is acknowledged by art professionals and is key to understanding how and why people may consider artworks appropriate for their private environment. The involved process of focused attention, causing a relaxing experience, a diversion of the main focus of the brain (on everyday problems of jobs, politics) and thus a sense of well-being, is likely the actual experience of works of art, not so much what the artwork represents. This kind of focus of attention, diverting the brain, is known to be fundamental for developing solutions and creating new ideas. An important lesson for any developer of technology and for any policy-maker regarding education is not to be blinded by new technological development in and of itself, but to check the usefulness of such developments in direct comparison to people's natural behavior.

Conclusions

In the immediate context of this book, *Re-designing Organizations—Concepts for the Connected Society*, for technology or the implementation of technology to be most effective, it is key for policy-makers to focus on speed for economic or geopolitical reasons. But it will be just as important for the future of human beings in a hybrid age to recognize the various human capabilities by which people can naturally handle and apply the speed of technologies.

Even in the early days of the first—mechanized—industrial production, maximizing the speed and efficiency of value-added chains was seen to be the only logical decision from an economic perspective. Workers were assumed to be extensions of the machines, merely adapting and hanging on. This attitude toward the relationship between technologies and human workers disrupted profitable production, causing many mistakes in production circles, as well as unhealthy conditions for workers.

Digitalization has the potential to improve business efficiencies, e-governmental services, and education to foster advancements in society. This can be successful when policymakers learn to understand the functioning of the human being and the natural behavioral constraints as set by biology. One has to search for a useful synthesis of human demands and human abilities with technological progress.

All thoughts and conclusions described in this essay are my own when no reference is mentioned in a footnote. My reading has been interdisciplinary, and I have also gathered information by interaction with many specialists from a wide range of professions. Although I have done my best to refer to the correct sources, I apologize when I have not mentioned sources I am not familiar with and will add appropriate acknowledgements in any future edition (Ilgen 2004, 2014).

References

Candy, L., & Edmonds, E. (2002). *Explorations in art and technology*. Berlin, Heidelberg: Springer Verlag. ISBN 1-85233-545-9. The revised and extended version is published 2019, ISBN 978-1-4471-7366-3.
Ilgen, F. (2004). *Art? No thing! Analogies between art, science, and philosophy*. Pro Foundation, The Netherlands, ISBN 90 9018543 7.
Ilgen, F. (2014). *ARTIST? The Hypothesis of Bodiness*. Tübingen, Wasmuth GmbH, Germany, ISBN 978-3–8030–3364–2.
Ilgen, F. (2017). *Art: For the professionals or for the art-viewer? Necessary involvement of the mind/body*. In Kate, L. (Ed.), *Fracturing conceptual art: The Asian turn*. Art Platform Asia, ISBN 979-11-961693-0-5.
Posner, M. I. (Ed.). (2012). *Cognitive neuroscience of attention* (2nd ed.). New York, London: The Guilford Press. ISBN 978-1-60918-985-3.

Fré Ilgen is based in Berlin, is not only a sculptor and a painter, but also a theorist and curator. His work is exhibited and owned widely in the United States, Europe, Asia, the Middle East and Australia. Works vary in size from small up to monumental size. His largest work is H5 × W7 × 40 m, in the main lobby of Heungkuk Life Insurance Building, a prominent office building, downtown Seoul. Ilgen's paintings, sculptures and mobiles depict a reality that is not a solid mass but a swirling movement of shifting relationships, using abstract as well as figurative forms. Ilgen combines features from Western and Eastern cultures and philosophies, like from the Baroque and Indian sunyata. His extensive interest in visual perception and his interest in neuroscience motivate him exploring artworks that are visually powerful, appealing in dynamic compositions, defying gravity and simulating continuous change. These purposely cause a pleasant bewilderment in the viewer, because his works do not offer any singular narrative besides a fusion of the positive and negative in life. Exhibitions include galleries, museums, corporations, foundations and art fairs around the world, and various biennials, including exhibition "Frontiers Reimagined" (Tagore Foundation), part of the official Collaterale Program of the 56th Venice Biennial. He moderates the award winning "Checkpoint Ilgen"—series, the idealistic art salon, which he and his wife Jacqueline host at their private apartment in Berlin. In 2011 Ilgen curated Mirrors of Continuous Change, a large exhibition of global art, for the Ilju and Seonhwa Foundations in Seoul. Monographies, published in Europe, USA and Asia, cover his creative output, including "Fré Ilgen: The Search", a retrospective book, 2001, and "To Be Free", 2012. His works and activities have been covered by many art magazines, newspapers, blogs and TV programs in various parts of the world. He also published the widely acknowledged major books "Art? No Thing! Analogies between Art, Science and Philosophy", 2004, and "ARTIST? The Hypothesis of Bodiness", Wasmuth Verlag, 2014, and "Press for Champagne - Inside the Art World", Studio Ilgen, 2019.

Chapter 27
The (Post-)Digital University

Markus Deimann

Introduction

We live in an age of change. This is the story we hear constantly. It is so big that every realm of society is part of the change. In a 2018 report, the World Economic Forum addresses *Future Scenarios and Implications for the Industry*:

> Incremental change is not an option any more in the construction industry. By redefining the ultimate frontier, leapfrogging innovations in construction will finally help address major societal challenges, from mass urbanization to climate change. The widespread adoption of game-changing innovations that consider a variety of possible futures is going to make a serious impact, socially, economically and environmentally.

In a similar vein, in 2017, the Organization for Economic Co-operation and Development (OECD) issued the report *Key Issues for Digital Transformation in the G20*, in which it is stated:

> As the cost of data collection, storage and processing continues to decline dramatically and computing power increases, social and economic activities are increasingly migrating to the Internet. Technologies, smart applications and other innovations in the digital economy can improve services and help address policy challenges in a wide range of areas, including health, agriculture, public governance, tax, transport, education, and the environment, among others. Information and communication technologies (ICTs) contribute not just to innovation in products, but also to innovation in processes and organizational arrangements.

In these and other reports (World Bank 2018), change is described as driven far and foremost by digital technology and the capitalist economy. They set the agenda for politics, which traditionally shapes social relations. Yet the agenda is more or less a hidden one as there are new values coming along with the change that are not based upon a general agreement but are set up by a small group of experts, entrepreneurs and evangelists as part of a new "digital feudalism" (Morozov 2016).

M. Deimann (✉)
Teaching, FernUniversität in Hagen, Hagen, Germany
e-mail: markus.deimann@fernuni-hagen.de

© Springer Nature Switzerland AG 2020
D. Feldner (ed.), *Redesigning Organizations*,
https://doi.org/10.1007/978-3-030-27957-8_27

357

There are parts of the society that seem especially appropriate for the new values circling around an entrepreneurial approach with respective metaphors such as disruptive innovation and methods like design thinking (Vinsel 2017). Other parts—in particular, education or the arts—have their own, inherent values that are about to clash with the new values.

With this brief preliminary remark, I intent to shed light on something that is not discussed much, but rather taken for granted. Technological innovations are inevitable for our society, and there is tremendous evidence from history that any society that wants to endure needs progress (see for example Smith 2003). This is even more apparent with the rise of Information and Communication Technologies (ICT) and the resulting narrative of digital transformation, which urges us to take matters into our hands. Every field of society is called upon to immediately update/upgrade its mechanisms and organize its principles according to the "digital logic". The prototype is described using the language of software engineering and versioning (1.0, 2.0, 4.0) to suggest that a new stage of development has been entered. Now we have reached the Fourth Industrial Revolution, which is characterized by a fusion of technologies and the blending of digital and biological spheres. Whereas the number of technological breakthroughs is undoubtful, the impact on culture and education has just begun to emerge. Therefore, to all appearances, Education 4.0 (or Bildung 4.0 in the German discourse) (Deimann 2017), is the term that is now used to signal that digitization is prevailing.

Yet the transfer from the logic of software engineering to education is so brute that it neglects and suppresses all the inner philosophies and cultural practices that have been developed over the last centuries. Even more bizarre, it is argued that the entire institution of the *university* has become obsolete thanks to the Internet and its digital offerings in the form of Massive Open Online Courses and others (Harden 2012). With all the content available on the web, there is no need anymore for a curated and cultivated knowledge production and dissemination system. This also holds true for certificates, which do not need to be issued via a university but via use of Blockchain technology (Kariuki 2018). However, Blockchain is also an expression for a lack of trust in authorities, governments and institutions (Baur and van Quaquebeke 2017).

There are other examples of such technology-focused debates that are put forward to shape the future of the university without taking care of social, cultural and educational factors. It is thus important to have a much more balanced examination if we want to sketch out the "Digital University". In the following sections, I intend to do so by unknotting dichotomies that are influential in current discussions (e.g. digital vs. analogue) with regard to certain functions of the university (e.g. teaching and learning). This can help to bring rivaling perspectives closer together. The perspective I want to suggest is informed by the concept of *postdigital*:

"The term 'Postdigital' is intended to acknowledge the current state of technology whilst rejecting the implied conceptual shift of the 'digital revolution'—a shift apparently as abrupt as the 'on/off,' 'zero/one' logic of the machines now pervading our daily lives" (Pepperell and Punt 2000). It can be conceived of as a state of being rather than the constant continuation of a process following distinct phases (1.0–4.0). Thus, there is no linear progression—such as the teleology that is ingrained in video and TV technology (e.g., from SD to HD to 4K)—but a mixture of subtle cultural shifts and ongoing mutations caused by digitalization and the global digital infrastructure (Cramer 2014). Postdigital allows a rethinking of the future of the university in a less heated way (beyond revolution and dystopia) but with a closer look at the power structures that emerge from the various entanglements between the digital and the non-digital. As Losh (2014) has written in *The War on Learning*, there is an immense pressure to recast the university as a product designed according to the demands from vendors of educational technologies.

Therefore, following the discussion on dichotomies, I will outline a line of thinking regarding the post-digital university.

Current Dichotomies in Relation to the University

One of the most prevailing dichotomies in current debates about the future of the university is analog versus digital. On a basic level, it refers to history: essentially, a very long time ago in Europe, some people had the idea to come together to collect, preserve and distribute knowledge. The gathering took place in a brick-and-mortar environment with the emblematic lecture hall. Therefore, the buildings were not only home to the academics, they were also the manifestation of the philosophy of scholarship.

With the advancement of technology, especially in the 20th century, this idea has come under criticism and new forms of distance education emerged. Television and video offered new possibilities for teachers and learners to exchange knowledge, even when they are not together in the same place at the same time. Yet, there were still buildings—for example, for the administration of education. Nowadays, this seems to be becoming obsolete thanks to digital infrastructures (i.e., the Internet) that offer a virtual space that contains all the necessary elements to study. Whereas in many distance education programs there is a residue of the traditional brick-and-mortar education, digital/virtual education attempts to eliminate that—for instance, Massive Open Online Courses (MOOCs) offered on virtual platforms such as Udacity and Coursera, with additional support and resources for online studying.

However, concepts that are used to signal a new model of education are oftentimes flawed, such as "Digitalization", which basically means a transformation from information presented in analog format to digital format with discrete and discontinuous values (1 and 0). In contrast to that, analog information contains varying values with a lot in betweennesses the 1 and 0. This allows the expression and storing of information in many different ways, such as punch cards.

Besides this material and the technical process of converting information, there is an additional meaning of the term digitalization, namely "(…) the way in which many domains of social life are restructured around digital communication and media infrastructures" (Brennen and Kreiss 2016). Digital does not necessarily have to be electronic, such as a mechanical typewriter. Conversely, "analog" does not mean non-computational or pre-computational, as there are analog computers. A similar confusion that exists pertains to "virtual and digital", which are used interchangeably. Instead of such subtle nuances, digitalization is typically defined as an open, dynamic, multiperspectival and unfinishable process that has become a significant configuration in people's lives—i.e., the lifeworld (Fors 2010). This is triggered by several convergences, such as regarding the infrastructure or devices. Smartphones are a telling example, as they not only physically consolidate many devices, but they also connect activities that were earlier linked to separate media (e.g. watching TV, reading the newspaper).

But even when the concepts are applied in a more accurate way, there are still dichotomies that are delusive. A prominent example is online versus offline, which refers to two separate states by which one is either connected to the Internet or not. Yet with the emergence of mobile computers and portable hardware, the digital and the physical are blending, and there is a rather constant status of being online. In addition to that, more and more physical devices (e.g., refrigerators) are using sensors to be (inter-)connected. They provide an additional layer to reality ("augmented reality") that also merges the digital and the physical world.

The constant drive for (inter-)connections between humans, machines and services sheds light on another dichotomy. Universities have—for a long time—been conceived of as single entities or monoliths that unite the basic functions of teaching and learning under one roof. Yet with the rise of science and technologies, the idea of a *Universitas magistrorum et scholarium* has become less important. Instead, there is a set of networks and communities that operate on their own and that are somewhat outside of the university. They have stronger links to the society and are aimed at a *Third Mission*, which is about opening scientific investigations and research projects to laymen. This is amplified by network technologies and Internet platforms (e.g. EU Citizen Science Platform) that offer a digital space to share ideas and to collaborate on projects generating a vast amount of resources. With these citizen scientists and other related activities, the monolithic construction of the university becomes questionable, particularly given the free flow of information on digital networks. One can argue that the monopoly of universities vanishes, because they are no longer the outstanding place for generating and distributing knowledge in the form of research and teaching. However, such claims are often put forward from the outside, neglecting the distinctiveness of the academic culture.

A third dichotomy that is revealed by constant technological advancement pertains to education as a "beautiful risk" (Biesta 2013) versus an algorithmically administered task. Based on a neoliberal agenda that seeks to make education more efficient, educational technologies are utilized to support/augment/replace teachers. Again, advocates are mostly from outside pedagogics and ignore the educational philosophy that is based on the fundamental belief that education is unpredictable, open-ended and risky. This is a challenging message for economists who want to create an educational system with a perfect match between input (costs) and output ("human capital") and technologists who want to "solve" alleged problems. A telling example for such a vision is presented by Knewton, a company specialized in adaptive learning technologies:

"We think of it like a robot tutor in the sky that can semi-read your mind and figure out what your strengths and weaknesses are, down to the percentile. (…) We can take the combined data power of millions of students—all the people who are just like you—[who] had to learn a particular concept before, that you have to learn today—to find the best pieces of content, proven most effective for people just like you, and give that to you every single time". (Westervelt 2015).

With the ongoing digitalization, a new line of conflict is emerging: On the one hand are the technocrats and "solutionists" (Morozov 2013) who want to redefine education based on the mandate of innovative technologies and the overall transformation of society. Opposing them, on the other hand are practitioners, scientists and philosophers of education who also attempt to redefine education, but from within education. Yet, besides these apparent discrepancies, there is also a commonality: Both follow essentialist and instrumentalist perspectives (Hamilton and Friesen 2013) for the integration of technology. As Hamilton and Friesen (2013) have argued, these positions are flawed and fail to capture the complex interplays between the social and the technical. The next section will provide an idea to overcome this.

How to Think About the Post-digital University

The prefix "post" is intended to give "digital" a meaning distinct from the one usually applied. Post-digital means to signal a new normality and to reject just another hype in the education business. As could be witnessed with the rise of the MOOCs, strong claims have been made with regard to the alleged changes for education and the university. Now, new trends like the Internet of Things or Blockchain, as well as renewed interest in Artificial Intelligence, emerge with similar patterns in media coverage. Also, in the past—such as with "E-Learning"—it has been a steady hyperbolic rhetoric.

From an educational standpoint, neither hype nor dystopia serves as a good adviser. Therefore, instead of disrupting or preserving the university, post-digital offers the possibility to continue the core mission (with a set of principles and values) in light of changed conditions. To give an example, learning management systems (LMS) have been commonplace for a decade, but they are also still talked about as something new. A post-digital perspective is based on this normality and focuses on new practices that can be derived from the interplay between technology (LMS) and education.

Furthermore, educational values (e.g., participation) can guide the selection of certain technologies, services and protocols (e.g., open source software) and can underpin the architecture of digital infrastructure. The project "Domains of One's Own" is an insightful example of such a system as it gives students the freedom to tinker with their own webspace and get acquainted with the logic of the web. In contrast to closed, proprietary platforms such as Facebook, the web is an open playground that offers new forms of collaboration and communication. Moreover, hosting your own domain sensitizes the learner to become more mindful about the backend of the Internet.

As could be witnessed with the so-called #PizzaGate—on December 6, 2016, a 28-year-old, heavily-armed father drove from Salisbury, North Carolina, to Washington, DC, to rescue children who he believed to be imprisoned at a small pizzeria as part of a large conspiracy including Hillary Clinton (Fisher et al. 2016)—network technologies and online behavior can be misused in a way that is in direct opposition to the earlier promises of the Internet as a technology of liberation. A lot of the developments that have led to #PizzaGate took place underneath the surface of the shining and soothing platforms and websites. They are providing an enormous challenge for universities and their mission of being a reflective entity for the current state and the future of humans and societies.

It is also a challenge that should not be answered with quick and repetitive solutions, such as integrating computer science in the compulsory education curriculum (Passey 2017), but rather with a balanced mix based on an informed understanding of educational philosophy, educational technology and the recent developments of Internet politics.

References

Baur, D., & van Quaquebeke, N. (2017, November 16). The blockchain does not eliminate the need for trust. Accessed from https://theconversation.com/the-blockchain-does-not-eliminate-the-need-for-trust-86481.

Biesta, G. (2013). *The beautiful risk of education*. Boulder: Paradigm Publishers.

Brennen, J. S., & Kreiss, D. (2016). Digitalization. In K. B. Jensen, E. W. Rothenbuhler, J. D. Pooley, & R. T. Craig (Hrsg.), *The international encyclopedia of communication theory and philosophy* (pp. 1–11). Hoboken, NJ, USA: Wiley & Sons, Inc. https://doi.org/10.1002/9781118766804. wbiect111.

Cramer, F. (2014). What is 'Post-digital'? In D. M. Berry & M. Dieter (Eds.), *Postdigital aesthetics: Art, computation and design* (pp. 12–26). New York: Palgrave Macmillan.

Deimann, M. (2017). Warum wir für Arbeit 4.0 nicht Bildung 4.0 brauchen. Held at the spring conference of the Deutsche Gesellschaft für Wissenschaftliche Weiterbildung und Fernstudien (German Society for Continuing Education and Distance Learning), Hannover. Accessed from https://docs.google.com/document/d/1IfwjRRiwcp0LsOJ_RkyYk0NsXB15yaz9gWBxg51cOSA/edit?usp=sharing.

Fors, A. C. (2010). The beauty of the beast: The matter of meaning in digitalization. *AI & Society, 25*(1), 27–33. https://doi.org/10.1007/s00146-009-0236-z.

Hamilton, E., & Friesen, N. (2013). Online education: A science and technology studies perspective (Éducation en ligne: Perspective des études en science et technologie). *Canadian Journal of Learning and Technology, 39*(2).

Harden, N. (2012). The end of the University as we know it. *The American Interest, 8*(3). Accessed from http://www.the-american-interest.com/articles/2012/12/11/the-end-of-the-university-as-we-know-it/.

Kariuki, D. (2018, February 8). *Digital certification of education degrees will be India's first blockchain project.* Accessed on June 19, 2018, from https://coinpedia.org/news/digital-certification-degrees-indias-blockchain-project/.

Losh, E. M. (2014). *The war on learning: Gaining ground in the digital University.* Cambridge, MA: The MIT Press.

Morozov, E. (2016, April 24). *Tech titans are busy privatising our data.* Accessed from https://www.theguardian.com/commentisfree/2016/apr/24/the-new-feudalism-silicon-valley-overlords-advertising-necessary-evil.

Morozov, E. (2013). *To save everything, click here: The folly of technological solutionism.* New York: PublicAffairs.

Passey, D. (2017). Computer science (CS) in the compulsory education curriculum: Implications for future research. *Education and Information Technologies, 22*(2), 421–443. https://doi.org/10.1007/s10639-016-9475-z.

Pepperell, R., & Punt, M. (2000). *The postdigital membrane.* Bristol: Intellect Books. https://doi.org/10.13140/2.1.4499.4241.

Smith, A. (2003). *The wealth of nations* (Bantam classic ed.). New York, NY: Bantam Classic.

Vinsel, L. (2017, December 6). *Design thinking is kind of like syphilis — It's contagious and rots your brains.* Accessed from https://medium.com/@sts_news/design-thinking-is-kind-of-like-syphilis-its-contagious-and-rots-your-brains-842ed078af29.

Westervelt, E. (2015, Oktober 13). *Meet the mind-reading robo tutor in the Sky.* Accessed from http://www.npr.org/sections/ed/2015/10/13/437265231/meet-the-mind-reading-robo-tutor-in-the-sky.

Studies

OECD. (2017). *Key issues for digital transformation in the G20.* https://www.oecd.org/g20/key-issues-for-digital-transformation-in-the-g20.pdf.

World Bank Group. (2018). World Development Report, Learning to Realize Education's Promise. https://openknowledge.worldbank.org/bitstream/handle/10986/28340/9781464810961.pdf.

World Economic Forum. (2018). *Future scenarios and implications for the industry.* https://www.weforum.org/reports/future-scenarios-and-implications-for-the-industry.

Web References

Domains of One's Own. https://www.wired.com/insights/2012/07/a-domain-of-ones-own/.

European Union Citizen Science Platform. http://digitalearthlab.jrc.ec.europa.eu/csp.

Fisher, M., Cox, J. W., & Hermann, P. (6, December 2016). https://www.washingtonpost.com/local/pizzagate-from-rumor-to-hashtag-to-gunfire-in-dc/2016/12/06/4c7def50-bbd4-11e6-94ac-3d324840106c_story.html?noredirect=on&utm_term=.1f397d15dfc1.

Dr. Markus Deimann currently is Deputy Professor for Media Didactics at the FU Hagen. He received his post-doctoral qualification ("Habilitation") from the FernUniversität in Hagen, his Ph.D. in educational science from the University of Erfurt, Germany and his MA in educational science and political science from the University of Mannheim, Germany. He has been a visiting scholar at Florida State University and Research Fellow at the Open Learning Network (Olnet) at The Open University. His research is located at the intersection of educational philosophy, media studies and ongoing socio-technological developments. Using the classical German conception of Bildung, he is attempting to translate its special meaning to a global and interconnected world and on the other hand to learn from digital transformation in order to update theories of Bildung. He is also a Member of the Hochschulforum Digitalisierung at the Stifterverband für die Deutsche Wissenschaft.

Chapter 28
Managing the Digital Change in Higher Education

Barbara Getto

Introduction

Within the past ten to fifteen years, digital media have come to play an increasing role—also in the higher education sector. The increasing digitization of life and work offers potential. Unlike in some other areas (e.g., like in the music or publishing industries), in education we are not experiencing any massive change in the "market" through digitization. Even if new actors and service providers take over tasks at various points, the system remains in place (see also Getto and Kerres 2018). Here we experience no disruption and no displacement through a new business model. The term "digital transformation" often refers to a "disruptive approach", and there is concern (or even hope?) that digitization would fundamentally change the system of education. But it seems more the case, that the role of digitization as a "driver" is often overestimated. Digitization will not magically revolutionize the system by itself. But it clearly does offer opportunities to change and rethink processes. The challenge, though, is not to create a new system from scratch, but to shape the existing system and, therefore, to drive changes that can harness the potential of digitization.

Shaping Digital Transformation in Higher Education

The systemic implications of e-learning for higher education are discussed by de Freitas and Oliver (2005), who refer to design options for the introduction of e-learning and criticize the frequently supported assumption of "Fordism" in the discussion (Clegg et al. 2003). The digitization of study and teaching is a comprehensive process of change that involves actors at various levels and, in addition to promoting

B. Getto (✉)
Learning Lab, University Duisburg-Essen, Essen, Germany
e-mail: barbara.getto@uni-due.de

© Springer Nature Switzerland AG 2020
D. Feldner (ed.), *Redesigning Organizations*,
https://doi.org/10.1007/978-3-030-27957-8_28

e-learning, affects all areas of the university. The universities must adapt to the changes and anchor digital technology as a tool for knowledge development, development and communication, but also face the new societal challenges in the content they are teaching. Since the beginning of the state funding programs at the end of the nineties, priorities have been identified for the sustainable anchoring of digital media in teaching and learning. From the development of the first technically focused teaching innovations in the "pioneer phase", to the phase of cross-university cooperation in the multimedia field, the question was whether e-learning would work at all, find acceptance and lead to the same or better learning outcomes. The later phases dealt with the question of the sustainable anchoring of digital media in universities across the board. Infrastructural questions, competence development, as well as personnel and organizational development were focused.

Actors in universities perceive the challenges of digitization differently, and they react differently. With reference to E. Rogers' model of a "technology adoption lifecycle", the diffusion of technological innovations in higher education has been described (e.g., by Euler and Seufert 2005). Ultimately, however, the process of change is considered to be the decision of individuals who are moving in a group-dynamic process. Therefore, their decision to turn to digital media depends on the experiences of others. Porter et al. (2014) applied the adoption model to institutional policies and structures of U.S. universities, distinguishing three phases that they studied at 11 higher education institutions. However, the special institutional framing of action at German universities—with their different levels of actors and constellations—is not systematically addressed. For this reason, it remains difficult to derive recommendations for higher education as a whole. Bogumil et al. (2013) outlined some of the specifics of higher education institutions in German-speaking countries, which include both the characteristics of an institution and an organization. Kehm (2012) describes universities as "special and incomplete organizations", because they lack the many possibilities of governance required for consistent strategy implementation. These two perspectives, briefly outlined here, have major implications for the question of how to tackle the challenge of digitization.

Opportunities for Digitization in the Area of Study and Teaching

Today, universities are increasingly adopting digital technologies for teaching. A lot of research has been devoted to analyzing the effects of educational technology on learning. For more than five decades, research on digital media in learning has studied the effects of educational technology on achievement in schools and higher education. Given the vast number of single studies on the topic, meta-analyses have been conducted to aggregate these findings. Since the seminal work of Kulik et al. (1980), the results of these meta-analyses demonstrate a remarkable stability: Digital technology can improve learning, especially when combined with traditional face-to-face instruction, with different strengths in various fields of education. With the proliferation of meta-analyses, Tamim et al. (2011) gathered these results into a meta-meta analysis, again proving a significant but small effect that, "the average student in a classroom where technology is used will perform 12 percentile points higher than the average student in the traditional setting that does not use technology to enhance the learning process" (p. 17). Therefore, the goal of educational reform in universities cannot be confined to the implementation or dissemination of digital technology in the classroom. It must center around the options digital technology offers for new teaching perspectives for the development of new curricula, new instructional methods and new ways to offer and to arrange learning opportunities for students.

This short summary of major findings in research on the impact of educational technology has substantial implications for the discussion in higher education. A university must provide a digital infrastructure for teaching and learning, starting with basic digital tools (email, etc.) and learning and teaching management systems, and ending with advanced and sophisticated technologies for exploring new methods of delivery and learning experiences. But given the outlined state of research on the impact of educational technology, it seems probable that the availability of digital tools for learning doesn't improve the quality of learning and—above all—does not change teaching and learning habits. It does even seem more plausible that teachers will continue the teaching routines they have developed over years—using digital technology. In the past, this scenario has repeatedly led to the observation that, despite heavy investments into digital technology, the daily routines of teaching in schools have not substantially changed. Therefore, many critics have questioned the usefulness of these investments.

From the perspective of educational technology, however, the successful implementation of digital media that leads to a substantial gain in the quality of learning requires rethinking the concepts of teaching and learning in a given school. For the administration, then, it becomes necessary to define the goals the institution wants to address with the use of digital technology, as well as the actions that are necessary to foster change. For the institution, the aim is not to implement digital technology; the aim is to find ways that digital technology can help to increase institutional goals.

What Strategic Goals Can Be Pushed with Digital Technology?

Many institutions are eagerly implementing various digital technologies and tools for learning; they seem to follow a general trend of modernizing university teaching. Following this understanding, digital transformation can be seen as a broader movement of modernization that all universities have to address in some way: a process that institutions are taking at different speeds and with different measures, but most probably with similar results in the end. From a strategic perspective, however, digital transformation has to be perceived as a chance for an organization to position itself in a competition. The management at universities must understand that digital transformation can be understood as a "general trend of modernization", but to some degree, it also opens up opportunities for a strategic positioning.

The public universities in Germany differ substantially in their background and goals: Small universities with only some thousand students, larger universities of more than 50,000 students, full research universities offering study programs that cover a broad range of topics, as well as universities with a specific focus (e.g., music and fine arts). Therefore, each university has to decide upon its strategy with regard to digital technology. Kerres and Getto (2018) have extracted central arguments and perspectives around the crucial question, "What are the strategic goals that can be pushed with digital technology?".

The results lead to the following conclusion: Basically, universities can aim toward several goals they want to achieve with the advancement of educational technology, but the major arguments can be clustered around the goal to …

- improve the quality of teaching and learning

    - by increasing the intensity of learning, supporting active learning (e.g., by providing materials for training and self-assessment)
    - by supporting individualized/personalized learning
    - by supporting online social learning

- optimize educational services (e.g., with digital registration, information, support, counseling, exams),
- improve outreach and to reach new target groups (e.g., lifelong learning, professional development, internationalization).

Digitization in higher education has implications on teaching on all levels. Key stakeholders promoting digitization in the field of teaching and learning on a strategic level are primarily found at the management level, where fundamental decisions on the design and focus of institutional strategic plans are made. They interact with stakeholders on political levels (federal and nationwide) and, within their own institution, with professors/chairs. We focus on stakeholders that impact strategic and political decision-making and their motives to give digitization a strategic significance:

- the chair/professor
- the institution (university, faculty or school)
- politics: (federal) ministry.

The key question is: Who can raise the profile and position themselves with digitization and what would be the main objectives? Professors and academic staff will have specific motives for using digital media in their teaching practice, depending on their view on digitization. If they perceive digitization merely as "modernization", they will act differently than when it's perceived as a chance to position themselves.

When modernization is the key motive, the focus lies on the hope of increases in efficiency in the organization of study and teaching. Crucial benefits are the online accessibility of study materials, an easier communication with students via digital media and more flexibility in the organization of study programs. If a professor wants to increase attention and raise awareness for his or her teaching practice, they will prefer to focus on innovative approaches to improve their educational content (Getto and Kerres 2017, p. 130).

On an institutional level, digitization can be seen as a modernization strategy or a profiling strategy. According to a study by Uhl (2013), universities will primarily focus on the following strategic topics as elements of differentiation:

– to increase outreach
– to improve quality in teaching and learning
– a substantial expansion of the study program.

If digitization is seen as a subject due to technical achievements, universities will follow a specific strategic approach focused on improvements (e.g., in their technological infrastructures, learning management systems or administration programs). In the digital age, these technological optimization measures cannot be seen as a specific feature of an institution; they are basic requirements for a modern operational procedure. Digitization is here an instrument for more efficiency. There is, therefore, no strategic focus on the digitization of teaching and learning.

Finding and defining a digital strategy for teaching and learning is not a trivial endeavor. Universities are quick to take measures to implement certain digital technologies. They invest in platforms, tools and applications. Often, substantial financial investments are made that lead to follow-up activities requiring even more long-term investments. Most of the actions are geared by the assumption that digitalization is a general trend a university has to follow. Investment decisions are not following a strategic plan but are based on daily micro-politics and the intra-organizational negotiations of stakeholders. Our workshops have pointed out that it is difficult for a university to develop a strategic plan for digital transformation in Higher Education. At the management level, discussion of strategic options, processes for the development of strategic plans and measures can be helpful in supporting these endeavors.

Final Remarks

Digitization permeates all areas of life. But we need to consider that digital technology does not inevitably lead to—positive or negative—changes in education. This may be disappointing for exuberant supporters, as well as critical skeptics of technology. Such technology determinism would fail to recognize the importance of actors in bringing about changes in education and a change in learning culture. At an institutional level, this means that we—the actors—must (and are able to!) continue the process of change. But beware: we will not live up to the massive changes if we pursue digitization as a "special topic" that runs parallel to other activities and aspirations. This may initially seem contradictory with regard to the discussion of digital strategies. And, yes, the buzzword "digitization" helps to mobilize and arouse interest.

To see digitization as a process of "technification" would be to greatly understate the potential. If we want to take advantage of the momentum, the design process will need special attention, and we will need the courage to tackle structural change, as well. Technical developments—such as artificial intelligence and robots, as well as increased networking and the use of big data—will increasingly shape the (educational) systems. But they will not replace the social learning processes. Algorithms have the potential to prepare decisions, but it is *we* who decide and *we* who design the path to education in the digital era.

References

Bogumil, J., Burgi, M., Heinze, R.G., Gerber, S., Gräf, I.-D., Jochheim, L., Schickentanz, M. (2013). Zwischen Selbstverwaltungs- und Managementmodell. In E. Grande, & D. Jansen, u.a. (G), *Reorganisation—externe Anforderungen—Medialisierung* (S. 49–71). Bielefeld: Transcript.

Clegg, S., Hudson, A., & Steel, J. (2003). The Emperor's new clothes: Globalisation and e-learning in higher education. *British Journal of Sociology of Education, 24*(1), 39–53. https://doi.org/10.1080/01425690301914.

de Freitas, S., & Oliver, M. (2005). Does e-learning policy drive change in higher education? A case study relating models of organisational change to e-learning implementation. *Journal of Higher Education Policy and Management, 27*(1), 81–96. https://doi.org/10.1080/13600800500046255.

Euler, D., & Seufert, S. (2005). *Nachhaltigkeit von eLearning-Innovationen: Fallstudien zu Implementierungsstrategien von eLearning als Innovationen an Hochschulen.* St. Gallen: SCIL Arbeitsbericht 4. Accessed from: http://www.scil.ch/publications/docs/2005-01-seufert-euler-nachhaltigkeit-elearning.pdf.

Getto B., & Kerres, M. (2017). Akteure der Digitalisierung im Hochschulsystem: Modernisierung oder Profilierung? In *Zeitschrift für Hochschulentwicklung* (Vol. 12, pp. 123–142). ZFHE-973-1-4416-1-10-20170325.pdf. Last accessed April 6, 2018.

Getto, B., & Kerres, M. (2018). Wer macht was? Akteurskonstellationen in der digitalen Hochschulbildung. In B. Getto, P. Hintze, & M. Kerres (Eds.), *Digitalisierung und Hochschulentwicklung. Proceedings zur 26. Tagung der Gesellschaft für Medien in der Wissenschaft e.V.* (p. 60–71). Münster: Waxmann. ISBN 978-3-8309-3868-2.

Kehm, B. (2012). Hochschulen als besondere und unvollständige Organisationen?—Neue Theorien zur ‚Organisation Hochschule'. In U. Wilkesmann, & C. Schmid (Eds.), *Hochschule als Organisation* (S. 17–25). Wiesbaden: VS Verlag für Sozialwissenschaften I Springer Fachmedien.

Kerres, M., & Getto, B. (2018). Developing strategies for digital transformation in higher education. In T. Bastians (ed.), *EdMedia.* Amsterdam: AACE. Abgerufen von. http://www.learntechlib.org/p/184240/.

Kulik, J. A., Kulik, C. C., & Cohen, P. A. (1980). Effectiveness of computer based college teaching: A meta-analysis of findings. *Review of educational research, 50,* 524–544.

Porter, W. W., Graham, C. R., Spring, K. A., & Welch, K. R. (2014). Blended learning in higher education: Institutional adoption and implementation. *Computers & Education, 75,* 185–195. Accessed from: http://www.sciencedirect.com/science/article/pii/S0360131514000451.

Tamim, R. M., Bernard, R. M., Borokhovski, E., Abrami, P. C., & Schmid, R. F. (2011). What forty years of research says about the impact of technology on learning. *Review of Educational Research, 81*(1), 4–28.

Uhl, V. (2013). *Virtuelle Hochschulen auf dem Bildungsmarkt: strategische Positionierung unter Berücksichtigung der Situation in Deutschland.* Österreich und England: Deutscher Universitätsverlag.

Dr. Barbara Getto is currently Teamleader "digitisation in higher education" at the Learning Lab of University Duisburg-Essen. She studied pedagogics and psychology at the RWTH Aachen and her Ph.D. thesis focuses on the sustainable establishment of learning-innovations in higher education. In 2005 Getto started at the Learning Lab as Research Assistant and her research focuses on development processes at universities with special reference to incentives, motivation, and e-learning. She advises universities on the development and implementation of e-learning processes. Her present research interests include digital strategies in higher education and the promotion of open education.

Chapter 29
The Shift from Stable Jobs to Dynamic Careers in Digital Manufacturing

Lina Huertas, Harald Egner and Martin Dury

Introduction

The recent exponential surge in digital solutions is not only changing the technological landscape, but also transforming businesses, at an accelerated pace. Increasingly, organizations would like to grasp the possible opportunities, but they are still wondering what digitalization means for them and how to get started on their journey. All around the world, there are examples of organizations investing in innovation projects that will enable them to extract the potential in digital transformation processes. Some have already captured significant benefits that have been reflected in financial rewards and, in the best cases, strategic global positioning.

Along with technology and businesses, the nature of work and jobs stands to be transformed, too. As innovative technology is introduced, processes are changing and so are human tasks related to those processes. For example, in the context of manufacturing, manual tasks are being replaced by more automated equipment and tools, and knowledge intensive tasks are being replaced by efficient solutions based on intelligent decision support systems.

While significant short-term savings are available, the transformation of jobs is happening organically rather than by design, hence, the outcomes are uncertain. For the current workforce, this level of uncertainty creates fear as their jobs may be at stake in the transition. This affects not only individuals, but society as a whole, as the

L. Huertas (✉)
Technology Strategy for Digital Manufacturing, The Manufacturing Technology Centre (MTC), Coventry, UK
e-mail: lina.huertas@the-mtc.org

H. Egner · M. Dury
The Manufacturing Technology Centre (MTC), Coventry, UK
e-mail: harald.egner@the-mtc.org

M. Dury
e-mail: martin.dury@the-mtc.org

© Springer Nature Switzerland AG 2020
D. Feldner (ed.), *Redesigning Organizations*,
https://doi.org/10.1007/978-3-030-27957-8_29

transformation is likely to lead to significant social changes and potentially social issues, as has happened in the past (Leondorf 2010). This can be seen as a threat or be transformed into an opportunity for society and organizations that want to engage in digitalization to realize its benefits.

The Decline of Stable Jobs

The elements of any transformation process are people, business and technology (Prodan 2015). Because of the nature of digital transformation—a process characterized by the sudden availability of numerous digitally enabled tools—there is a trend for digital transformation processes to be focused mainly on technology, followed by changes in businesses and then consequences for people within the organization, as shown in Fig. 29.1. This leads to changes where technology is introduced to generate changes in the business that lead to financial benefits. As the transformation process progresses, those changes in the business eventually result in the transformation in jobs.

Changes in the human side of organizations as a consequence of digital transformation often happen organically, and actions are taken reactively, rather than strategically and proactively. As a result, the process can be perceived as a threat to the current workforce. This perception can easily become reality when transformation processes lead to layoffs. The perspectives of employees can be represented on a decline down the Maslow Hierarchy of Needs (Maslow 1943), as depicted in Fig. 29.2.

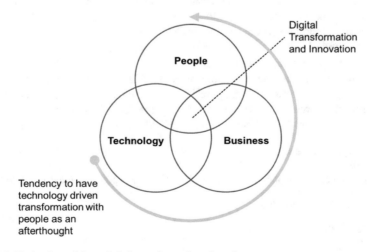

Fig. 29.1 Technology driven digital transformation. *Own Source*

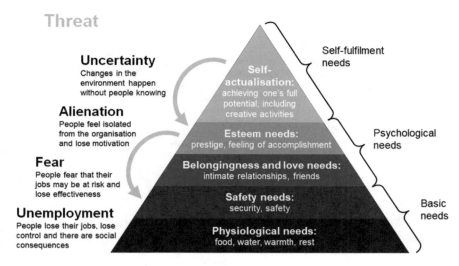

Fig. 29.2   The downward movement in the Maslow pyramid. *Own Source*

From an employee perspective, the process starts with uncertainty about their future created by sudden changes in their work environment, along with ubiquitous media headlines about the destruction of jobs brought by digital technologies and the dangers of automation—such as "Automation to take 1 in 3 jobs in UK's northern centres" (The Guardian 2018) or "Robot automation will 'take 800 million jobs by (2030)'" (BBC 2017). Feelings of "digital uncertainty" are reinforced and amplified by other uncertainties, such as global warming, migration, political instabilities and populist trends. If employees are not made participants of transformation processes early on, feelings of uncertainty become alienation, as people start feeling isolated from the organization and lose motivation.

When the transformation process and the uncertainty close in on an individual's job, it is usually too late to get them involved, since their feelings have turned into fear that their jobs may be at risk, leading to a lack of effectiveness and accelerating the downfall. The feeling of fear is intensified by questions about their own ability to adapt to changes and risks related to status and income, financial stability and family security. In the worst case, the journey finishes with the bleak prospect of unemployment.

In addition to the obvious impact of unemployment for the individual and the well-researched societal impact of unemployment, there are further consequences for the organization and the transformation process itself, as shown in Fig. 29.3. As the workforce loses motivation and effectiveness, alongside an organic digitalization process that has not been well communicated, opportunities for the existing workforce are eroded, and organizations create requirements for new skills that are not immediately available.

As a result, while technological introduction may not stop, the transformation process itself slows down and is weakened. This is a potentially declining cycle of

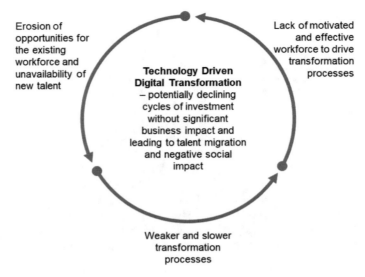

Erosion of opportunities for the existing workforce and unavailability of new talent

Lack of motivated and effective workforce to drive transformation processes

Technology Driven Digital Transformation
– potentially declining cycles of investment without significant business impact and leading to talent migration and negative social impact

Weaker and slower transformation processes

Fig. 29.3 Technology-driven declining transformation. *Own Source*

investment, where business returns may be reduced. Outside the organization, significant negative psychological and social impacts—such as poverty, homelessness and the breakdown of the family—could be driven (McBride 1999).

A Shift in Paradigm

A shift in paradigm, based on cultural change and a strategic approach, is required for successful digital transformation processes. These types of processes should be based on an integrated vision of people, business and technology—with organizations developing and actively engaging existing workforce and talent to generate value through business change—enabled by digital tools, as shown in Fig. 29.4. Future organizational cultures need to appreciate the power of human talent and put people at the center of transformation processes. Strategies should recognize the professional and personal development of individuals as a key mechanism to drive transformation processes. Ultimately, traditional stable jobs will be overtaken by modern dynamic careers that have the potential to propel the evolution of the organization.

Personal and professional development, lifelong learning and career management need to be at the core of the strategy and the culture, based on an understanding of the roles of the future and the development of an associated integrated competency framework for the whole organization. Embedding those competency frameworks in a wider framework that provides transparency for potential progression and career paths is a major step toward changing organizational culture and behavior, and toward getting people involved, making new career opportunities visible to employees. This

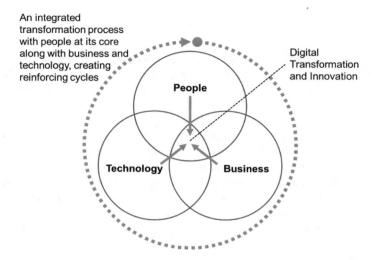

Fig. 29.4 Integrated approach to digital transformation. *Own Source*

puts people in the driving seat, giving them choice and control over their career. Behavioral models, such as the one presented by Michie et al. (2011), show how this type of mechanism can change behaviors and, therefore, the culture of an organization.

Competency Frameworks

Competency frameworks capture industrial requirements in terms of knowledge, skills and behavior (McLaren 2017). The process should be based on exhaustive consultation and a collaborative approach whenever possible, to make sure that all the key stakeholders have an input and share the ownership. This is key for two main reasons. First, it is essential to have the input of a variety of stakeholders, including: state of the art experts (e.g., universities, RTOs, R&D representatives) who can ensure an informed understanding of the future; current practitioners who can highlight key elements of current practice; and organizational leaders who can ensure alignment with business strategies. Second, it is key that there is wide consensus to achieve accreditation. Accreditation is key to ensure continuous professional development (CPD), so that careers can progress not only inside organizations, but also across different organizations. The latter element will become more relevant as careers become more dynamic and talent turnover becomes ubiquitous.

Implementation of Retraining Agenda

The definition of an integrated competency framework will show a vision of the future organizational structure of a business as the backdrop of the future shape of business processes and technologies to be used. These will provide people with opportunities for the future and the possibility of retraining. However, the definition is not enough. There are also significant challenges with the implementation of a retraining agenda.

One of the challenges is the increasing competency/skills gap. A recent survey in the UK concluded that employers want to cap off the job training at a maximum of 20% of working time to avoid loss of productivity. While a lot of current curricula are 20 years old or more and include the training of skills that will not necessarily be required in the future, technologies keep changing at an accelerated pace, and the competency gap (competency demand vs. current competency available in organizations) is ever increasing. This means that retraining needs in the future will be significantly larger than those currently being considered, making the skills gap grow at a faster rate.

Businesses face potential loss of productivity and revenue with high rates of required retraining. However, the demand for change in retraining needs is not coming from the employer side only, as employees also expect better integration between work and personal life. Bespoke training, eLearning, virtual classrooms, peer discussions and short courses are some of the approaches designed to solve this challenge. High levels of flexibility, adaptability and novel training methodologies are required to balance retraining and productivity. Models with increased flexibility—such as "roll on, roll off" training—are required, where students can step in and out and adapt the pace and intensity of the training according to changes in individual personal circumstances. In addition to this, approaches that use the job as a reinforcing element of retraining—providing a platform to practice and accelerate learning—are becoming increasingly relevant. Pertinent emerging terms include "learn while you earn" and "learning by doing".

Organizations must be aware that there might be limitations related to motivation and ability. According to Fogg (2018), an individual's action toward a change in career relies on a balance between motivation and ability. If motivation or ability are significantly low, the transition is too difficult. It is important to consider that the motivation of an individual might change over a career life cycle, that some of the root causes may be outside the control of the organization and that, in certain cases, there may be "natural" limitations in ability. As a minimum, the resources (time and budget) must be provided by the organization to stimulate action. Even with sufficient motivation and ability, stimuli are still required—for example, in the form of clear signals from leadership to prompt activities.

The Role of Policy and Training Offerings as National Enablers

Finally, it must be recognized that, while industry should be the champions of the shift in paradigm and the development of competency framework-based agendas, government and training providers are key enabling agents of a successful shift.

Governments play a key role in the development of policy that provides a strategy with clear direction and guidance, supported by funding. This should boost collaboration between the right stakeholders to achieve solutions of national impact. Similarly, training providers need to work toward developing innovative methodologies that are aligned with the requirements described, and appropriate accreditation is required to ensure that cross-organizational career management is provided for efficient management of talent at a national level.

To summarize, an effective paradigm shift to drive more successful digital transformation processes should be based on:

- A co-developed organizational strategy and culture, developed collaboratively, providing meaningful career options toward higher value jobs;
- An appropriate implementation process to ensure that careers can be fast-tracked to keep up with the pace of technological change;
- Policy funding and training offerings to make the shift possible.

The Opportunity for Dynamic Careers

A shift in paradigm as described above can put people in control and make them drivers of a digital transformation process. It also alters the expectation of change and creates a new breed of dynamic, motivated and empowered employees. Only with this shift in paradigm can humans become agents instead of victims and the threat perceived by individuals turn into opportunity for them, as well as for organizations and society as a whole. Such a shift will result in a positive change of mindset from the employee perspective, as well as a journey in the opposite direction—upwards—in Maslow's Hierarchy of Needs, as shown in Fig. 29.5.

When the appropriate strategy and culture is established—based on an implementable and enabled retraining agenda, supported by policy and providers—the workforce in an organization can start to flourish. As they are involved in defining a range of future meaningful career opportunities, they are instantly, positively empowered to evolve their jobs, transforming their careers. Motivation is created by the visibility of "tangible" and realistic opportunities for development and achievement. This process elevates their natural human talent, which when developed, can lead to the fulfilment of new, higher value roles that are better rewarded. Individuals are rewarded by a sense of achievement and the recognition that they are the

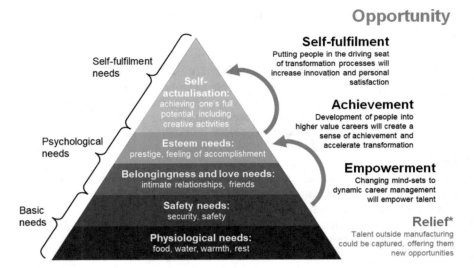

Fig. 29.5 The upward journey on the Maslow Pyramid. *Own Source*

accelerating agents of a digital transformation process. Most workers, with the right resources and support, should be able to get to this level.

During the process, the exceptional talent will become visible and those individuals will have the opportunity to further develop their careers by changing from a delivery role to a transformational role where the mission changes from delivering a business function to delivering—and even defining—the transformation process itself, by leveraging and stimulating their curiosity, creativeness and critical thinking. From the individual's perspective, this will lead to an even higher level of self-fulfillment and satisfaction.

Captivating and motivating exceptional internal talent into roles that drive digitalization will strengthen the process, accelerate it and increase the amount of innovation in the process, opening the door to higher benefits and even the opportunity to leapfrog in the market. In turn, more successful digitalization processes lead to the creation of new opportunities for existing talent in the organization or new talent outside the organization. When talent is attracted from outside manufacturing—for example, from labor markets with less potential—relief is created in society as a whole, as indicated with the asterisk in Fig. 29.5. This is a self-reinforcing cycle of positive impact in businesses and society, enabled by technology introductions, illustrated in Fig. 29.6.

For organizations, this also results in capitalization of talent when workers decide to commit to their organization. Even when individuals eventually decide to leave an organization, empowered and fulfilled talent has the potential to extend and strengthen strategic partnerships with other organizations. For society, the wider impacts ultimately translate into prosperity through metrics related to employment.

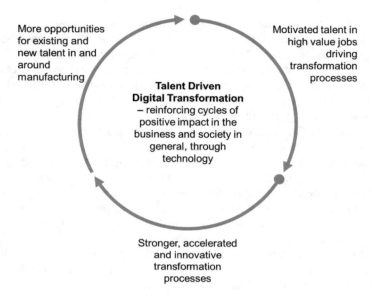

More opportunities
for existing and
new talent in and
around
manufacturing

Motivated talent in
high value jobs
driving
transformation
processes

**Talent Driven
Digital Transformation**
– reinforcing cycles of
positive impact in the
business and society in
general, through
technology

Stronger, accelerated
and innovative
transformation
processes

Fig. 29.6 People orientated, self-reinforcing driven declining transformation. *Own Source*

The Example of Additive Manufacturing in the UK

The transformation of manufacturing organizations through the use of additive manufacturing technologies is akin to digital transformation. Given the reduced scope of additive manufacturing in comparison to digital manufacturing (which covers a wide variety of technologies and applications across different areas of manufacturing businesses), some aspects of additive manufacturing are more advanced than they are in digital. In the UK, one of those aspects is training, providing a representative example to illustrate the methods and insights presented in this article.

As with digital, additive manufacturing transformation processes have tended to be driven from the technology side. The most ambitious organizations have charged ahead and implemented the technology successfully for specific components. In spite of this progress, organizations are now realizing that retraining is essential to continue a successful transformation process. For other smaller organizations in the supply chain, the transformation process has been truncated in its infancy due to cost barriers and the lack of people who have the right competencies to understand the potential impact of the technology, implement it and use it.

The Advanced Manufacturing Training Centre (AMTC) and the National Centre for Additive Manufacturing (NCAM) (both MTC organizations) have taken proactive action to support adoption of additive manufacturing, unlocking its benefits at a national level. Over the last three years, they have completed a comprehensive exercise to help the UK additive manufacturing community develop an integrated vision of the future of businesses operating with additive manufacturing. As a result, they have also developed the associated integrated competency framework, including all the key future manufacturing roles that can drive and deliver additive manufacturing.

Using this framework, AMTC has developed a suite of flexible training offerings, including eLearning, Virtual Classrooms and short professional courses. This offer is already available in the market. This includes eight competency frameworks, eight curricula and eighty courses that will be delivered collaboratively by universities, commercial technology providers and the AMTC at the MTC.

NCAM and AMTC have covered the three elements of the proposed shift: an integrated competency framework capturing the vision of an organization embarked in the transformation process; an implementable training plan; and supporting policy and commercial offerings that help deliver the plan. As a result, some of the partners of NCAM are already well into their journey, and others are looking forward to joining, to make the most of the potential benefits of additive manufacturing. From a societal point of view, it is already foreseen that the successful implementation of additive manufacturing—in conjunction with the development of talent—will help anchor the supply chain for high value manufacturing sectors in the UK. This will also generate increased economic benefits and sustained social benefits through the generation of high value jobs (AM UK 2017).

Conclusion

Digital transformation processes are transforming jobs organically, which may lead to unintended consequences. When jobs are not transformed strategically and proactively, transformation processes may be ineffective and even truncated for organizations. For the current workforce of an organization undergoing such a process, this may mean a psychological downward spiral, resulting in unemployment and the attrition of stable jobs. The negative consequences for individuals and for society could be significant.

A shift in paradigm is proposed in this article, where strategy and culture are developed based on an integrated vision of the future of organizations, covering people, business and technology. This can be captured in the format of an integrated competency framework and should be accompanied by an implementable training plan that aligns with the needs of businesses and individuals. Finally, policy and the availability of commercial training offerings are key to enable the shift.

From the employee perspective, such a shift in paradigm has the potential to transform the perceived threat of a decline of stable jobs into an optimistic journey of developing a dynamic career that provides long-term safety in employment, a sense of achievement in meaningful roles and, potentially, the self-fulfilling opportunity to become the drivers of digital transformation processes themselves. This prospect has the potential to strengthen and accelerate digital transformation processes, increasing the benefits to organizations and creating opportunities for new talent that can further increase the overall social impact.

Additive manufacturing poses similar challenges and opportunities to digital manufacturing. The Manufacturing Technology Centre in the UK has already implemented the shift proposed for additive manufacturing, showing positive results and encouraging signs for the proposed recommendations.

Acknowledgements The authors would like to acknowledge the personal contributions of Paul Rowlett (Managing Director, AMTC), Ross Trepleton (Chief Engineer, Component Manufacturing Technology MTC) and Paul Shakespeare (Consultant—Education and Skills, HVMC), as well as the support of the National Centre for Additive Manufacturing, the Advanced Manufacturing Training Centre and the High Value Manufacturing Catapult.

References

Automation to Take 1 in 3 Jobs in UK's Northern Centres, Report Finds. (2018, January 29). The Guardian. Retrieved from: https://www.theguardian.com/technology/2018/jan/29/automation-to-take-1-in-3-jobs-in-uks-northern-centres-report-finds.

Fogg, B. J. (2018, September 28). BJ Fogg's behavior model. Behaviour Design Lab, viewed. https://www.behaviormodel.org/.

Leondorf, W. (2010). The social, economic and political impact of technology: An historical perspective. In *American Society for Engineering Education, Annual Conference & Exposition.* Retrieved from http://www.asee.org/search/proceedings.

Maslow, H. (1943). The theory of human motivation. *Psychological Review, 50*(4), 370–396.

McBride, S. (1999). Towards permanent insecurity: The social impact of unemployment. *Journal of Canadian Studies, 34*(2), 13–30.

McLaren, RTPI Scotland. (2017). *Developing Skills, Behaviours and Knowledge to Deliver Outcomes.* Retrieved from: https://beta.gov.scot/binaries/content/documents/govscot/publications/corporate-report/2017/04/planning-review-developing-skills-behaviours-knowledge-report/documents/18458cdb-c8f7-43a1-95ea-8ae9a9b2d329/18458cdb-c8f7-43a1-95ea-8ae9a9b2d329/govscot:document/?inline = true/.

Michie, S., van Stralen, M., & West, R. (2011). The behaviour change wheel: A new method for characterising and designing behaviour change interventions. *Implementation Science, 6*, 42.

Prodan, C. (2015). *Three new dimensions to people, process, technology improvement model.* Switzerland: Springer International Publishing.

Robot Automation will 'Take 800 Million Jobs by 2030'—Report. (2017, November 29). *BBC News.* Retrieved from: https://www.bbc.co.uk/news/world-us-canada-42170100.

Dr. Lina A. Huertas in her current role focuses on driving the creation of value in the UK through the digitalization of UK industry. This include shaping the UK national strategy and policy, leading the relevant activities at High Value Manufacturing Catapult and running the MTC corporate strategy on Digital Manufacturing. In recent years she has formed and led different engineering and technology teams that offer unique capabilities to develop and mature technological solutions for the UK manufacturing sector. Her experience includes a Ph.D. in the application of simulation and informatics solutions to support manufacturing, and 12 years' experience delivering industrial projects that generated business benefits on the back of digital technologies. Lina is a Fellow of the IMechE and has a first class honours degree and a master's degree in Mechanical Engineering.

Dr. Harald Egner's current role is with the Manufacturing Technology Centre (MTC), a technology and innovation centre in the UK which is part of the High Value Manufacturing Catapult (HVMC). His responsibilities are in developing international and research partnerships, networks and ecosystems. With his arrival at the MTC he brought along the ideas and developments of "Industrie 4.0" in Germany which have been adapted early in the MTC. He also is an active member of various EC work groups in Digital Manufacturing and a strategic initiative to great a national stakeholder platform in the UK. After a career in Stuttgart including the role of a Deputy Director and a Fraunhofer Advisory Board position Harald moved with his family to the UK in 2002 to gain new experience on international level and a focus to build new European business partnerships and networks. From the time Harald came to the UK he was involved in the UK innovation infrastructure as a board member with Faradays and KTNs. Furthermore, he was involved in the "Hauser Review" and developments following on from there which led to the set-up of the Catapult infrastructure in the UK. Harald graduated from the University of Stuttgart with a master's degree in mechanical engineering and joined Fraunhofer IPA in Stuttgart. His work there took him from the early stages of automation in the early 80s to design for automation and into product development. His focus in the 90s was very much on customer driven product development and relevant methodologies such as QFD. All through his professional career he was working very much on the application and implementation of technology, particularly for SME.

Martin Dury's current role is Learning Design Manager at the Lloyds Bank Advanced manufacturing Training centre. Martin is an accomplished learning and development specialist, who has implemented innovative, effective and efficient learning strategies for a number of global engineering and manufacturing companies. Martin has 20 years' experience working in the education and skills sector, typically producing manufacturer specific technical competency frameworks, nationally accredited curricula and modular commercial training programmes. Previously, as Learning Design Manager for Volkswagen Group (VWG) UK, Martin was responsible for creating, designing and implementing an innovative blended learning strategy that transformed the speed to competence of VWG's 25,000 learners across the UK. All learning interventions embraced new delivery technologies which significantly reduced the pressure on the National Learning Centre. And as all learning provided the business with a measurable return on investment, it achieved the buy-in of the franchise businesses and significant improvements to behaviour change across the UK network. Martin is an advocate of the Training Foundations curriculum in learning and development and their approach to skill-based practical assessment in training design and delivery, of which he has achieved their Diploma in Learning Design. Martin has also enjoyed significant success internationally (Europe, Middle East, Australasia) in the design and delivery of both technical and soft-skills programmes for other large clients, such as Jaguar Land Rover, Aston Martin, Bentley, Honda and Volvo. Martin is passionate about creating positive behaviour change as a result of his learning design and delivery strategies, which have a tangible return on investment.

Printed in the United States
by Baker & Taylor Publisher Services